QUANTUM
OPTICS
FOR ENGINEERS

Quantum Optics

Optics

for Engineers

F.J. Duarte

CRC Press
Taylor & Francis Group
Boca Raton London New York

CRC Press is an imprint of the
Taylor & Francis Group, an **informa** business

CRC Press
Taylor & Francis Group
6000 Broken Sound Parkway NW, Suite 300
Boca Raton, FL 33487-2742

First issued in paperback 2017

© 2014 by Taylor & Francis Group, LLC
CRC Press is an imprint of Taylor & Francis Group, an Informa business

No claim to original U.S. Government works
Version Date: 20131009

ISBN 13: 978-1-138-07754-6 (pbk)
ISBN 13: 978-1-4398-8853-7 (hbk)

Library of Congress Cataloging-in-Publication Data

Duarte, F. J. (Frank J.)
   Quantum optics for engineers / F.J. Duarte.
     pages cm
   Includes bibliographical references and index.
   ISBN 978-1-4398-8853-7 (hbk. : alk. paper)
   1. Quantum optics. I. Title.

QC446.2.D83 2014
535'.15--dc23   2013037851

Visit the Taylor & Francis Web site at
http://www.taylorandfrancis.com

and the CRC Press Web site at
http://www.crcpress.com

In memory of my father Luis Enrique (1921–2011).

Contents

List of Figures

List of Tables

Preface

Optics plays a crucial role in the success of today's science and technology. This role is certainly destined to increase significantly in the future. Within the discipline of optics, *quantum optics* is rapidly becoming a dominant asset.

The aim of this book is to make available the principles of quantum optics directly to engineers and engineering students. This is done by presenting the subject, and associated elements of coherent optics, from a pragmatic and forthright perspective using Dirac's quantum notation.

It is indeed a pleasure to offer, to the scientific and engineering communities, this first edition of *Quantum Optics for Engineers*.

F.J. Duarte
Rochester, New York

MATLAB® is a registered trademark of The MathWorks, Inc. For product information, please contact:

The MathWorks, Inc.
3 Apple Hill Drive
Natick, MA, 01760-2098 USA
Tel: 508-647-7000
Fax: 508-647-7001
E-mail: info@mathworks.com
Web: www.mathworks.com

Author

F.J. Duarte is a research physicist with Interferometric Optics, Rochester, New York, and adjunct professor at the Electrical and Computer Engineering Department, University of New Mexico. He graduated with first-class honors in physics from Macquarie University (Sydney, Australia), where he was also awarded a PhD in physics for his research on optically pumped molecular lasers. At Macquarie, he was a student of the well-known quantum physicist J. C. Ward. Duarte is the author of the generalized multiple-prism dispersion theory, has made various unique contributions to the physics and architecture of tunable laser oscillators, and has pioneered the use of Dirac's quantum notation in classical optics. These contributions have found applications in the design of laser resonators, laser pulse compression, imaging, microscopy, medicine, and the nuclear industry. He is the author and editor of *Dye Laser Principles*, *High-Power Dye Lasers*, *Selected Papers on Dye Lasers*, *Tunable Lasers Handbook*, and *Tunable Laser Applications*. He is also the author of *Tunable Laser Optics*. Dr. Duarte received the Engineering Excellence Award from the Optical Society of America and is a fellow of both the Australian Institute of Physics and the Optical Society of America.

1

Introduction

1.1 Introduction

Perhaps no other subject in the history of physics has captured the human imagination more than quantum mechanics has. This captivation has extended beyond physics, and science, and well into the realm of popular culture. This is because quantum mechanics, also known as quantum physics, correctly describes the microworld and the nanoworld in a mathematical way that appears to be mysterious to us, the inhabitants of the classical world.

Perhaps one of the best, and most succinct, descriptions of quantum mechanics has been given by the well-known quantum and particle physicist John Clive Ward: *"The inner mysteries of quantum mechanics require a willingness to extend one's mental processes into a strange world of phantom possibilities, endlessly branching into more and more abstruse chains of coupled logical networks, endlessly extending themselves forward and even backwards in time"* (Ward, 2004).

1.2 Brief Historical Perspective

Quantum mechanics came to light via the work of Max Planck, published in 1901. In that contribution, Planck used concepts of thermodynamics to explain the energy distribution of light sources as a function of wavelength (Planck, 1901). In doing so he introduced, without derivation, an equation where the energy of the emission was a function of frequency ν, that is,

$$E = h\nu \tag{1.1}$$

where
   the units of the energy E is the joule (J)
   the units of the frequency ν is the Hz
   h is known as Planck's constant ($h = 6.62606957 \times 10^{-34}$ J s).

That was the birth of quantum mechanics. It was born from the experiment; it was an empirical birth.

Another important experimental observation relevant to the development of quantum mechanics was the photoelectric effect (Hertz, 1887). This effect, of fundamental significance to modern photomultipliers, and photo detectors in general, means that when a surface composed of charged particles is irradiated with light of frequency v, there is a probability that electrons will be emitted from that surface. An explanation to the photoelectric effect was provided by Einstein (1905) via the relationship

$$E = \hbar\omega - W \tag{1.2}$$

where W is defined as the work function or energy required to emit an electron from the irradiated surface. In this contribution, Einstein also proposed that light behaves as a stream of localized units of energy that he called *lightquanta*.

A few years later, Bohr (1913) postulated that electrons in an atom can only populate well-defined orbits at discrete energies W_n. When the electron jumps from one orbit of energy W_n to another one at W_{n+1}, it does so emitting radiation at a frequency v, so that (Bohr, 1913)

$$W_n - W_{n+1} = hv \tag{1.3}$$

The developments introduced earlier were the preamble to the 1925–1927 revolution that yielded the *quantum mechanics* we know today. Heisenberg (1925), Born and Jordan (1925), and Born et al. (1926) introduced the quantum mechanics in matrix form. Schrödinger (1926) introduced his quantum wave equation. Dirac (1925) first established that there was a correspondence between Heisenberg's non-commuting dynamical variables and the Poisson bracket (Dirac, 1925). Then he discovered that there was an equivalence between the Born–Jordan formulation and Schrödinger's equation (Dirac, 1926). Further, he demonstrated that there was a direct correspondence between the Heisenberg–Dirac quantum mechanics and Schrödinger's wave mechanics (Dirac, 1927).

In addition to the three formulations just mentioned, Dirac (1939) further introduced his *bra–ket* notation, also known as the Dirac notation, which is the preferred formulation of quantum mechanics used in this book.

Further approaches to quantum mechanics include the Feynman formulation via *integral paths* (Feynman and Hibbs, 1965) and the phase-space formulation (Moyal, 1949). There are also other lesser-known formulations.

Post quantum mechanical developments include *quantum electrodynamics* (Tomonaga, 1946; Schwinger, 1948; Dyson, 1949; Feynman, 1949), *renormalization theory* (Ward, 1950), *Feynman diagrams* (Feynman, 1949), and the *standard*

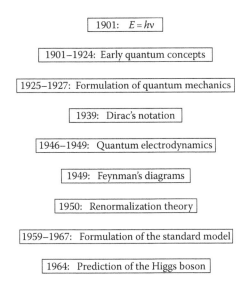

FIGURE 1.1
Time line depicting important developments in the quantum era.

model of particle physics (see, e.g., Salam and Ward, 1959, 1964; Glashow, 1961; Weinberg, 1967). The Higgs boson was theorized in 1964 (Higgs, 1964).

Figure 1.1 provides a time line of important developments in the quantum era.

1.3 Principles of Quantum Mechanics

The Principles of Quantum Mechanics is the landmark book written by one of the creators of quantum mechanics Paul Adrien Maurice Dirac. The first edition of this masterpiece was published in 1930, the second edition in 1935, and the third edition in 1947. The fourth edition was released in 1958, and it is this edition that gives origin to the 1978 version, its ninth revised printing, used as the standard reference in this book.

An interesting aspect of this book is that the Dirac *bra–ket* notation was introduced in its third edition (1947). This is explained by the Australian particle physicist R. H. Dalitz (known of the *Dalitz plot* and the *Dalitz pair*) whom in 1947 was taking lectures from Dirac in Cambridge (Dalitz, 1987).

The Principles of Quantum Mechanics, third and fourth editions, are the vehicles by which the Dirac notation was introduced to physicists although Dirac first disclosed the notation in a paper entitled *A new notation for quantum mechanics* (Dirac, 1939). This paper, in a fairly mechanistic style, limits itself to introduce the new notation and to provide a correspondence between it and the "old notation." The paper does not explain how Dirac discovered or created the new notation. Nor does he explain it in the book. At one time

I did ask Dick Dalitz if Dirac had explained in his lectures how he created, or discovered, his *bra–ket* notation, and his reply was "no."

Here we should make a necessary point: albeit we use as reference in this book, a revised version of the fourth edition of *The Principles of Quantum Mechanics*, we should be very much aware that the first edition was published in 1930 and that the Dirac notation was incorporated in 1947. Thus, given Dirac's famous precision as a communicator, we should assume that our version of this masterpiece goes back to 1947.

Dirac's book, *The Principles of Quantum Mechanics*, includes 12 chapters. The most relevant of those chapters to our immediate interest are

The principle of superposition
Dynamical variables and observables
Representations
The quantum conditions
The equations of motion
Perturbation theory
Systems containing several similar particles
Theory of radiation

Throughout the book he does use his *bra–ket* notation extensively albeit it is not the only type of notation he utilizes.

Besides issues of notation, the Dirac book is remarkable in that it provides probably the very first discussion of optics in a quantum context. It does so via a brilliant and prophetic discussion of interferometry. He begins by considering a beam of "roughly monochromatic light" and continues by referring to this beam of light as "consisting of a large number of photons," and the beam is then "split up into two components of equal intensity" (Dirac, 1978). In today's terminology Dirac's discussion applies perfectly well to a high-power narrow-linewidth laser beam undergoing interference in a Mach–Zehnder interferometer (Duarte, 1998). This discussion qualifies Dirac as the father of quantum optics and laser optics (Duarte, 2003).

It is also apparent that *The Principles of Quantum Mechanics* served as inspiration to Feynman for his lectures on physics not only on the central topic of the Dirac notation but also on the fundamental ideas on interference and other various mathematical formalisms.

1.4 *The Feynman Lectures on Physics*

Volume III of *The Feynman Lectures on Physics* (Feynman et al., 1965) offers a brilliant discussion of quantum mechanics via the Dirac notation. From a fascinating discussion of the two-slit interference thought experiments, using electrons, to practical applications of two-state systems, and beyond, this book is a physics treasure.

At a basic level, *The Feynman Lectures on Physics*, Volume III, is a valued introduction to the use and practice of the Dirac notation in quantum mechanics.

At this stage it is also instructive to mention that in his 1965 book on the path integral approach to quantum mechanics, Feynman applies quantum mechanics directly to the classical problem of diffraction (Feynman and Hibbs, 1965). It is necessary to make this observation for the benefit of some practitioners that insist in imposing the use of classical tools only to describe macroscopic diffraction and interference.

1.5 Photon

In this section first we explore the opinion on this subject given by a few luminaries of quantum physics: Dirac, Feynman, Haken, and Lamb. Then, our own opinion on the subject is examined.

Dirac (1978): "Quantum mechanics is able to effect a reconciliation of the wave and corpuscular properties of light. The essential point is the association of each of the translational states of a photon with one of the wave functions of ordinary wave optics… the wave function gives information about the probability of one photon being in a particular place and not the probable number of photons in that place."

Feynman (1965): "Newton thought that light was made up of particles, but then it was discovered that it behaves like a wave…." We say: "It is like *neither.*"

Haken (1981): "In quantum mechanics we attribute an infinite extended wave to a freely moving particle with momentum p so that $\lambda = h/p$. The wave must be of infinite extent, otherwise it would not have a definite wavelength."

Lamb (1995): "Photons cannot be localized in any meaningful manner, and they do not behave at all like particles, whether described by a wave function or not." Indeed, the nonlocality of the photon is intuitive to experimentalists experienced in optics: "All the indistinguishable photons illuminate the array of N slits, or grating, simultaneously. If only one photon propagates, at any given time, then that individual photon illuminates the whole array of N slits simultaneously" (Duarte, 2003).

So, those are some the written definitions of a photon. As can be observed, there is no complete conceptual convergence on the meaning of a photon. Here, rather than offering yet another language-based concept of the photon, we shall examine in a pragmatic approach what we know about the basic features of the photon:

1. A single photon moves, in vacuum, at the speed of light c.
2. A single photon has a wavelength λ, which is related to its frequency ν by

$$\lambda = \frac{c}{\nu} \tag{1.4}$$

3. A single photon exhibits a quantum energy of

$$E = h\nu$$

   or

$$E = \hbar\omega \tag{1.5}$$

4. A single photon exhibits a quantum momentum of

$$p = \hbar k \tag{1.6}$$

5. A single photon is associated with the wave functions of ordinary wave optics, such as

$$\psi(x,t) = \psi_0 e^{-i(\omega t - kx)} \tag{1.7}$$

6. Single photons are nonlocal and can exhibit *enormous coherence lengths* as described by

$$\Delta x \approx \frac{c}{\Delta \nu} \tag{1.8}$$

Under these premises we can attempt a conceptual description of a photon as a unique entity that can be mathematically described using the wave functions of ordinary wave optics (Dirac, 1978) while exhibiting a quantum energy $E = h\nu$ and a quantum momentum of $p = \hbar k$. As of now, limitations in the existing language prevent us from a more definite description other than this tenuous outline. Therefore, when we refer to a photon, or quanta, we refer to a unique energetic entity, which is the basic component of light.

Notice that in the case of emission resulting in the generation of an ensemble of coherent photons, as in the case of a narrow-linewidth lasers, a refinement on the wave description of Haken (1981) should refer to a *near-infinite* wave since the wavelength, in practice, would always exhibit a measurable linewidth; in other words the wavelength would be $\lambda \pm \Delta\lambda$ and not just λ.

One further point of extreme importance is the following: quantum mechanically indistinguishable photons *are the same photon.* In other words, two photons coming from different narrow-linewidth lasers with energies $E = h\nu_1$ and $E = h\nu_1$ are the same photon and will interfere precisely as described by Dirac (1978) even though they originate from different sources. Thus, a seventh item to be added to the list earlier becomes

7. Ensembles of indistinguishable photons exhibiting very narrow-linewidth $\Delta\nu$ originating from nearly monochromatic sources, such as narrow-linewidth lasers, approximates the behavior of a single photon.

Finally, in this terminology a monochromatic source is an ideal laser source with an emission linewidth, which is extremely narrow, so that $\Delta\nu \rightarrow 1$ Hz or less.

1.6 Quantum Optics

As described previously, the first known discussion of quantum optics was provided by Dirac in his book. He did so via interference. Furthermore, and very importantly, he did so considering a macroscopic interferometric experiment involving a beam of "roughly monochromatic light" and continues by referring to this beam of light as "consisting of a large number of photons," and the beam is then "split up into two components of equal intensity" (Dirac, 1978). In other words, Dirac applies his quantum concepts directly to a *macroscopic interferometric experiment.*

The use of quantum physics in macroscopic optics is not unique to Dirac. In 1965, Feynman used his path integrals to describe divergence and diffraction resulting from the passage through a Gaussian slit (Feynman and Hibbs, 1965). Even further, Feynman in his *Feynman Lectures on Physics* (problem book to Feynman et al., 1965) gives credit to Hanbury Brown and Twiss (1956) as performing an early experiment in quantum optics.

Hanbury Brown and Twiss collected light from the star Sirius in two separate detectors, and the signals from these detectors are then made to interfere. Building on Feynman's description of double-slit electron interference, the Dirac quantum notation was applied to N-slit interference (Duarte, 1991).

In addition to applications to macroscopic interference, a clear and intrinsic quantum physics development was the derivation of probability amplitude equations associated with counterpropagating photons with entangled polarizations (Pryce and Ward, 1947; Snyder et al., 1948; Ward, 1949)

$$| s \rangle = \frac{1}{\sqrt{2}} \left(| x \rangle_1 | y \rangle_2 - | y \rangle_1 | x \rangle_2 \right) \tag{1.9}$$

and the subsequent experimental confirmation provided via the measurements of polarization correlations by Wu and Shaknov (1950). A development directly related to photon entanglement was the introduction of Bell's inequality (1964). All-optical experiments on polarization entanglement were reported by Aspect et al. (1982).

A further development in quantum optics was the introduction of quantum cryptography (Bennett, 1992). An advance directly related to the physics of entanglement is quantum teleportation (Bennett et al., 1993). Figure 1.2 highlights the time line of important developments in quantum optics while emphasizing the application of the Dirac notation.

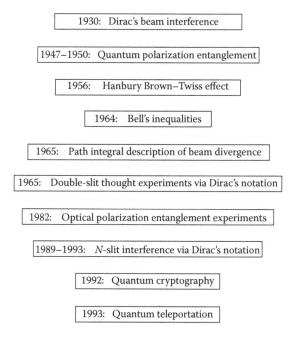

FIGURE 1.2
Time line depicting important developments in quantum optics while emphasizing the application of the Dirac notation.

1.7 Quantum Optics for Engineers

Quantum Optics for Engineers is designed as a textbook, primarily utilizing the Dirac quantum notation, to describe optics in a unified and coherent approach. The emphasis is practical. This approach uses a minimum of mathematical sophistication. In other words, the reader should be able to use the tools provided primarily with the knowledge of first-year courses in calculus and algebra.

The subject matter is contained in Chapters 1 through 21, while a set of companion Appendices A–K provide additional necessary information relevant to the chapter material. The concept here is to offer the student a self-contained book, thus minimizing the need to refer to additional texts except for those who would like to expand their knowledge of a particular subject.

The reader will also notice that some of the equations, and figures, in this book are reproduced in several of the chapters. In other words, they are repeated. This has been done quite deliberately to avoid having to go back in the text to find a particular equation and then forward again to continue the work. Besides highlighting the importance of some concepts, this approach should facilitate remembering those equations and easing the lecture process. Hopefully, this will enhance the learning process according to the old Roman saying *repetitio est mater studiorum* (approximately translated as "repetition is the mother of learning").

References

Aspect, A., Dalibard, J., and Roger, G. (1982). Experimental test of Bell's inequalities using time-varying analyzers. *Phys. Rev. Lett.* **49**, 1804–1807.

Bell, J. S. (1964). On the Einstein-Podolsky-Rosen paradox. *Physics* **1**, 195–200.

Bennett, C. H. (1992). Quantum cryptography using any two nonorthogonal states. *Phys. Rev. Lett.* **68**, 3121–3124.

Bennett, C. H., Brassard, G., Crépeau, C., Jozsa, R., Peres, A., and Wootters, W. K. (1993). Teleporting an unknown quantum state via dual classical and Einstein-Podolsky-Rosen channels. *Phys. Rev. Lett.* **70**, 1895–1899.

Bohr, N. (1913). On the constitution of atoms and molecules. *Phil. Mag.* **26**, 857–875.

Born, M. and Jordan, P. (1925). Zur quantenmechanik. *Z. Phys.* **34**, 858–888.

Born, M., Heisenberg, W., and Jordan, P. (1926). Zur quantenmechanik II. *Z. Phys.* **35**, 557–617.

Dalitz, R. H. (1987). Another side to Paul Dirac. In *Paul Adrien Maurice Dirac* (Kursunoglu, B. N. and Wigner E. P., eds.). Cambridge University, Cambridge, U.K., Chapter 10.

Dirac, P. A. M. (1925). The fundamental equations of quantum mechanics. *Proc. Roy. Soc. London A* **109**, 642–653.

Dirac, P. A. M. (1926). On the theory of quantum mechanics. *Proc. Roy. Soc. London A* **112**, 661–677.

Dirac, P. A. M. (1927). The physical interpretation of the quantum dynamics. *Proc. Roy. Soc. London A* **113**, 621–641.

Dirac, P. A. M. (1939). A new notation for quantum mechanics. *Math. Proc. Cambridge Phil. Soc.* **35**, 416–418.

Dirac, P. A. M. (1978). *The Principles of Quantum Mechanics*, 4th edn. Oxford, London, U.K.

Duarte, F. J. (1991). Dispersive dye lasers. In *High Power Dye Lasers* (Duarte, F. J., ed.). Springer-Verlag, Berlin, Germany, pp. 7–43.

Duarte, F. J. (1998). Interference of two independent sources. *Am. J. Phys.* **66**, 662–663.

Duarte, F. J. (2003). *Tunable Laser Optics*, Elsevier-Academic, New York.

Dyson, F. J. (1949). The S matrix in quantum electrodynamics. *Phys. Rev.* **75**, 1736–1755.

Einstein, A. (1905). Uber einen erzeugung und verwandlung des lichtes betreffenden. *Ann. Phys.* **17**, 132–148.

Feynman, R. P. (1949). Space–time approach to quantum electrodynamics. *Phys. Rev.* **76**, 769–789.

Feynman, R. P., Leighton, R. B., and Sands, M. (1965). *The Feynman Lectures on Physics*, Vol. III, Addison-Wesley, Reading, MA.

Feynman, R. P. and Hibbs, A. R. (1965). *Quantum Mechanics and Path Integrals*, McGraw-Hill, New York.

Glashow, S. L. (1961). Partial-symmetries of weak interactions. *Nucl. Phys.* **22**, 579–588.

Haken, H. (1981). *Light*, North-Holland, Amsterdam, the Netherlands.

Hanbury Brown, R. and Twiss, R. Q. (1956). A test of a new type of stellar interferometer on Sirius, *Nature* **178**, 1046–1048.

Heisenberg, W. (1925). Uber quantenthoretische umdeutung kinematischer und mechanischer beziehungen. *Z. Phys.* **33**, 879–893.

Hertz, H. (1887). Uber einen einfluss des ultravioletten lichtes auf die elektrische entladung. *Ann. Phys.* **31**, 983–1000.

Higgs, P. W. (1964). Broken symmetries and the masses of gauge bosons. *Phys. Rev. Lett.* **13**, 508–509.

Lamb, W. E. (1995). Anti-photon. *Appl. Phys. B* **60**, 77–84.

Moyal, J. E. (1949). Quantum mechanics as a statistical theory. *Proc. Cambridge Phil. Soc.* **45**, 99–124.

Planck, M. (1901). Ueber das gesetz der energieverteilung im normal spectrum. *Ann. Phys.* **309**(3), 553–563.

Pryce, M. H. L. and Ward, J. C. (1947). Angular correlation effects with annihilation radiation. *Nature* **160**, 435.

Salam, A. and Ward, J. C. (1959). Weak and electromagnetic interactions. *Nuovo Cimento* **11**, 568–577.

Salam, A. and Ward, J. C. (1964). Electromagnetic and weak interactions. *Phys. Lett.* **13**, 168–171.

Schrödinger, E. (1926). An undulatory theory of the mechanics of atoms and molecules. *Phys. Rev.* **28**, 1049–1070.

Schwinger, J. (1948). On quantum-electrodynamics and the magnetic moment of the electron. *Phys. Rev.* **73**, 416–417.

Snyder, H. S., Pasternack, S., and Hornbostel, J. (1948). Angular correlation of scattered annihilation radiation. *Phys. Rev.* **73**, 440–448.

Tomonaga, S. (1946). On a relativistically invariant formulation of the quantum theory of wave fields. *Prog. Theo. Phys.* **1**, 27–42.

Ward, J. C. (1949). *Some Properties of the Elementary Particles*. D. Phil Thesis, Oxford University, Oxford.

Ward, J. C. (1950). An identity in quantum electrodynamics. *Phys. Rev.* **78**, 182.

Ward, J. C. (2004). *Memoirs of a Theoretical Physicist*, Optics Journal, New York.

Weinberg, S. (1967). A model of leptons. *Phys. Rev. Lett.* **19**, 1264–1266.

Wu, C. S. and Shaknov, I. (1950). The angular correlation of scattered annihilation radiation. *Phys. Rev.* **77**, 136.

2

Planck's Quantum Energy Equation

2.1 Introduction

In his work on light sources and black body radiation, Max Planck, around 1900, was confronted with experimental data that could not be explained with the prevailing theoretical concepts of the time. The problem was to express the energy distribution of light emission as a function of wavelength. Using Planck's own notation (Planck, 1901), the number of electromagnetic modes per unit volume can be expressed as

$$u = \frac{8\pi v^2}{c^3} U \tag{2.1}$$

where
  v is the frequency of the emission, in Hz or cycles per second
  c is the speed of light ($c = 2.99792458 \times 10^8$ ms^{-1})

Then the question becomes how to define the energy distribution U. Planck approaches this problem using thermodynamic arguments related to the entropy. However, immediately prior to doing that, he makes the crucial step of introducing, without any derivation whatsoever, the energy expression (Planck, 1901):

$$E = hv \tag{2.2}$$

All he writes prior to this equation is that the "energy is proportional to the frequency v." Once he unveils this enormous discovery, Planck proceeds with his thermodynamics argument providing an explicit expression for the entropy of the system:

$$S = k\left(\left(1+\frac{U}{hv}\right)\ln\left(1+\frac{U}{hv}\right)-\frac{U}{hv}\ln\left(\frac{U}{hv}\right)\right) \tag{2.3}$$

and then using $(1/T) = dS/dU$, he proceeds to differentiate Equation 2.3 and arrives at

$$U = h\nu(e^{h\nu/kT} - 1)^{-1} \tag{2.4}$$

which, using Equation 2.1, leads directly to Planck's distribution (Planck, 1901):

$$u = \frac{8\pi h\nu^3}{c^3}(e^{h\nu/kT} - 1)^{-1} \tag{2.5}$$

In the previous equations, $k = 1.3806488 \times 10^{-23}$ J K^{-1} is Boltzmann's constant, T is the absolute temperature, and $h = 6.62606957 \times 10^{-34}$ J s is of course Planck's constant.

Planck's quantum equation, also known as Planck's relation, $E = h\nu$, is one of the most fundamental principles of quantum mechanics and one of the most important equations in physics. The fact that the energy of the emission depends on the frequency of the emission ν, according to the elegant relation $E = h\nu$, is a fundamental quantum law that was arrived to from *macroscopic observations* performed on a *classical experiment*. Planck's contribution represents one of the greatest triumphs of a physicist's intuition in the history of physics. Around 1900 when Planck discovered this empirical law, the physics world was classical and entirely macroscopic.

2.2 Planck's Equation and Wave Optics

From Planck's quantum energy equation

$$E = h\nu$$

one can use special relativity's $E = mc^2$ (de Broglie, 1923) to arrive at

$$p = h\frac{\nu}{c} \tag{2.6}$$

which, using $\lambda = c/\nu$, leads directly to

$$p = \hbar k \tag{2.7}$$

where $k = 2\pi/\lambda$ is known as the wave number. This momentum expression in waveform, $p = \hbar k$, is known as de Broglie's equation and plays a significant role in developing further concepts in quantum optics.

In Chapter 3, the Heisenberg uncertainty principle (Heisenberg, 1927; Feynman et al., 1965; Dirac, 1978) is developed from interferometric principles. The approach is based on interferometric principles that lead to the identity (Duarte, 2003)

$$\Delta\lambda\,\Delta x \approx \lambda^2 \tag{2.8}$$

which can also be expressed as

$$\Delta\nu\,\Delta x \approx c \tag{2.9}$$

Using the expression for momentum $p = \hbar k$, which is based on $E = h\nu$, one arrives directly to the Heisenberg uncertainty principle:

$$\Delta p\,\Delta x \approx h \tag{2.10}$$

This simple exercise is useful in exposing the order of fundamental concepts in quantum mechanics. This order places $E = h\nu$, and interferometric principles, at the very foundations of quantum physics.

References

de Broglie, L. (1923). Waves and quanta. *Nature* **112**, 540.

Dirac, P. A. M. (1978). *The Principles of Quantum Mechanics*, 4th edn. Oxford, London, U.K.

Duarte, F. J. (2003). *Tunable Laser Optics*, Elsevier-Academic, New York.

Feynman, R. P., Leighton, R. B., and Sands, M. (1965). *The Feynman Lectures on Physics*, Vol. III, Addison-Wesley, Reading, MA.

Heisenberg, W. (1927). Uber den anschaulichen inhalt der quantentheoretischen kinematic und mechanic. *Z. Phys.* **43**, 172–198.

Planck, M. (1901). Ueber das gesetz der energieverteilung im normal spectrum. *Ann. Phys.* **309**(3), 553–563.

3

Uncertainty Principle

3.1 Heisenberg Uncertainty Principle

Perhaps no other topic in physics is surrounded with more awe and mystique as Heisenberg's succinct and beautiful uncertainty principle. Besides its beauty, the uncertainty principle is central to the most successful physical theory ever discovered by mankind: quantum mechanics. Richard Feynman, by far, physics' most successful communicator, said: "The uncertainty principle protects quantum mechanics. Heisenberg recognized that if it were possible to measure the momentum and the position simultaneously with a greater accuracy, quantum mechanics would collapse. So he proposed that it must be impossible… Quantum mechanics maintains its perilous but still correct existence" (Feynman et al., 1965).

Heisenberg (1927) introduced his famed *uncertainty principle* as

$$p_1 q_1 \sim h \tag{3.1}$$

where
p_1 refers to the momentum uncertainty
q_1 refers to the position uncertainty
$h = 6.62606957 \times 10^{-34}$ J s is Planck's constant, one of the most fundamental constants in the whole of nature

Dirac in his book (Dirac, 1978) expresses the Heisenberg uncertainty principle as

$$\Delta q \Delta p \approx h \tag{3.2}$$

Similarly, Feynman et al. (1965) describes the uncertainty relation as

$$\Delta y \Delta p_y \approx h \tag{3.3}$$

In general, for the three space coordinates, we have

$$\Delta x \Delta p_x \approx h, \quad \Delta y \Delta p_y \approx h, \quad \Delta z \Delta p_z \approx h \tag{3.4}$$

Considering the uncertainty principle relevant to the x coordinate

$$\Delta x \Delta p_x \approx h \tag{3.5}$$

it should be mentioned that the uncertainties in x and p_x are intimately related. In other words, a series of measurements on these quantities yield $(x \pm \Delta x)$ and $(p_x \pm \Delta p_x)$ with Δx and Δp_x directly related via $\Delta x \Delta p_x \approx h$. The larger the value of Δx, the smaller the value of Δp_x. In other words, the more accurately we can measure the position of a particle, the less accurately we can determine its momentum and *vice versa*. In quantum physics, x and p_x are said to be non-commuting observables (Dirac, 1978). Uncertainty and errors are essential to physical measurements and have been part of physics since the dawn of physics. Already Newton in his *Principia* (Newton, 1687) dealt with measurement errors and uncertainties. In this regard, a measurement of x with $\Delta x = 0$ is not possible in physics. Similarly, a measurement of p_x with $\Delta p_x = 0$ is physically impossible. There is always an uncertainty, no matter how small, no matter how infinitesimal. A similar observation was made, in the quantum context, by Dirac in 1930 (Dirac, 1978) and, as we shall see later, this observation is crucial to interpretational issues of quantum mechanics.

In his lectures, Feynman also relates the uncertainty principle to the double-slit experiment and hence to interferometry. He did so conceptually, in reference to the impossibility of determining the path of the electron without disturbing the interference pattern (Feynman et al., 1965). In this chapter we offer further elucidation on the link between the uncertainty principle and N-slit interferometry. In fact, we show that it is possible to obtain the old uncertainty principle from interferometric principles, thus indicating that interferometry is of fundamental importance to quantum mechanics, even more fundamental than the Heisenberg uncertainty principle.

The approach to the Heisenberg uncertainty principle, disclosed here, is from a physics perspective while avoiding the use of unnecessary mathematics, or preestablished analytical techniques, which might obscure the essence of the physics.

3.2 Wave–Particle Duality

The quantum energy of a wave, of frequency ν, is given by Planck's quantum energy equation:

$$E = h\nu \tag{3.6}$$

Equating this to the relativistic energy of a particle $E = mc^2$ and using the identity $\lambda = c/v$, an expression for the momentum is given as

$$p = \frac{h}{\lambda} \tag{3.7}$$

which, using the identity

$$k = \frac{2\pi}{\lambda} \tag{3.8}$$

can also be expressed as

$$p = \hbar k \tag{3.9}$$

This momentum equation was applied to particles, such as electrons, by de Broglie (Haken, 1981). Thus, wave properties such as frequency and wavelength were attributed to the motion of particles. Notice that the momentum expression can be rewritten as

$$p\lambda = h \tag{3.10}$$

which already embodies the dimensionality described in the Heisenberg uncertainty principle. As it will be seen in the exposition given next, this expression is crucial in the approximate derivation of the uncertainty principle.

3.3 Feynman Approximation

In his book on *Quantum Mechanics and Path Integrals*, Feynman makes use of the two-slit experiment to provide an approximate description of the physics behind the uncertainty principle (Feynman and Hibbs, 1965). Here, Feynman's description is outlined using a slightly different notation. Feynman observes that in the two-slit experiment (see Figure 3.1), the separation of the center of the slits, a, divided by the intra-interferometric distance (distance from the slits to the interferometric plane) D, that is, a/D, is approximately equal to the ratio of the wavelength λ to the distance from the central maxima to the first secondary maxima at the interferometric plane Δx. In other words,

$$\frac{a}{D} \approx \frac{\lambda}{\Delta x} \tag{3.11}$$

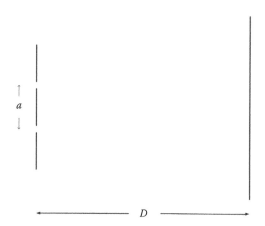

FIGURE 3.1

Two-slit interferometer used in the Feynman approximation. The separation between the center of the slits is a, while the intra-interferometric distance from the slits to the interferometric plane is D.

It should be noted that this geometrical identity can be either obtained directly from experimental results or using the interferometric equation (Duarte, 1993) described in the next section. Then, using the same two-slit experiment for electrons, Feynman postulates that passage through the slits induces a change in momentum and that the ratio $(\Delta p/p)$ is approximately equal to a/D, so that

$$\frac{\Delta p}{p} \approx \frac{a}{D} \tag{3.12}$$

then, using $p\lambda = h$, we get

$$\Delta x \Delta p \approx h \tag{3.13}$$

3.3.1 Example

In Figure 3.1 the basic geometry of a two-slit interferometer is illustrated. The separation of the center of the slits is a, and the intra-interferometric distance from the slits to the interferometric plane is D. For two 570 μm slits, separated by 570 μm, $a \approx 1140$ μm. Thus, for $D \approx 7.235$ m, the ratio $(a/D) \approx 1.58 \times 10^{-4}$. The corresponding two-slit interferogram for He–Ne laser illumination at $\lambda \approx 632.8$ nm is shown in Figure 3.2. From this interferogram, given that each pixel is ~20 μm wide, we get $(\lambda/\Delta x) \approx 1.62 \times 10^{-4}$, thus corroborating the Feynman approximation $(a/D) \approx (\lambda/\Delta x)$. The slight difference between the two ratios is well within the experimental uncertainties involved, which are not included here in order to keep the exposition simple.

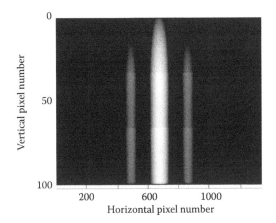

FIGURE 3.2
Measured double-slit interferogram generated by two slits 570 μm wide separated by 570 μm. The intra-interferometric distance (from the slits to the interferometric plane) is $D \approx 7.235$ m and the laser wavelength from the He–Ne laser is $\lambda \approx 632.8$ nm.

3.4 Interferometric Approximation

As mentioned at the instruction, Feynman relates the uncertainty principle to the double-slit electron experiment and hence to interferometry. Here, the link between the uncertainty principle and generalized N-slit interferometry is described in detail. The following exposition is based on the approach given in Duarte (2003).

The generalized 1D interferometric equation derived using the Dirac notation is given by (Duarte, 1991, 1993)

$$|\langle x|s\rangle|^2 = \sum_{j=1}^{N} \Psi(r_j)^2 + 2\sum_{j=1}^{N} \Psi(r_j)\left(\sum_{m=j+1}^{N} \Psi(r_m)\cos(\Omega_m - \Omega_j)\right) \quad (3.14)$$

The interference term in this equation, $\cos(\Omega_m - \Omega_j)$, can be expressed as (Duarte, 1997)

$$\cos\left((\theta_m - \theta_j) \pm (\phi_m - \phi_j)\right) = \cos\left(|l_m - l_{m-1}|k_1 \pm |L_m - L_{m-1}|k_2\right) \quad (3.15)$$

and from this equation the diffraction grating equation can be obtained (see Chapter 5):

$$d_m(\sin\Theta_m \pm \sin\Phi_m) = m\lambda \quad (3.16)$$

where $m = 0, \pm1, \pm2, \pm3\ldots$. For a grating utilized in the reflection domain, in Littrow configuration, $\Theta_m = \Phi_m = \Theta$ so that the grating equation reduces to

$$2d \sin\Theta = m\lambda \tag{3.17}$$

Using Equation 3.17 and considering an expanded light beam incident on a reflection grating, as illustrated in Figure 3.3, and allowing for an infinitesimal change in wavelength,

$$\lambda_1 = \frac{2d}{m}\left(\frac{\Delta x_1}{l}\right) \tag{3.18}$$

$$\lambda_2 = \frac{2d}{m}\left(\frac{\Delta x_2}{l}\right) \tag{3.19}$$

where
  l is the grating length
  Δx is the path difference

Subtracting Equation 3.18 from 3.19, it follows that

$$\Delta\lambda = \frac{l\lambda}{\Delta x}\left(\frac{\Delta x_1 - \Delta x_2}{l}\right) \tag{3.20}$$

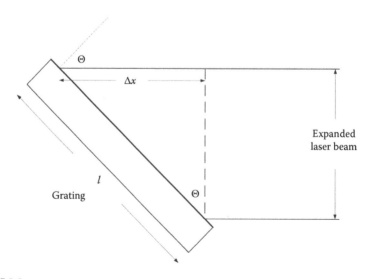

FIGURE 3.3
Path differences in a diffraction grating of the reflective class in Littrow configuration. From the geometry, $\sin\Theta = (\Delta x/l)$.

In order to distinguish between a maximum and a minimum, the *difference* in path differences should be equal to a single wavelength, so that

$$(\Delta x_1 - \Delta x_2) \approx \lambda \tag{3.21}$$

Hence, Equation 3.20 reduces to the well-known and important diffraction identity

$$\Delta \lambda \approx \frac{\lambda^2}{\Delta x} \tag{3.22}$$

which leads to

$$\Delta v \approx \frac{c}{\Delta x} \tag{3.23}$$

Considering $p = \hbar k$ for two slightly different wavelengths, one can write

$$p_1 - p_2 = h \frac{(\lambda_1 - \lambda_2)}{\lambda_1 \lambda_2} \tag{3.24}$$

and since the difference between λ_1 and λ_2 is infinitesimal, then we have

$$\Delta p \approx h \frac{\Delta \lambda}{\lambda^2} \tag{3.25}$$

Substitution of Equation 3.22 into 3.25 leads directly to

$$\Delta p \Delta x \approx h \tag{3.26}$$

which is known as the *Heisenberg uncertainty principle* (Dirac, 1978). This approach describes the interferometric foundation of the uncertainty principle and hints toward interference as a more fundamental principle.

3.5 Minimum Uncertainty Principle

The paths described here have used approximate optical and interferometric methods to arrive to the uncertainty principle

$$\Delta p \Delta x \approx h$$

where Δx and Δp are outlined as the *uncertainties* in displacement x and momentum p, respectively. From a physics perspective, this is quite alright since an "exact" derivation of the uncertainty principle appears to be a contradiction.

An alternative, more restrictive, version of the uncertainty principle is given by Feynman as

$$\Delta p \Delta x \approx \frac{\hbar}{2} \tag{3.27}$$

Feynman arrives at this expression using a probability density approach and states: "for any other form of a distribution in x or p, the product $\Delta p \Delta x$ cannot be smaller than the one we have found here" (Feynman et al., 1965). Thus, we call this the minimum uncertainty principle.

The literature offers several approaches to this definition. Here, we briefly describe the approach of Landau and Lifshitz (1976). These authors begin by defining the uncertainties via the standard of deviation:

$$(\delta x)^2 = (x - \bar{x})^2 \tag{3.28}$$

$$(\delta p_x)^2 = (p_x - \bar{p}_x)^2 \tag{3.29}$$

and then, they consider the inequality

$$\int_{-\infty}^{\infty} |\alpha x \psi + (d\psi/dx)|^2 dx \geq 0 \tag{3.30}$$

where
  α is an arbitrary real constant
  ψ is an ordinary wave function

Evaluation of this integral leads to

$$\delta x \delta p_x \geq \frac{\hbar}{2} \tag{3.31}$$

which is known as the least possible value of the $\delta x \delta p_x$ product or as the *minimum uncertainty product*. Thus, the minimum product is 4π smaller than the approximate expression derived from the physics. Landau and Lifshitz (1976) write that "the least possible value" of the uncertainty product is $(\hbar/2)$.

3.6 Generalized Uncertainty Principle

Here, we mainly refer to a generalization of the Heisenberg uncertainty principle provided by Robertson in 1929. In that short paper Roberson states: "This principle, as formulated by Heisenberg for two conjugate quantum mechanical variables states that the accuracy with which two such variables can be measured simultaneously is subject to the restriction that the product of the uncertainties in the two measurements is at least of order h" (Robertson, 1929). Robertson then explains the desirability to extend the principle to two variables, which are not conjugate.

To this effect, Robertson defines a mean value (A_0) of a Hermitean operator A (see Chapter 14), in a system described by the wave function ψ, as

$$A_0 = \int \psi^* A \psi d\tau \tag{3.32}$$

The uncertainty in A, that is, ΔA, is defined "in accordance with statistical usage" as (Robertson, 1929)

$$(\Delta A)^2 = \int \psi^*(A - A_0)^2 \psi d\tau \tag{3.33}$$

likewise, we can write

$$(\Delta B)^2 = \int \psi^*(B - B_0)^2 \psi d\tau \tag{3.34}$$

Then, Robertson introduces the Schwarzian inequality

$$\left[\int \left(f_1 f_1^* + f_2 f_2^* \right) d\tau \right] \left[\int \left(g_1 g_1^* + g_2 g_2^* \right) d\tau \right] \ge \left| \int (f_1 g_1 + f_2 g_2) d\tau \right|^2 \tag{3.35}$$

and defines

$$f_1^* = (A - A_0)\psi = f_2 \tag{3.36}$$

$$g_1 = (B - B_0)\psi = -g_2^* \tag{3.37}$$

Using these definitions it follows that

$$f_1 = (A - A_0)\psi^* = f_2^* \tag{3.38}$$

$$g_1^* = (B - B_0)\psi^* = -g_2 \tag{3.39}$$

substituting into the left-hand side of Equation 3.35, we get

$$\left[2\int\psi^*(A-A_0)^2\psi d\tau\right]\left[2\int\psi^*(B-B_0)^2\psi d\tau\right]\geq\left|\int(f_1g_1+f_2g_2)d\tau\right|^2$$

Next, reducing the $(f_1g_1+f_2g_2)$ term, within the integral on the right-hand side of Equation 3.35, and using Equations 3.33 and 3.34 (on the left-hand side), leads to Robertson's result:

$$4(\Delta A)^2(\Delta B)^2\geq\left|\int\psi^*(AB-BA)\psi d\tau\right|^2$$

$$\Delta A\Delta B\geq\frac{1}{2}\left|\int\psi^*(AB-BA)\psi d\tau\right|$$

(3.40)

This result is also reproduced in the recent literature as (Erhart et al., 2012)

$$\sigma(A)\sigma(B)\geq\frac{1}{2}|\langle\psi|[A,B]|\psi\rangle|$$

(3.41)

where

$$[A,B]=AB-BA$$

(3.42)

Revisiting Robertson's result here teaches us that this generalization flows mainly from a mathematical technique, that is, the application of the Schwarzian inequality.

Recent work has led to further generalized formulations of the Heisenberg uncertainty principle. This work has been carried out motivated by concerns that the original Heisenberg version of the uncertainty principle only applies to a limited array of measurement apparatuses (Ozawa, 2004). Very briefly, and without further discussion, a generalized version of the uncertainty principle has been put forward for a measurable A and an observable B (Erhart et al., 2012):

$$\left(e(A)\eta(B)+e(A)\sigma(B)+\sigma(A)\eta(B)\right)\geq\frac{1}{2}|\langle\psi|[A,B]|\psi\rangle|$$

(3.43)

where $e(A)$ is the root mean square deviation of an output operator O_A, while the disturbance $\eta(B)$ is defined as the root mean square of the change in the observable B during the measurement. The second and third terms, of the left-hand side, flow from the non-commutability of A and B (Ozawa, 2004).

Notice that the right-hand side of the inequality is the same as Robertson's, and it is on the right-hand side that the physics resides. Here we should mention that in Chapter 17, on the subject of entanglement, an analogous expression to $[A,B] = AB-BA$ is developed in the form of $(|x\rangle|y\rangle - |y\rangle|x\rangle)$.

Interest in generalized forms of the uncertainty principle has gained recent attention due to activity in the area of successive quantum measurements.

3.7 Additional Versions of the Heisenberg Uncertainty Principle

The Heisenberg uncertainty principle, $\Delta x \Delta p \approx h$, can be expressed in several useful versions. Assuming an independent derivation of $\Delta p \Delta x \approx h$, and using $p = \hbar k$, it can be expressed in its wavelength-spatial form:

$$\Delta \lambda \approx \frac{\lambda^2}{\Delta x} \tag{3.44}$$

This is a widely used identity in interferometry utilized to express linewidth in wavelength units (m). We also know that the interferometric identity given in Equation 3.44 can be expressed in its frequency-spatial version:

$$\Delta v \approx \frac{c}{\Delta x} \tag{3.45}$$

which is also widely used in interferometry to express linewidth in frequency units (Hz).

Using $E = mc^2$, the uncertainty principle can also be expressed in its energy-time form:

$$\Delta E \, \Delta t \approx h \tag{3.46}$$

which, using the quantum energy $E = hv$, can be transformed to its frequency-time version

$$\Delta v \, \Delta t \approx 1 \tag{3.47}$$

This succinct and beautiful expression is a crucial result for the field of pulsed lasers and in particular for femtosecond (fs) and attosecond (at) lasers. In fact, $\Delta v \Delta t \approx 1$ means that, for a laser working optimally at the limit established by the uncertainty principle, the time duration of the pulses Δt can be determined from its spectral profile.

From Equation 3.47 we can write directly an expression for the time segment:

$$\Delta t \approx \frac{1}{\Delta \nu} \tag{3.48}$$

which is also known as the *coherence time*. From this time the *coherence length* can be defined as

$$\Delta x \approx \frac{c}{\Delta \nu} \tag{3.49}$$

which is an alternative form of Equation 3.45.

One final observation: The highly practical identities $\Delta \lambda \approx \lambda^2/\Delta x$ and $\Delta \nu \approx c/\Delta x$ are routinely applied in the field of interferometry to express measured linewidths in either frequency or wavelength units (Duarte, 2003). Note that all these expressions are simply based on $\Delta x \Delta p \approx h$ and not in its more restrictive minimum product version.

3.7.1 Example

An optimized multiple-prism grating solid-state organic dye laser (Duarte, 1999) yields kW pulses tunable in the $565 \leq \lambda \leq 603$ nm range. Its smooth temporal pulse is indicative of single-longitudinal-mode oscillation and is shown in Figure 3.4. The duration of this pulse at FWHM is $\Delta t \approx 3$ ns. The corresponding Fabry–Perot interferogram from this single-longitudinal-mode emission is shown in Figure 3.5. The half-width of the rings is measured to be $\Delta \nu \approx 350$ MHz. Thus, it can be directly established that the product

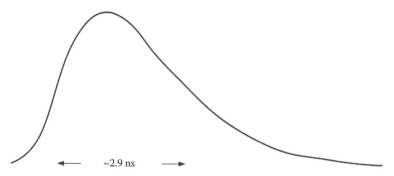

~ 2.9 ns

FIGURE 3.4
Temporal profile of a laser pulse from an optimized multiple-prism grating solid-state dye laser. The temporal scale is 1 ns/div. (Reproduced from Duarte, F.J., *Appl. Opt.* 38, 6347, 1999, with permission from the Optical Society of America.)

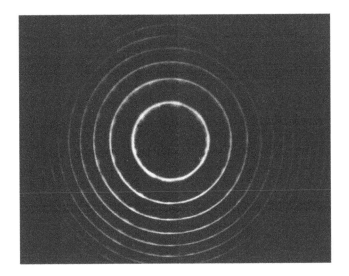

FIGURE 3.5
Corresponding Fabry–Perot interferogram of the single-longitudinal-mode oscillation from the optimized multiple-prism grating solid-state dye laser. (Reproduced from Duarte, F.J., *Appl. Opt.* 38, 6347, 1999, with permission from the Optical Society of America.)

$\Delta\nu\Delta t \approx 1.06$ for this narrow-linewidth pulsed laser emission, which is near the limit established frequency-time version of the Heisenberg uncertainty principle $\Delta\nu\Delta t \approx 1$.

3.8 Applications of the Uncertainty Principle in Optics

The uncertainty principle is widely applied in optics. It applies to interferometry, linewidth measurements, and beam divergence measurements. Here we focus on the uncertainty principle and beam divergence. Applications of these concepts to astronomy are also mentioned.

3.8.1 Beam Divergence

The Heisenberg uncertainty principle can be used to derive some useful identities in optics and interferometry. Starting from

$$\Delta p \, \Delta x \approx h$$

and substituting for Δp using $p = \hbar k$, yields

$$\Delta k \, \Delta x \approx 2\pi \tag{3.50}$$

which leads directly to

$$\Delta\lambda \approx \frac{\lambda^2}{\Delta x}$$

For a diffraction-limited beam traveling in the z direction, $k_x = k \sin \theta$. Thus, for a very small angle θ,

$$k_x \approx k\theta \tag{3.51}$$

so that

$$\Delta k_x \approx k \, \Delta\theta \tag{3.52}$$

Using $\Delta k_x \, \Delta x \approx 2\pi$ and Equation 3.52, it is readily seen that the beam has an angular divergence given by

$$\Delta\theta \approx \frac{\lambda}{\Delta x} \tag{3.53}$$

which is a succinct equation for angular beam divergence and in essence an additional manifestation of the uncertainty principle. Equation 3.53 indicates that the angular spread of a propagating beam of wavelength λ is inversely proportional to its original width. The narrower the beam, the larger its divergence. This equation also states that light of shorter wavelength experiences less beam divergence, which is a well-known experimental fact in laser physics. This implies that the beam divergence can be controlled using geometrical (Δx) as well as physical means (λ).

Equation 3.53 has the same form of the return-pass beam divergence equation derived from classical principles (Duarte, 1990), namely,

$$\Delta\theta = \frac{\lambda}{\pi w}\left[1+\left(\frac{L_R}{B}\right)^2+\left(\frac{A\,L_R}{B}\right)^2\right]^{1/2} \tag{3.54}$$

where
  w is the beam waist
  $L_R = (\pi w^2/\lambda)$ is known as the Rayleigh length
  A and B are geometrical-spatial matrix propagation parameters defined in Duarte (2003) and explained in Appendix C. For an optimized design,

$$\left(1+\left(\frac{L_R}{B}\right)^2+\left(\frac{AL_R}{B}\right)^2\right)^{1/2} \approx 1$$

and Equation 3.54 reduces to

$$\Delta\theta \approx \frac{\lambda}{\pi w} \qquad (3.55)$$

This minimized beam divergence is known as the *diffraction limit*. The equivalence of Equations 3.53 and 3.55 is self-evident.

In summary, the generalized interference equation, that is, Equation 3.14, leads to the interferometric identity $\Delta\lambda \approx \lambda^2/\Delta x$, which, in conjunction with the uncertainty principle $\Delta p\Delta x \approx h$, leads to the expression of the diffraction-limited beam divergence $\Delta\theta \approx \lambda/\Delta x$.

3.8.2 Beam Divergence and Astronomy

An important application of the uncertainty principle manifests itself in calculations of the angular resolution limit of telescopes used in astronomical observations. Reflection telescopes such as the Newtonian and Cassegrain telescopes are depicted in Figure 3.6 and discussed further in Duarte (2003). The angular resolution that can be achieved with these telescopes, under ideal conditions, is approximately quantified by the diffraction limit of the beam divergence given in Equation 3.53. That is,

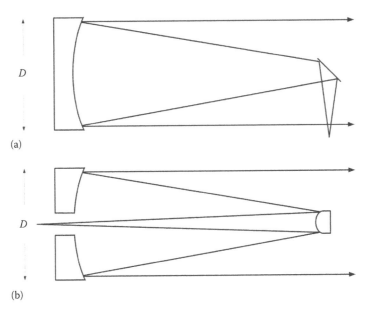

(a)

(b)

FIGURE 3.6
Reflection telescopes used in astronomical observations. (a) Newtonian telescope. (b) Cassegranian telescope. The diameter D of the main mirror determines the angular resolution of the telescope.

TABLE 3.1

Angular Resolutions for Newtonian
and Cassegrain Terrestrial Telescopes

Diameter (m)	Area (m²)	$\Delta\theta$ (rad)
10	25π	$(1/\pi) \times 10^{-7}$
50	625π	$(2/\pi) \times 10^{-8}$
100	2500π	$(1/\pi) \times 10^{-8}$
$1000/\pi$	$(25/\pi) \times 10^4$	10^{-9}

the smallest angular discrimination, or resolution limit, of a telescope with a diameter $D = 2w$ is given by

$$\Delta\theta \approx \frac{2\lambda}{\pi D} \tag{3.56}$$

This equation teaches us that the two alternatives to improve the angular resolution of a telescope are either to observe at shorter wavelengths (λ) or to increase the diameter (D) of the telescope. This equation explains the quest toward the building of large, very large, and extremely large terrestrial telescopes. Previously, we estimated the angular resolution for telescope diameters of 10 and 100 m (Duarte 2003). For $\lambda = 500$ nm the angular resolution for large and very large telescopes, at various diameters, are listed in Table 3.1. In addition to better angular resolutions, large aperture telescopes provide increased signal since the area of collection increases substantially. The construction of very large telescopes, with diameters greater than 50 m, should be feasible in the future via the use of very light segmented mirrors.

The subject of laser beam divergence for *laser guide star* in astronomy, using narrow-linewidth oscillators emitting at $\lambda \approx 589$ nm, is discussed by Duarte (2003).

3.8.3 Uncertainty Principle and the Cavity Linewidth Equation

In this section, the nexus between the uncertainty principle and the single-pass cavity linewidth equation is outlined: the generalized 1D interferometric equation (Duarte, 1991, 1992)

$$|\langle x | s \rangle|^2 = \sum_{j=1}^{N} \Psi(r_j)^2 + 2\sum_{j=1}^{N} \Psi(r_j) \left(\sum_{m=j+1}^{N} \Psi(r_m)\cos(\Omega_m - \Omega_j) \right)$$

is used in Chapter 9 to derive the cavity linewidth equation (Duarte, 1992):

$$\Delta\lambda \approx \Delta\theta \left(\frac{\partial\theta}{\partial\lambda} \right)^{-1} \tag{3.57}$$

which is also expressed as

$$\Delta\lambda \approx \Delta\theta(\nabla_\lambda\theta)^{-1} \tag{3.58}$$

where $\nabla_\lambda\theta = (\partial\theta/\partial\lambda)$. This equation is used extensively to determine the emission linewidth in high-gain pulsed narrow-linewidth dispersive laser oscillators (Duarte, 1990). As indicated, this equation originates from the generalized N-slit interference equation and incorporates the beam divergence expression $\Delta\theta$ whose diffraction-limited value

$$\Delta\theta \approx \frac{\lambda}{\Delta x}$$

can be derived from the uncertainty principle $\Delta p\Delta x \approx h$, as previously illustrated.

In addition to the explicit equations for beam divergence given here, it is also important to indicate that the beam profile can be generated directly from the generalized N-slit interferometric equation (Equation 3.14) and the beam divergence calculated from the history of the beam profiles. In other words, the interferometric equation does contain the correct information on beam divergence, which is not surprising since it can also be used to derive the uncertainty principle as we have just seen.

Equation 3.54 also has a geometrical origin (Robertson, 1955), thus illustrating, once again, the synergy between classical and quantum physics.

3.8.4 Tuning Laser Microcavities

A fine-tuning technique applicable to microelectromechanical system (MEMS)-driven miniature laser cavities consists simply in changing the cavity length as illustrated in Figure 3.7. This approach exploits the very fact that the free spectral range (FSR) of the cavity is a function of Δx. Here, we

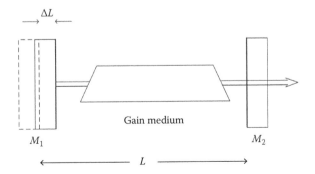

FIGURE 3.7
Wavelength tuning by changing the length of the cavity L. This is accomplished via the displacement of one of the mirror of the resonator.

examine this approach to wavelength tuning following the approach of Duarte (2003, 2009). Going back to the interferometric identity

$$\delta\lambda \approx \frac{\lambda^2}{\Delta x} \tag{3.59}$$

one can write for an initial wavelength λ_1

$$\delta\lambda_1 \approx \frac{\lambda_1^2}{2L} \tag{3.60}$$

and for a subsequent wavelength λ_2

$$\delta\lambda_2 \approx \frac{\lambda_2^2}{2(L \pm \Delta L)} \tag{3.61}$$

Also, it is convenient to define the number of longitudinal modes in each case as

$$N_1 = \frac{\Delta\lambda_1}{\delta\lambda_1} \tag{3.62}$$

and

$$N_2 = \frac{\Delta\lambda_2}{\delta\lambda_2} \tag{3.63}$$

where $\Delta\lambda_1$ and $\Delta\lambda_1$ are the corresponding laser linewidths. If the laser linewidth, during this ΔL change, is maintained so that $\Delta\lambda_1 \approx \Delta\lambda_2$, then taking the ratio of Equations 3.60 and 3.61 leads to (Duarte, 2003)

$$\lambda_2 \approx \lambda_1 \left(\frac{N_1}{N_2}\right)^{1/2} \left(1 \pm \frac{\Delta L}{L}\right)^{1/2} \tag{3.64}$$

For $N_1 \approx N_2$, or single-longitudinal-mode oscillation, this equation reduces to (Duarte 2003)

$$\lambda_2 \approx \lambda_1 \left(1 \pm \frac{\Delta L}{L}\right)^{1/2} \tag{3.65}$$

Uenishi et al. (1996) report on experiments using the $\Delta L/L$ method to perform wavelength tuning in a MEMS-driven semiconductor laser cavity.

In that experiment they observed wavelength tuning, in the absence of mode-hopping, as long as the change in wavelength did not exceed $\lambda_2 - \lambda_1 \approx$ 1 nm. Using their graphical data for the scan initiated at $\lambda_1 \approx 1547$ nm, it is established that $\Delta L \approx 0.4$ µm, and using $L \approx 305$ µm, Equation 3.65 yields $\lambda_1 \approx 1548$ nm, which approximately agrees with the authors' observations (Uenishi et al., 1996) In this regard, it should be mentioned that Equation 3.65 was implicitly derived with the assumption of a wavelength scan obeying the condition $\delta\lambda_1 \approx \delta\lambda_2$. Albeit here, we use the term microcavity this approach should also apply to cavities in the submicrometer regime or nanocavities.

3.8.5 Sub-Microcavities

The longitudinal-mode spacing in a cavity of length $L = \Delta x/2$ is known as the *FSR* of the cavity and can be designated as

$$\delta\lambda \approx \frac{\lambda^2}{2L} \tag{3.66}$$

or

$$\delta v \approx \frac{c}{2L} \tag{3.67}$$

For very short cavities, with cavity lengths in the submicrometer range, or the nanometer range, the longitudinal-mode spacing becomes rather large. For example, for $L \approx 300$ nm at $\lambda \approx 540$ nm, the longitudinal-mode spacing becomes $\delta\lambda \approx 486$ nm, which is an enormous separation. This means that a measured linewidth in the tens of nm easily meets the criteria for single-longitudinal-mode emission. Thus, the challenge with nanocavities lies in restricting the emission to a single transverse mode since, as established by Equation 3.14, a very short cavity length requires an infinitesimal aperture size.

Sub-microcavities are of interest not just because of their size but also because they exhibit some interesting characteristics such as lasing with extremely low thresholds (de Martini and Jakobovitz, 1980). Spatial and spectral coherent emission from an electrically driven laser-dye-doped organic semiconductor nanocavity ($L \approx 300$ nm), in the pulsed regime, has been reported by Duarte et al. (2005, 2008). A directional beam with a near-Gaussian spatial distribution was obtained by using an extracavity double-interferometric filter configuration. The single transverse mode thus selected was determined to have a linewidth of $\Delta\lambda \approx 10$ nm.

Problems

3.1 Show that $\Delta x \Delta p \approx h$ can be expressed as $\Delta \lambda \approx \lambda^2 / \Delta x$.

3.2 Show that $\Delta x \Delta p \approx h$ can be expressed as $\Delta \nu \approx c / \Delta x$.

3.3 Show that $\Delta x \Delta p \approx h$ can be expressed as $\Delta \nu \Delta t \approx 1$.

3.4 Show that $(f_1 g_1 + f_2 g_2) = \psi^*(AB - BA)\psi$.

3.5 Calculate the diffraction-limited beam divergence, at full-width-half-maximum (FWHM), for (a) a laser beam with a 150 μm radius at $\lambda = 590$ nm and (b) a laser beam with a 500 μm radius at $\lambda = 590$ nm.

3.6 Repeat the calculations of the previous problem for the excimer laser (XeCl) wavelength $\lambda = 308$ nm. Comment.

3.7 Calculate the dispersive cavity linewidth for a high-power tunable laser yielding a diffraction-limited beam divergence, 150 μm in radius, at $\lambda = 590$ nm. Assume that an appropriate beam expander illuminates a 3300 lines mm^{-1} grating deployed in the first order. The grating has a 50 mm length perpendicular to the grooves. Assume a fully illuminated grating deployed in Littrow configuration in its first order.

3.8 (a) For a pulsed laser delivering a 350 MHz laser linewidth, at the limit established by the Heisenberg uncertainty principle, estimate its shortest possible pulse width. (b) For a laser emitting 1 fs pulses, estimate its broadest possible spectral width in nanometers centered around $\lambda = 500$ nm.

3.9 For a cavity with a length $L = 100$ μm, calculate the change in wavelength for $\Delta L = 1.0$ μm, given that the initial wavelength is $\lambda = 1500$ nm.

References

De Martini, F. and Jakobovitz, J. R. (1980). Anomalous spontaneous-emission-decay phase transition and zero-threshold laser action in a microscopic cavity. *Phys. Rev. Lett.* **60**, 1711–1714.

Dirac, P. A. M. (1978). *The Principles of Quantum Mechanics*, 4th edn. Oxford, London, U.K.

Duarte, F. J. (1990). Narrow-linewidth pulsed dye laser oscillators. In *Dye Laser Principles* (Duarte, F. J. and Hillman, L. W., eds.) Academic, New York, Chapter 4.

Duarte, F. J. (1991). Dispersive dye lasers. In *High Power Dye Lasers* (Duarte, F. J. ed.) Springer-Verlag, Berlin, Germany. Chapter 2.

Duarte, F. J. (1992). Cavity dispersion equation $\Delta \lambda \approx \Delta \theta \ (\partial \theta / \partial \lambda)^{-1}$: A note on its origin. *Appl. Opt.* **31**, 6979–6982.

Duarte, F. J. (1993). On a generalized interference equation and interferometric measurements. *Opt. Commun.* **103**, 8–14.

Duarte, F. J. (1997). Interference, diffraction, and refraction, via Dirac's notation. *Am. J. Phys.* **65**, 637–640.

Duarte, F. J. (1999). Multiple-prism grating solid-state dye laser oscillator: Optimized architecture. *Appl. Opt.* **38**, 6347–6349.

Duarte, F. J. (2003). *Tunable Laser Optics*, Elsevier-Academic, New York.

Duarte, F. J. (2008). Coherent electrically excited semiconductors: Coherent or laser emission? *Appl. Phys. B.* **90**, 101–108.

Duarte, F. J. (2009). Broadly tunable dispersive external-cavity semiconductor lasers. In *Tunable Laser Applications* (Duarte, F. J. ed.). CRC, New York, Chapter 5.

Duarte, F. J., Liao, L. S., and Vaeth, K. M. (2005). Coherent characteristics of electrically excited tandem organic light emitting diodes. *Opt. Lett.* **30**, 3072–3074.

Erhart, J., Sponar, S., Suluok, G., Badurek, G., Ozawa, M., and Hasegawa, Y. (2012). Experimental demonstration of a universally valid error–disturbance uncertainty relation in spin measurements. *Nat. Phys.* **8**, 185–189.

Feynman, R. P. and Hibbs, A. R. (1965). *Quantum Mechanics and Path Integrals*, McGraw-Hill, New York.

Feynman, R. P., Leighton, R. B., and Sands, M. (1965). *The Feynman Lectures on Physics*, Vol. III, Addison-Wesley, Reading, MA.

Haken, H. (1981). *Light*, North Holland, Amsterdam, the Netherlands.

Heisenberg, W. (1927). Über den anschaulichen inhalt der quantentheoretischen kinematik und mechanic. *Z. Phys.* **43**, 172–198.

Landau, L. D. and Lifshitz, E. M. (1976). *Quantum Mechanics*, Pergamon, New York.

Newton, I. (1687). *Principia Mathematica*, Royal Society, London, U.K.

Ozawa, M. (2004). Uncertainty relations for noise and disturbance in generalized quantum measurements. *Ann. Phys.* **311**, 350–416.

Robertson, H. P. (1929). The uncertainty principle. *Phys. Rev.* **34**, 163–164.

Robertson, J. K. (1955). *Introduction to Optics: Geometrical and Physical*, Van Nostrand, New York.

Uenishi, Y., Honna, K., and Nagaola, S. (1996). Tunable laser diode using a nickel micromachined external mirror. *Electron. Lett.* **32**, 1207–1208.

4

Dirac Quantum Optics

4.1 Dirac Notation in Optics

The Dirac notation is one of the mathematical avenues that can be used to describe nature quantum mechanically. This mathematical notation was invented by Dirac in 1939 and is particularly well suited to describe quantum optics. In this chapter, we introduce the reader to the basics of the Dirac notation and apply the notation to the generalized description of the fundamental phenomenon of interference that, as it will be seen, is crucial to quantum physics itself. This description is based on topics and elements of a review given by Duarte (2003).

In *Principles of Quantum Mechanics*, first published in 1930, Dirac discusses the essence of interference as a one-photon phenomenon. Albeit his discussion is qualitative, it is also profound. In 1965, Feynman discussed electron interference via a two-slit thought experiment using probability amplitudes and the Dirac notation as a tool (Feynman et al., 1965). In 1989, inspired on Feynman's discussion, the Dirac notation was applied to the propagation of coherent light in an N-slit interferometer (Duarte and Paine, 1989; Duarte, 1991, 1993).

The ideas of the notation invented by Dirac (1939) can be explained by considering the propagation of a photon from plane s to plane x, as illustrated in Figure 4.1. According to the Dirac concept, there is a probability amplitude, denoted by $\langle x|s \rangle$, that quantifies such propagation. Historically, Dirac introduced the nomenclature of *ket vectors*, denoted by $|\ \rangle$, and *bra vectors*, denoted by $\langle\ |$, which are mirror images of each other. Thus, the probability amplitude is described by the *bra–ket* $\langle x|s \rangle$, which is a *complex number*.

In the Dirac notation, the propagation from s to x is expressed in reverse order by $\langle x|s \rangle$. In other words, the *starting* position is at the *right* and the *final* position is at the *left*. If the propagation of the photon is not directly from plane s to plane x but involves the passage through an intermediate plane j, as illustrated in Figure 4.2, then the probability amplitude describing such propagation becomes

$$\langle x|s \rangle = \langle x|j \rangle\langle j|s \rangle \tag{4.1}$$

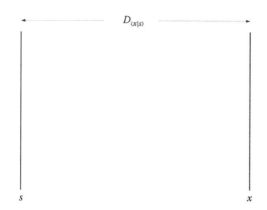

FIGURE 4.1
Propagation from s to the interferometric plane x is expressed as the probability amplitude $\langle x|s\rangle$. $D_{\langle x|s\rangle}$ is the physical distance between the two planes.

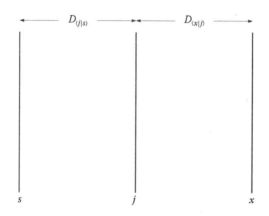

FIGURE 4.2
Propagation from s to the interferometric plane x via an intermediate plane j is expressed as the probability amplitude $\langle x|s\rangle = \langle x|j\rangle\langle j|s\rangle$. $D_{\langle j|s\rangle}$ and $D_{\langle x|j\rangle}$ are the distances between the designated planes.

If the photon from s must also propagate through planes j and k in its trajectory to x, that is, $s{\rightarrow}j{\rightarrow}k{\rightarrow}x$, as illustrated in Figure 4.3, then the probability amplitude is given by

$$\langle x|s\rangle = \langle x|k\rangle\langle k|j\rangle\langle j|s\rangle \tag{4.2}$$

If an additional intermediate plane l is added, so that the propagation, from plane to plane, proceeds as $s{\rightarrow}j{\rightarrow}k{\rightarrow}l{\rightarrow}x$, then the probability amplitude is given by

$$\langle x|s\rangle = \langle x|l\rangle\langle l|k\rangle\langle k|j\rangle\langle j|s\rangle \tag{4.3}$$

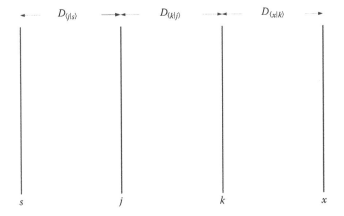

FIGURE 4.3
Propagation from s to the interferometric plane x via two intermediate planes j and k is expressed as the probability amplitude $\langle x|s \rangle = \langle x|k \rangle \langle k|j \rangle \langle j|s \rangle$. $D_{\langle j|s \rangle}$, $D_{\langle k|j \rangle}$, and $D_{\langle x|k \rangle}$, are the distances between the designated planes.

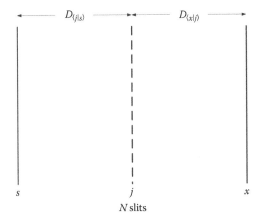

FIGURE 4.4
Propagation from s to the interferometric plane x via an array of N slits positioned at the intermediate plane j. $D_{\langle x|j \rangle}$ is the distance between the N-slit array and the x plane.

When at the intermediate plane, in Figure 4.2, a number of N alternatives are available to the passage of the photon, as depicted in Figure 4.4, then the overall probability amplitude must consider *every possible alternative*, which is expressed mathematically by a summation over j in the form of

$$\langle x|s \rangle = \sum_{j=1}^{N} \langle x|j \rangle \langle j|s \rangle \tag{4.4}$$

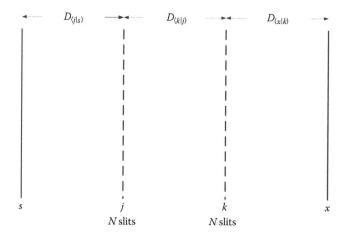

FIGURE 4.5
Propagation from s to the interferometric plane x via an array of N slits positioned at the intermediate plane j and via an additional array of N slits positioned at k. $D_{\langle k|j\rangle}$ is the distance between the N-slit arrays.

Consideration of *every possible alternative N* in the computation of probability amplitude, as described in Equation 4.4, is an essential and crucial quantum feature.

For the case of an additional intermediate plane with N alternatives, as illustrated in Figure 4.5, the probability amplitude is written as

$$\langle x\,|\,s\rangle = \sum_{k=1}^{N}\sum_{j=1}^{N}\langle x\,|\,k\rangle\langle k\,|\,j\rangle\langle j\,|\,s\rangle \tag{4.5}$$

And for the case including three intermediate N-slit arrays, the probability amplitude becomes

$$\langle x\,|\,s\rangle = \sum_{l=1}^{N}\sum_{k=1}^{N}\sum_{j=1}^{N}\langle x\,|\,l\rangle\langle l\,|\,k\rangle\langle k\,|\,j\rangle\langle j\,|\,s\rangle \tag{4.6}$$

The addition of further intermediate planes, with N alternatives, can then be systematically incorporated in the notation. The Dirac notation albeit originally applied to the propagation of single particles (Feynman et al., 1965; Dirac, 1978) also applies to describe the propagation of ensembles of coherent, or indistinguishable, photons (Duarte, 1991, 1993, 2004). This observation is compatible with the postulate that indicates that the principles of quantum mechanics are applicable to the description of macroscopic phenomena that are not perturbed by observation (van Kampen, 1988).

4.2 Dirac Quantum Principles

The principles established by Dirac in 1939, associated to his description of quantum mechanics, via the Dirac notation, can be described rather succinctly (Feynman et al., 1965). The first principle stipulates that any state, such as ψ, can be described in terms of a set of *base states*. The amplitude to transition from any state to another state can be written as a sum of products, such as

$$\langle x \,|\, s \rangle = \sum_{i=1}^{N} \langle x \,|\, i \rangle \langle i \,|\, s \rangle \tag{4.7}$$

The base states are orthogonal. This means that the amplitude to be in one if you are in the other is zero, or

$$\langle i \,|\, j \rangle = \delta_{ij} \tag{4.8}$$

Furthermore, the amplitude to get from one state to another directly is the complex conjugate of the reverse

$$\langle x \,|\, s \rangle^* = \langle s \,|\, x \rangle \tag{4.9}$$

As a matter of formality, it should be mentioned that the space of *bra–ket* vectors, when the vectors are restricted to a finite length and finite scalar products, is called a *Hilbert space* (Dirac, 1978). However, Dirac himself points out that *bra–ket* vectors form a more general space than a Hilbert space (Note: a Hilbert space is a generalized Euclidean space).

4.3 Interference and the Interferometric Equation

The Dirac notation offers a natural avenue to describe the propagation of particles from a source to a detection plane, via a pair of slits. This was done by Feynman in a thought experiment using electrons and two slits. The Feynman approach was extended to the description of indistinguishable *photon propagation* from a source *s* to an interferometric plane *x*, via a transmission grating *j* comprised by *N* slits, as illustrated in Figure 4.6, by Duarte (1989, 1991).

In the interferometric architecture of Figure 4.6, an expanded laser beam, from a single-transverse-mode narrow-linewidth laser, becomes the

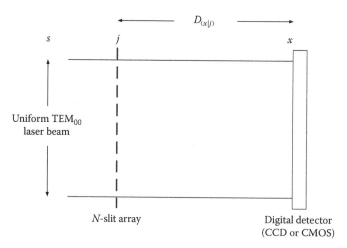

FIGURE 4.6

N-slit laser interferometer: A near-Gaussian beam from a single transverse mode (TEM$_{00}$), narrow-linewidth laser is preexpanded in 2D telescope and then expanded in one dimension (parallel to the plane of incidence) by an MPBE (Duarte, 1987). This expanded beam can be further transformed into a nearly uniform illumination source (s) (Duarte, 2003). Then, a uniform light source (s) illuminates an array of N slits at j. Interaction of the coherent emission with the slit array produces interference at the interferometric plane x (Duarte, 1993).

radiation source (s) and illuminates an array of N slits or transmission grating (j). The interaction of the coherent radiation with the N-slit array (j) produces an interference signal at x. A crucial point here is that all the indistinguishable photons illuminate the array of N slits, or grating, simultaneously. If only one photon propagates, at any given time, then that individual photon illuminates the whole array of N slits simultaneously (Duarte, 2003). The probability amplitude that describes the propagation from the source (s) to the detection plane (x), via the array of N slits (j), is given by (Duarte, 1991, 1993)

$$\langle x \,|\, s \rangle = \sum_{j=1}^{N} \langle x \,|\, j \rangle \langle j \,|\, s \rangle$$

According to Dirac (1978), the probability amplitudes can be represented by *wave functions of ordinary wave optics*. Thus, following Feynman et al. (1965),

$$\langle j \,|\, s \rangle = \Psi(r_{j,s}) e^{-i\theta_j} \tag{4.10}$$

$$\langle x \,|\, j \rangle = \Psi(r_{x,j}) e^{-i\phi_j} \tag{4.11}$$

where θ_j and ϕ_j are the phase terms associated with the incidence and diffraction waves, respectively. Using Equations 4.10 and 4.11, for the probability amplitudes, the propagation probability amplitude

$$\langle x \,|\, s \rangle = \sum_{j=1}^{N} \langle x \,|\, j \rangle \langle j \,|\, s \rangle$$

can be written as

$$\langle x \,|\, s \rangle = \sum_{j=1}^{N} \Psi(r_j) e^{-i\Omega_j} \qquad (4.12)$$

where

$$\Psi(r_j) = \Psi(r_{x,j})\Psi(r_{j,s}) \qquad (4.13)$$

and

$$\Omega_j = (\theta_j + \phi_j) \qquad (4.14)$$

Next, the propagation probability is obtained by expanding Equation 4.12 and multiplying the expansion by its complex conjugate, or

$$|\langle x \,|\, s \rangle|^2 = \langle x \,|\, s \rangle \langle x \,|\, s \rangle^* \qquad (4.15)$$

Expansion of the probability amplitude and multiplication with its complex conjugate, following some algebra, lead to

$$|\langle x \,|\, s \rangle|^2 = \sum_{j=1}^{N} \Psi(r_j) \sum_{m=1}^{N} \Psi(r_m) e^{i(\Omega_m - \Omega_j)} \qquad (4.16)$$

Expanding Equation 4.16 and using the identity

$$2\cos(\Omega_m - \Omega_j) = e^{-i(\Omega_m - \Omega_j)} + e^{i(\Omega_m - \Omega_j)} \qquad (4.17)$$

lead to the explicit form of the generalized propagation probability in one dimension (Duarte and Paine, 1989; Duarte, 1991):

$$|\langle x \,|\, s \rangle|^2 = \sum_{j=1}^{N} \Psi(r_j)^2 + 2\sum_{j=1}^{N} \Psi(r_j) \left(\sum_{m=j+1}^{N} \Psi(r_m)\cos(\Omega_m - \Omega_j) \right) \qquad (4.18)$$

This equation is the 1D generalized interferometric equation. The reader should keep in mind that it is completely equivalent to Equation 4.16.

4.3.1 Examples: Double-, Triple-, Quadruple-, and Quintuple-Slit Interference

Expanding Equation 4.16 for two slits ($N = 2$), as applicable to double-slit interference, we get

$$|\langle x|s\rangle|^2 = \Psi(r_1)\left(\Psi(r_1)e^{i(\Omega_1-\Omega_1)} + \Psi(r_2)e^{i(\Omega_2-\Omega_1)}\right)$$

$$+ \Psi(r_2)\left(\Psi(r_1)e^{i(\Omega_1-\Omega_2)} + \Psi(r_2)e^{i(\Omega_2-\Omega_2)}\right)$$

$$|\langle x|s\rangle|^2 = \Psi(r_1)\left(\Psi(r_1) + \Psi(r_2)e^{i(\Omega_2-\Omega_1)}\right) + \Psi(r_2)\left(\Psi(r_1)e^{i(\Omega_1-\Omega_2)} + \Psi(r_2)\right)$$

$$|\langle x|s\rangle|^2 = \Psi(r_1)^2 + \Psi(r_2)^2 + \left(\Psi(r_1)\Psi(r_2)e^{i(\Omega_2-\Omega_1)} + \Psi(r_1)\Psi(r_2)e^{i(\Omega_1-\Omega_2)}\right)$$

$$|\langle x|s\rangle|^2 = \Psi(r_1)^2 + \Psi(r_2)^2 + \left(\Psi(r_1)\Psi(r_2)e^{i(\Omega_2-\Omega_1)} + \Psi(r_1)\Psi(r_2)e^{-i(\Omega_2-\Omega_1)}\right)$$

$$|\langle x|s\rangle|^2 = \Psi(r_1)^2 + \Psi(r_2)^2 + 2\Psi(r_1)\Psi(r_2)\cos(\Omega_2-\Omega_1) \tag{4.19}$$

Expanding Equation 4.18 for three slits ($N = 3$), applicable to triple-slit interference, we get

$$|\langle x|s\rangle|^2 = \Psi(r_1)^2 + \Psi(r_2)^2 + \Psi(r_3)^2 + 2\left(\Psi(r_1)\Psi(r_2)\cos(\Omega_2-\Omega_1)\right.$$

$$\left. + \Psi(r_1)\Psi(r_3)\cos(\Omega_3-\Omega_1) + \Psi(r_2)\Psi(r_3)\cos(\Omega_3-\Omega_2)\right) \tag{4.20}$$

Expanding Equation 4.18 for four slits ($N = 4$), applicable to quadruple-slit interference, we get

$$|\langle x|s\rangle|^2 = \Psi(r_1)^2 + \Psi(r_2)^2 + \Psi(r_3)^2 + \Psi(r_4)^2 + 2\left(\Psi(r_1)\Psi(r_2)\cos(\Omega_2-\Omega_1)\right.$$

$$+ \Psi(r_1)\Psi(r_3)\cos(\Omega_3-\Omega_1) + \Psi(r_1)\Psi(r_4)\cos(\Omega_4-\Omega_1)$$

$$+ \Psi(r_2)\Psi(r_3)\cos(\Omega_3-\Omega_2) + \Psi(r_2)\Psi(r_4)\cos(\Omega_4-\Omega_2)$$

$$\left. + \Psi(r_3)\Psi(r_4)\cos(\Omega_4-\Omega_3)\right) \tag{4.21}$$

Expanding Equation 4.18 for five slits ($N = 5$), applicable to quintuple-slit interference, we get

$$|\langle x \,|\, s \rangle|^2 = \Psi(r_1)^2 + \Psi(r_2)^2 + \Psi(r_3)^2 + \Psi(r_4)^2 + \Psi(r_5)^2 + 2\big(\Psi(r_1)\Psi(r_2)\cos(\Omega_2 - \Omega_1)$$
$$+ \Psi(r_1)\Psi(r_3)\cos(\Omega_3 - \Omega_1) + \Psi(r_1)\Psi(r_4)\cos(\Omega_4 - \Omega_1)$$
$$+ \Psi(r_1)\Psi(r_5)\cos(\Omega_5 - \Omega_1) + \Psi(r_2)\Psi(r_3)\cos(\Omega_3 - \Omega_2)$$
$$+ \Psi(r_2)\Psi(r_4)\cos(\Omega_4 - \Omega_2) + \Psi(r_2)\Psi(r_5)\cos(\Omega_5 - \Omega_2)$$
$$+ \Psi(r_3)\Psi(r_4)\cos(\Omega_4 - \Omega_3) + \Psi(r_3)\Psi(r_5)\cos(\Omega_5 - \Omega_3)$$
$$+ \Psi(r_4)\Psi(r_5)\cos(\Omega_5 - \Omega_4)\big) \tag{4.22}$$

and so on. Besides the explicit interferometric expressions for $N = 2$, $N = 3$, $N = 4$, and $N = 5$, Equation 4.18 can be programmed to include sextuple ($N = 6$), septuple ($N = 7$), octuple ($N = 8$), nonuple ($N = 9$), or any number of slits, and in practice it has been used to do calculations, and comparisons with measurements, in the $2 \leq N \leq 2000$ range (Duarte, 1993, 2009).

4.3.2 Geometry of the *N*-Slit Interferometer

The relevant geometry, and geometrical parameters, at the transmission grating (j) and the plane of interference (x) are illustrated in Figures 4.6 through 4.8. According to the geometry, the phase difference term in Equations 4.16 and 4.18 can be expressed as (Duarte, 1997)

$$\cos\big((\theta_m - \theta_j) \pm (\phi_m - \phi_j)\big) = \cos\big(|\, l_m - l_{m-1}\,|\, k_1 \pm |\, L_m - L_{m-1}\,|\, k_2\big) \tag{4.23}$$

where

$$k_1 = \frac{2\pi n_1}{\lambda_v} \tag{4.24}$$

and

$$k_2 = \frac{2\pi n_2}{\lambda_v} \tag{4.25}$$

are the wave numbers of the two optical regions defined in Figure 4.8. Here, $\lambda_1 = \lambda_v/n_1$ and $\lambda_2 = \lambda_v/n_2$ where λ_v is the vacuum wavelength, while n_1 and n_2 are the corresponding indexes of refraction (Wallenstein and Hänsch, 1974; Born and Wolf, 1999). The phase differences are expressed *exactly* via the following geometrical equations (Duarte, 1993):

$$|\, L_m - L_{m-1}\,| = \frac{2\xi_m d_m}{|\, L_m + L_{m-1}\,|} \tag{4.26}$$

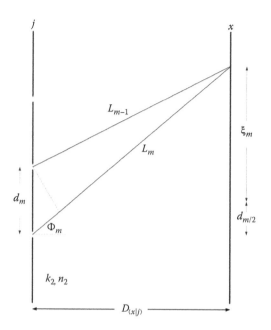

FIGURE 4.7
The N-slit array, or transmission grating, plane (j) and the interferometric plane (x) (not to scale) illustrating the path difference and the various parameters involved in the exact description of the geometry.

$$L_m^2 = D_{\langle x|j\rangle}^2 + \left(\xi_m + \frac{d_m}{2}\right)^2 \tag{4.27}$$

$$L_{m-1}^2 = D_{\langle x|j\rangle}^2 + \left(\xi_m - \frac{d_m}{2}\right)^2 \tag{4.28}$$

In this notation ξ_m is the lateral displacement, on the x plane, from the projected median of d_m to the interference plane, and $D_{\langle x|j\rangle}$ is the intra-interferometric distance from the j plane to the x plane. Accurate representation of the exact geometry is important when writing software to generate numerical interferograms based on Equation 4.18.

4.3.3 Diffraction Grating Equation

In the phase term equation

$$\cos\left((\theta_m - \theta_j) \pm (\phi_m - \phi_j)\right) = \cos\left(|\,l_m - l_{m-1}\,|\,k_1 \pm |\,L_m - L_{m-1}\,|\,k_2\right)$$

the corresponding path differences are $|\,l_m - l_{m-1}\,|$ and $|\,L_m - L_{m-1}\,|$. Since *maxima* occur at

$$\left(|\,l_m - l_{m-1}\,|\,n_1 \pm |\,L_m - L_{m-1}\,|\,n_2\right)\frac{2\pi}{\lambda_v} = M\pi \tag{4.29}$$

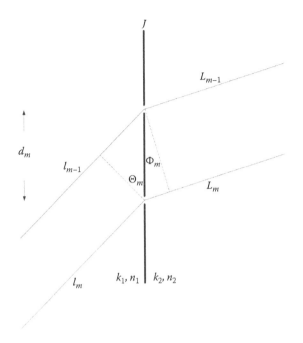

FIGURE 4.8
Close up of the *N*-slit array, or transmission grating, plane (*j*) illustrating the path length difference and the angles of incidence (Θ_m) and diffraction (Φ_m) for the condition $D_{\langle x|j\rangle} \gg d_m$ as described. (Reproduced from Duarte, F.J., *Am. J. Phys.* 65, 637, 1997, with permission from the American Association of Physics Teachers.)

where $M = 0, \pm2, \pm4, \pm6 \cdots$, it can be shown that

$$d_m\left(n_1 \sin\Theta_m \pm n_2 \sin\Phi_m\right)\frac{2\pi}{\lambda_v} = M\pi \tag{4.30}$$

which, for $n_1 = n_2 = 1$ and $\lambda = \lambda_v$, reduces to

$$d_m(\sin\Theta_m \pm \sin\Phi_m)k = M\pi \tag{4.31}$$

that can be expressed as the well-known grating equation

$$d_m(\sin\Theta_m \pm \sin\Phi_m) = m\lambda \tag{4.32}$$

where $m = 0, \pm1, \pm2, \pm3 \cdots$. For a grating utilized in the reflection domain, in *Littrow configuration*, $\Theta_m = \Phi_m = \Theta$ so that the grating equation reduces to

$$2d \sin\Theta = m\lambda \tag{4.33}$$

These diffraction equations are reconsidered, in a more general form, with an extra sign alternative in Chapter 5.

4.3.4 *N*-Slit Interferometer Experiment

The *N*-slit interferometer is illustrated in Figure 4.9. In practice this interferometer can be configured with a variety of lasers including tunable lasers. However, one requirement is that the laser to be utilized must emit in the narrow-linewidth regime and in a single transverse mode (TEM_{00}) with a near-Gaussian profile. Ideally the source should be a single-longitudinal-mode laser (see Chapter 9). The reason for this requirement is that narrow-linewidth lasers yield sharp well-defined interference patterns close to those predicted theoretically for a single wavelength.

One particular configuration of the *N*-slit laser interferometer (NSLI), described by Duarte (1993), utilizes a TEM_{00} He–Ne laser ($\lambda \approx 632.82$ nm) with a beam 0.5 mm in diameter as the illumination source. This class of laser yields a smooth near-Gaussian beam profile and narrow-linewidth emission ($\Delta\nu \approx 1$ GHz). The laser beam is then magnified, in two dimensions, by a Galilean telescope. Following the telescopic expansion, the beam is further expanded, in one dimension, by a multiple-prism beam expander (MPBE). This class of optical architecture can yield an expanded smooth near-Gaussian beam approximately 50 mm wide. An option is to insert a convex lens prior to the multiple-prism expander thus producing an extremely elongated near-Gaussian beam (Duarte, 1987, 1993). The beam propagation through this system can be accurately characterized using ray transfer matrices as discussed in Duarte (2003) (see Appendix C). Also, as an option, at the exit of the MPBE, an aperture, a few mm wide, can be deployed.

The beam profile thus produced can be neatly reproduced by the interferometric equation as illustrated later in this chapter. Thus, the source *s* can be either the exit prism of the MPBE or the wide aperture. For the results discussed in this chapter, on the detection side, the interference screen at *x* is a digital detector comprised of a photodiode array with individual pixels each 25 μm in width.

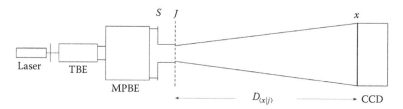

FIGURE 4.9

Top view schematics of the *N*-slit interferometer. A neutral density filter follows the TEM_{00} narrow linewidth. The laser beam is then magnified in two dimensions by a telescope beam expander (TBE). The magnified beam is then expanded in one dimension by an MPBE. A wide aperture then selects the central part of the expanded beam to illuminate the *N*-slit array (*j*). The interferogram then propagates via the intra-interferometric path $D_{\langle x|j \rangle}$ in its ways toward the interference plane *x*. Detection of the interferogram at *x* can either be performed by a silver halide film or a digital array such as a CCD or CMOS detector. (Reproduced from Duarte, F.J. et al., *J. Opt.* 12, 015705, 2010, with permission from the Institute of Physics.)

Now, we consider a series of cases that demonstrate the measurement capability of the NSLI and the ability of the generalized interferometric equation to either predict or reproduce the measurement. The first case considered is the well-known double-slit experiment also known as Young's interference experiment. For ($N = 2$) with slits 50 μm in width, separated by 50 μm, the elongated Gaussian beam provides a nearly plane illumination. That is also approximately the case even if a larger number of slits, of these dimensions, are illuminated. For the particular case of a two-slit experiment involving 50 μm slits separated by 50 μm and a grating to screen distance $D_{\langle x|j\rangle}$ = 10 cm, the interference signal is displayed in Figure 4.10a. The calculated interference, using Equation 4.18, and assuming plane wave illumination are given in Figure 4.10b.

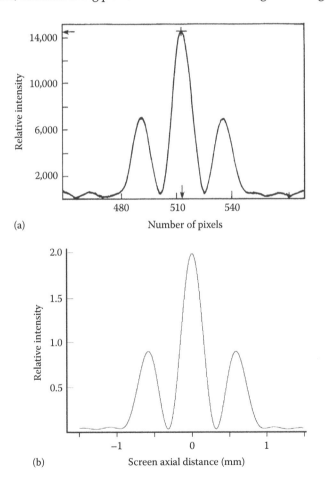

FIGURE 4.10
(a) Measured interferogram resulting from the interaction of coherent laser emission at λ = 632.82 nm and two slits ($N = 2$) 50 μm wide, separated by 50 μm. The j to x distance is $D_{\langle x|j\rangle}$ = 10 cm. Each pixel is 25 μm wide. (b) Corresponding theoretical interferogram from Equation 4.18.

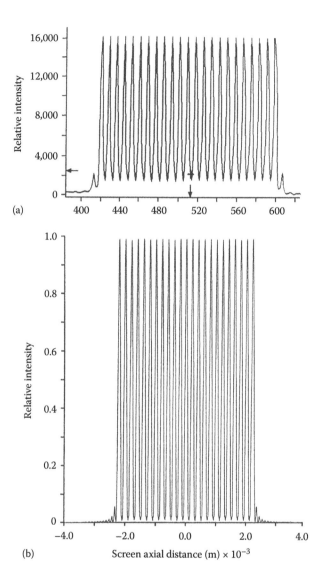

FIGURE 4.11

(a) Measured interferogram, in the near field, resulting from the interaction of coherent laser emission at $\lambda = 632.82$ nm and $N = 23$ slits, 100 μm wide, separated by 100 μm. Here, $D_{(x|j)} = 1.5$ cm. (b) Corresponding near-field theoretical interferogram from Equation 4.18. (Reproduced from Duarte, F.J., *Opt. Commun.* 103, 8, 1993, with permission from Elsevier.)

For an array of $N = 23$ slits, each 100 μm in width and separated by 100 μm, the measured and calculated interferograms are shown in Figure 4.11. Here the grating to digital detector distance is $D_{(x|j)} = 1.5$ cm. This is a *near field* result and corresponds entirely to the interferometric regime.

For an array of $N = 100$ slits, each 30 μm in width and separated by 30 μm, the measured and calculated interferograms are shown in Figure 4.12.

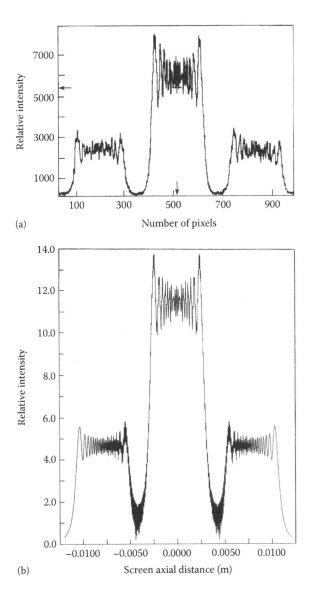

FIGURE 4.12

(a) Measured interferogram resulting from the interaction of coherent laser emission at $\lambda = 632.82$ nm and $N = 100$, slits 30 μm wide, separated by 30 μm. Here, $D_{\langle x|j\rangle} = 75$ cm. (b) Corresponding theoretical interferogram from Equation 4.18. (Reproduced from Duarte, F.J., *Opt. Commun.* 103, 8, 1993, with permission from Elsevier.)

Here the grating to digital detector distance is $D_{\langle x|j\rangle} = 75$ cm. In Figure 4.12 the ±1 diffraction orders are present.

In practice, the transmission gratings are not perfect and offer an uncertainty in the dimension of the slits. The uncertainty in the slit dimensions of the grating, incorporating the 30 μm slits, used in this experiments was

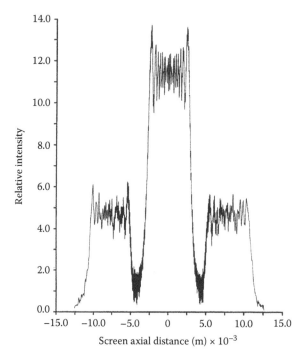

FIGURE 4.13

Theoretical interferometric/diffraction distribution using a ≤2% uncertainty in the dimensions of the 30 μm slits. In this calculation, $N = 100$ and $D_{\langle x|j \rangle} = 75$ cm. A deterioration in the spatial symmetry of the distribution is evident. (Reproduced from Duarte, F.J., *Opt. Commun.* 103, 8, 1993, with permission from Elsevier.)

measured to be ≤2%. The theoretical interferogram for the grating comprised by $N = 100$ slits, each 30.0 ± 0.6 μm wide and separated by 30.0 ± 0.6 μm, is given in Figure 4.13. Notice the symmetry deterioration.

When a wide slit is used to select the central portion of the elongated Gaussian beam, the interaction of the coherent laser beam with the slit results in diffraction prior to the illumination of the transmission grating. The interferometric Equation 4.18 can be used to characterize this diffraction. This is done by dividing the wide slit in hundreds of smaller slits. As an example a 4 mm wide aperture is divided into 800 slits each 4 μm wide and separated by a 1 μm interslit distance (Duarte, 1993). The calculated near- field diffraction pattern, for a distance of $D_{\langle x|j \rangle} = 10$ cm, is shown in Figure 4.14. Using this as the radiation source to illuminate the $N = 100$ slit grating, comprised of 30 μm slits with an interslit distance of 30 μm (for $D_{\langle x|j \rangle} = 75$ cm), yields the theoretical interferogram displayed in Figure 4.15. This is a *cascade interferometric technique* in which the interferometric distribution in one plane is used to illuminate an N-slit array in the immediately following plane and is applied further in the results discussed in Chapter 20.

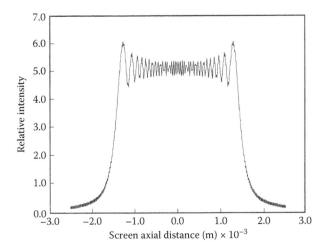

FIGURE 4.14
Theoretical near-field diffraction distribution produced by a 4 mm aperture illuminated at $\lambda = 632.82$ nm, and $D_{\langle x|j\rangle} = 10$ cm. (Reproduced from Duarte, F.J., *Opt. Commun.* 103, 8, 1993, with permission from Elsevier.)

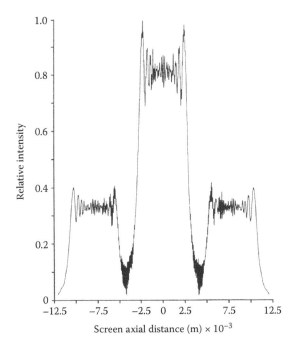

FIGURE 4.15
Theoretical interferometric distribution incorporating diffraction-edge effects in the illumination. In this calculation, the width of the slits in the array is 30 μm, separated by 30 μm, $N = 100$, and $D_{\langle x|j\rangle} = 75$ cm. The aperture-grating distance is 10 cm. (Reproduced from Duarte, F.J., *Opt. Commun.* 103, 8, 1993, with permission from Elsevier.)

4.4 Coherent and Semicoherent Interferograms

The interferometric equation

$$|\langle x \mid s \rangle|_\lambda^2 = \sum_{j=1}^{N} \Psi(r_j)_\lambda^2 + 2 \sum_{j=1}^{N} \Psi(r_j)_\lambda \left(\sum_{m=j+1}^{N} \Psi(r_m)_\lambda \cos(\Omega_m - \Omega_j) \right) \qquad (4.34)$$

was originally derived to account for single-photon propagation only (Duarte, 1993, 2004). This is illustrated by adding a single-wavelength subscript to Equation 4.18, as made explicit now in Equation 4.34. Thus, this equation is intrinsically related to monochromatic and/or highly coherent emission. In practice it has also been found that it accounts for the propagation of ensembles of indistinguishable photons or narrow-linewidth emission as available from narrow-linewidth laser sources (Duarte, 1993, 2003). The question then arises on the applicability of Equation 4.34 to the case of semicoherent, partially coherent, or broadband emission.

Equation 4.34 provides an interferogram for a single wavelength and in practice for an ensemble of indistinguishable photons. These interferograms are narrow and spatially sharp and well defined. For broadband emission, or semicoherent emission, the sharpness of the interferogram diminishes and the interferometric pattern becomes broad and less defined. This is how this occurs: each wavelength has a unique interferometric signature defined by Equation 4.34. A detector registers that signature. If the emission is broadband or semicoherent, a multitude of different interferograms are generated, and the detector (either digital or a photographic plaque) provides an integrated picture of a composite interferogram produced by the array of wavelengths involved in the emission.

Thus, for broadband emission Equation 4.34 is modified to include a sum over the wavelength range involved, so that

$$\sum_{\lambda=\lambda_1}^{\lambda_n} |\langle x \mid s \rangle|_\lambda^2 = \sum_{\lambda=\lambda_1}^{\lambda_n} \left(\sum_{j=1}^{N} \Psi(r_j)_\lambda^2 + 2 \sum_{j=1}^{N} \Psi(r_j)_\lambda \left(\sum_{m=j+1}^{N} \Psi(r_m)_\lambda \cos(\Omega_m - \Omega_j) \right) \right) \qquad (4.35)$$

The concept just described has been previously outlined by Duarte (2007, 2008) and is further illustrated next. In Figure 4.16 the double-slit interferogram produced with narrow-linewidth emission from the $3s_2 - 2p_{10}$ transition of a He–Ne laser, at $\lambda \approx 543.3$ nm, is displayed. The visibility of this interferogram is calculated using (Michelson, 1927)

$$\mathcal{V} = \frac{I_1 - I_2}{I_1 + I_2} \qquad (4.36)$$

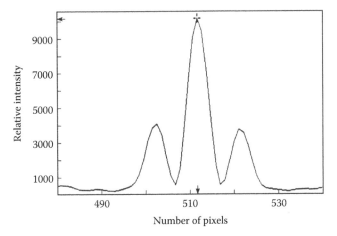

FIGURE 4.16
Measured double-slit interferogram generated with He–Ne laser emission from the $3s_2 - 2p_{10}$ transition at $\lambda \approx 543.3$ nm. Here, $N = 2$ for a slit width of 50 μm. The intra-interferometric distance is $D_{\langle x|j \rangle} = 10$ cm. (Reproduced from Duarte, F.J., *Opt. Lett.* 32, 412, 2007, with permission from the Optical Society of America.)

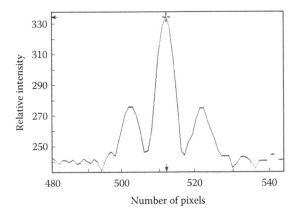

FIGURE 4.17
Measured double-slit interferogram generated with emission from an organic semiconductor interferometric emitter at $\lambda \approx 540$ nm. Here, $N = 2$ for a slit width of 50 μm. The intra-interferometric distance is $D_{\langle x|j \rangle} = 10$ cm. (Reproduced from Duarte, F.J., *Opt. Lett.* 32, 412, 2007, with permission from the Optical Society of America.)

to be $V \approx 0.95$. In Figure 4.17 the double-slit interferogram produced, under identical geometrical conditions, but with the emission from an electrically excited coherent organic semiconductor interferometric emitter, at $\lambda \approx 540$ nm, is displayed. Here the visibility is lower $V \approx 0.90$. A comparison between the two interferograms reveals that the second interferogram has slightly broader spatial features relative to the interferogram produced with illumination from the $3s_2 - 2p_{10}$ transition of the He–Ne laser. The differences in spatial distributions between these two interferograms have been used to estimate the linewidth of the emission from the interferometric emitter (Duarte, 2008).

A double-slit interferogram produced, under identical geometrical conditions, but with the emission from a broadband semicoherent source, centered around $\lambda \approx 540$ nm, yields a visibility of $\mathcal{V} \approx 0.55$. A survey of measured double-slit interferogram visibilities from relevant semicoherent, or partially coherent, sources reveals a visibility range of $0.4 \leq \mathcal{V} \leq 0.65$. On the other hand, the visibility range for double-slit interferograms originating from various laser sources is $0.85 \leq \mathcal{V} \leq 0.99$ (Duarte, 2008).

4.5 Interferometric Equation in Two and Three Dimensions

The 2D interferometric case can be described considering a diffractive grid, or 2D N-slit array, as depicted in Figure 4.18. Photon propagation takes place from s to the interferometric plane x via a 2D transmission grating j_{zy}, that is, j is replaced by a grid comprised of j components in the y direction and j components in the z direction. Note that in the 1D case only the y component of j is present, which is written simply as j. The plane configured by the j_{zy} grid is orthogonal to the plane of propagation. Hence,

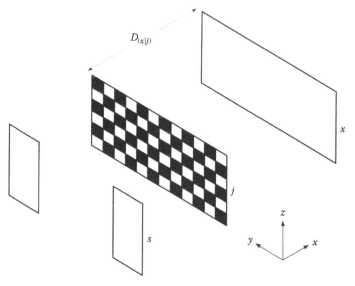

FIGURE 4.18
Two-dimensional depiction of the interferometric system $\langle x|j\rangle\langle j|s\rangle$. (Adapted from Duarte, F.J., Interferometric imaging, In: *Tunable Laser Applications*, Duarte, F. J. (ed.), Marcel Dekker, New York, 1995.)

for photon propagation from s to x, via j_{zy}, the probability amplitude is given by (Duarte, 1995)

$$\langle x \mid s \rangle = \sum_{j_z=1}^{N} \sum_{j_y=1}^{N} \langle x \mid j_{zy} \rangle \langle j_{zy} \mid s \rangle \qquad (4.37)$$

Now, if the j is abstracted from j_{zy}, then Equation 4.37 can be expressed as

$$\langle x \mid s \rangle = \sum_{z=1}^{N} \sum_{y=1}^{N} \Psi(r_{zy}) e^{-i\Omega_{zy}} \qquad (4.38)$$

and the corresponding probability is given by (Duarte, 1995)

$$|\langle x \mid s \rangle|^2 = \sum_{z=1}^{N} \sum_{y=1}^{N} \Psi(r_{zy}) \sum_{q=1}^{N} \sum_{p=1}^{N} \Psi(r_{pq}) e^{i(\Omega_{qp} - \Omega_{zy})} \qquad (4.39)$$

For a 3D transmission grating, it can be shown that

$$|\langle x \mid s \rangle|^2 = \sum_{z=1}^{N} \sum_{y=1}^{N} \sum_{x=1}^{N} \Psi(r_{zyx}) \sum_{q=1}^{N} \sum_{p=1}^{N} \sum_{r=1}^{N} \Psi(r_{qpr}) e^{i(\Omega_{qpr} - \Omega_{zyx})} \qquad (4.40)$$

It is important to emphasize that the equations described here apply either to the propagation of single photons or to the propagation of ensembles of *coherent, indistinguishable,* or *monochromatic* photons. For broadband emission, as described in the previous section, an interferometric summation over the emission spectrum is necessary.

The application of quantum principles to the description of propagation of a large number of monochromatic, or indistinguishable, photons was already advanced by Dirac in his discussion of interference (Dirac, 1978; Duarte, 1998).

4.6 Classical and Quantum Alternatives

Increasingly the field of optics has seen a transition from a classical description to a quantum description. Some phenomena is purely quantum and cannot be described classically. Other phenomena, such as polarization and interference, can be described either classically or quantum mechanically. In the case of interference, and diffraction, the beauty is that the quantum mechanical description also applies to the description of ensembles of indistinguishable photons. Also, when we describe interference and diffraction using quantum mechanical tools, all we are doing is following the steps of giants such as Dirac (1978) and Feynman (see, Feynman and Hibbs, 1965) (Note: Dirac's description comes from the 1930s, but the book edition we are using is 1978).

As we shall see in Chapter 5, there is another powerful reason to describe interference quantum mechanically: This description provides a unified avenue to the whole of optics in a succinct inverse hierarchy that goes

INTERFERENCE → DIFFRACTION → REFRACTION → REFLECTION

(Duarte, 1997). By contrast the situation from a traditional classical perspective is rather disjointed as can be observed by perusing any good book on classical optics. There, the description goes like reflection, refraction, diffraction, and interference in a non-cohesive manner. Classically, there is no mathematical coherence in the description.

Furthermore, in our quantum description, a single equation is used to describe interference, and interference–diffraction phenomena, in the near and the far field in a unified manner (Duarte, 1993).

Problems

4.1 Show that substitution of Equations 4.10 and 4.11 into Equation 4.4 leads to Equation 4.12.

4.2 Show that Equation 4.16 can be expressed as Equation 4.18.

4.3 From the geometry of Figure 4.7, derive Equations 4.26 through 4.28.

4.4 Write an equation for $|\langle x|s \rangle|^2$ in the case relevant to $N = 3$ starting from Equation 4.16.

4.5 Write an equation for $|\langle x|s \rangle|^2$ using the probability amplitude given in Equation 4.5.

References

Born, M. and Wolf, E. (1999). *Principles of Optics*, 7th edn. Cambridge, New York.
Dirac, P. A. M. (1939). A new notation for quantum mechanics. *Math. Proc. Cam. Phil. Soc.* **35**, 416–418.
Dirac, P. A. M. (1978). *The Principles of Quantum Mechanics*, 4th edn. Oxford, London, U.K.
Duarte, F. J. (1987). Beam shaping with telescopes and multiple-prism beam expanders. *J. Opt. Soc. Am. A* **4**, p30.

Duarte, F. J. (1991). Dispersive dye lasers. In *High Power Dye Lasers*, (Duarte, F. J., ed.). Springer-Verlag, Berlin, Germany, Chapter 2.

Duarte, F. J. (1993). On a generalized interference equation and interferometric measurements. *Opt. Commun.* **103**, 8–14.

Duarte, F. J. (1995). Interferometric imaging. In *Tunable Laser Applications*, (Duarte, F. J., ed.). Marcel Dekker, New York, Chapter 5.

Duarte, F. J. (1997). Interference, diffraction, and refraction, via Dirac's notation. *Am. J. Phys.* **65**, 637–640.

Duarte, F. J. (1998). Interference of two independent sources. *Am. J. Phys.* **66**, 662–663.

Duarte, F. J. (2003). *Tunable Laser Optics*. Elsevier-Academic, New York.

Duarte, F. J. (2004). Comment on "Reflection, refraction and multislit interference". *Eur. J. Phys.* **25**, L57–L58.

Duarte, F. J. (2007). Coherent electrically-excited organic semiconductors: Visibility of interferograms and emission linewidth, *Opt. Lett.* **32**, 412–414.

Duarte, F. J. (2008). Coherent electrically excited organic semiconductors: Coherent or laser emission? *Appl. Phys. B* **90**, 101–108.

Duarte, F. J. (2009). Interferometric imaging. In *Tunable Laser Applications*, 2nd edn. (Duarte, F. J., ed.). CRC, New York, Chapter 12.

Duarte, F. J. and Paine, D. J. (1989). Quantum mechanical description of N-slit interference phenomena, in *Proceedings of the International Conference on Lasers'88* (Sze, R. C. and Duarte, F. J., eds.). STS Press, McLean, VA, pp. 42–47.

Duarte, F. J., Taylor, T. S., Clark, A. B., and Davenport, W. E. (2010). The N-slit interferometer: An extended configuration. *J. Opt.* **12**, 015705.

Feynman. R. P. and Hibbs, A. R. (1965). *Quantum Mechanics and Path Integrals*, Mc Graw-Hill, New York.

Feynman, R. P., Leighton, R. B., and Sands, M. (1965). *The Feynman Lectures on Physics*, Vol. III, Addison-Wesley, Reading, MA.

Michelson, A. A. (1927). *Studies in Optics*, University of Chicago, Chicago, IL.

van Kampen, N. G. (1988). Ten theorems about quantum mechanical measurements. *Physica A* **153**, 97–113.

Wallenstein, R. and Hänsch, T. W. (1974). Linear pressure tuning of a multielement dye laser spectrometer. *Appl. Opt.* **13**, 1625–1628.

5

Interference, Diffraction, Refraction, and Reflection via the Dirac Notation

5.1 Introduction

In this chapter we use the generalized 1D interferometric equation to introduce the concepts of diffraction, refraction, and reflection, cohesively and unified in that order. Thus, we have an equation derived using the probabilistic concept of single-photon propagation, via the Dirac quantum notation, explaining in a unified manner the major concepts of classical optics.

In the original exposition of this united and cohesive approach to optics (Duarte, 1997, 2003), only positive refraction was considered. Here, both, *positive* and *negative refraction*, are incorporated (Duarte, 2006). Subsequently, the brief exposition of generalized prismatic dispersion also encompasses the positive and negative alternatives.

5.2 Interference and Diffraction

Feynman, in his usual style, stated that *"no one has ever been able to define the difference between interference and diffraction satisfactorily"* (Feynman et al., 1965).

In the discussion related to Figure 5.1, and its variants, reference, up to now, was only made to interference. However, what we really have is interference in three diffraction orders, that is, the 0th, or central order, and the ±1, or secondary orders. In other words, there is an interference pattern associated with each diffraction order.

Physically, however, it is part of the same phenomenon. The interaction of coherent light with a set of slits, in the near field, gives rise to an interference pattern. As the intra- interferometric distance $D_{\langle x|j \rangle}$, from j to x, increases, the central interference pattern begins to give origin to secondary patterns that gradually separate from the central order at lower intensities.

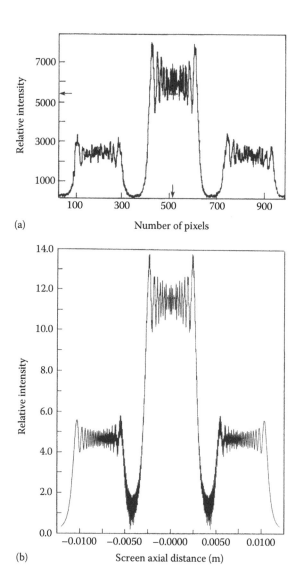

(a)

(b)

FIGURE 5.1

(a) Measured interferogram resulting from the interaction of coherent laser emission at $\lambda = 632.82$ nm and 100 slits 30 μm wide, separated by 30 μm. The j to x distance is 75 cm. (b) Corresponding theoretical interferogram from Equation 5.1. (Reproduced from Duarte, F.J., *Opt. Commun.* 103, 8, 1993, with permission from Elsevier.)

These are the ±1 diffraction orders. This physical phenomenon, as one goes from the near to the far field, is clearly illustrated in Figures 5.2 through 5.5. One of the beauties of the Dirac description of optics is the ability to continuously describe the evolution of the interferometric distribution, as it moves from the *near* to the *far* field, with a single mathematical equation.

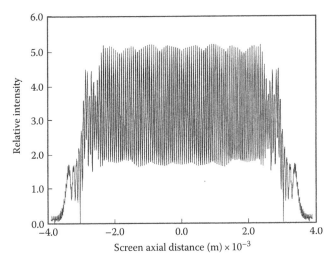

FIGURE 5.2

Interferogram at a grating to screen distance of $D_{(x|j)} = 5$ cm. The interferometric distribution, in the *near field*, is mainly part of a single order. At the boundaries there is an incipient indication that of emerging orders. Slit width is 30 μm and slits are separated by 30 μm, $N = 100$, and $\lambda = 632.8$ nm. (From Duarte, F.J., *Tunable Laser Optics*, Elsevier-Academic, New York, 2003.)

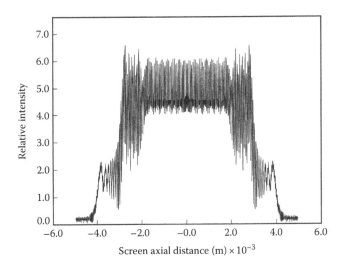

FIGURE 5.3

Interferogram at a grating to screen distance of $D_{(x|j)} = 10$ cm. The presence of the emerging (±1) orders is more visible. Slit width is 30 μm and slits are separated by 30 μm, $N = 100$, and $\lambda = 632.8$ nm. (From Duarte, F.J., *Tunable Laser Optics*, Elsevier-Academic, New York, 2003.)

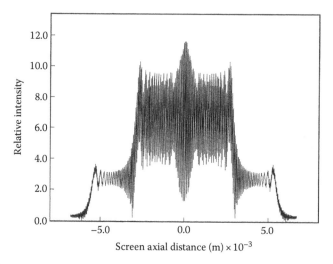

FIGURE 5.4
Interferogram at a grating to screen distance of $D_{\langle x|j\rangle} = 25$ cm. The emerging (±1) orders give rise to an overall distribution with clear "shoulders." Slit width is 30 μm and slits are separated by 30 μm, $N = 100$, and λ = 632.8 nm. (From Duarte, F.J., *Tunable Laser Optics*, Elsevier-Academic, New York, 2003.)

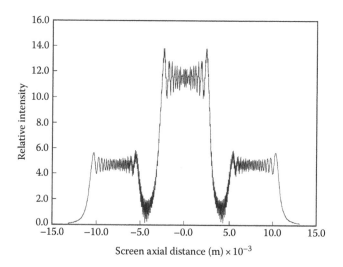

FIGURE 5.5
Interferogram at a grating to screen distance of $D_{\langle x|j\rangle} = 75$ cm. The −1, 0, and +1 diffraction orders are clearly established. Notice the increase in the width of the distribution as the *j* to *x* distance increases from 5 to 75 cm. Slit width is 30 μm and slits are separated by 30 μm, $N = 100$, and λ = 632.8 nm. (From Duarte, F.J., *Tunable Laser Optics*, Elsevier-Academic, New York, 2003.)

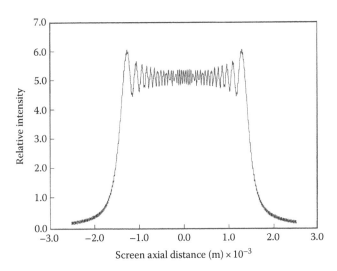

FIGURE 5.6
Theoretical near field diffraction distribution produced by a 4 mm aperture illuminated at $\lambda = 632.82$ nm. The j to x distance is 10 cm. (Reproduced from Duarte, F.J., *Opt. Commun.* 103, 8, 1993, with permission from Elsevier.)

The second interference–diffraction entanglement refers to the fact that our generalized interference equation can be naturally applied to describe a diffraction pattern produced by a single wide slit as previously shown in Figure 5.6. Under those circumstances the wide slit is mathematically represented by a multitude of subslits.

5.2.1 Generalized Diffraction

The intimate relation between interference and diffraction has its origin in the interferometric equation itself (Duarte, 2003):

$$|\langle x|s \rangle|^2 = \sum_{j=1}^{N} \Psi(r_j)^2 + 2\sum_{j=1}^{N} \Psi(r_j)\left(\sum_{m=j+1}^{N} \Psi(r_m)\cos(\Omega_m - \Omega_j) \right) \qquad (5.1)$$

for it is the $\cos(\Omega_m - \Omega_j)$ term that gives rise to the different diffraction orders.

Here, we revisit the geometry at the N-slit plane j and illustrate what is obviously seen in Figures 5.1 through 5.5: up on arrival to a slit, diffraction occurs symmetrically toward both sides as illustrated in Figures 5.7 through 5.10. Figure 5.7 depicts the usual description associated with incidence below the normal (–) and diffraction above the normal (+).

Figure 5.8 illustrates incidence above the normal (+) and diffraction above the normal (+) (Duarte, 2006). For completeness we also include the case of incidence below the normal (–) followed by diffraction below the normal (–)

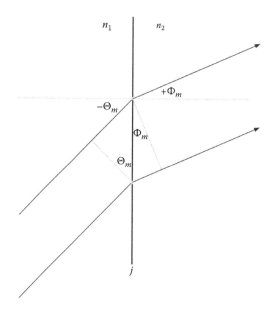

FIGURE 5.7
Outline of the *j* plane in a transmission grating showing incidence below the normal (–) and diffraction above the normal (+) consistent with the convention leading to positive refraction. (Reproduced from Duarte, F.J., *Appl. Phys. B*, 82, 35, 2006, with permission from Springer Verlag.)

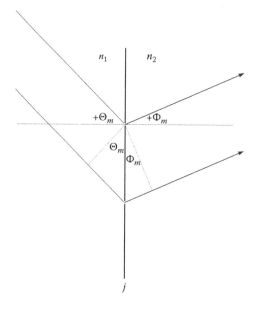

FIGURE 5.8
Outline of the *j* plane in a transmission grating showing incidence above the normal (+) and diffraction above the normal (+) consistent with the convention leading to negative refraction. (Reproduced from Duarte, F.J., *Appl. Phys. B*, 82, 35, 2006, with permission from Springer Verlag.)

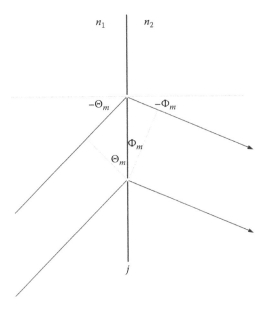

FIGURE 5.9
Outline of the j plane in a transmission grating showing incidence below the normal (–) and diffraction below the normal (–) consistent with the convention leading to negative refraction.

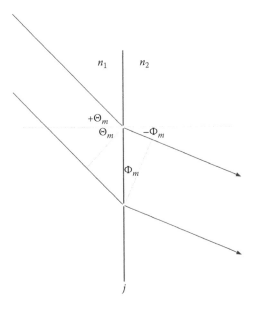

FIGURE 5.10
Outline of the j plane in a transmission grating showing incidence above the normal (+) and diffraction below the normal (–) consistent with the convention leading to positive refraction.

and incidence above the normal (+) followed by diffraction below the normal (−) (Figures 5.9 and 5.10).

Thus, the equations describing the geometry (Duarte, 1997) are slightly modified to account for all the ± alternatives

$$\cos\left(\pm(\theta_m - \theta_j) \pm (\phi_m - \phi_j)\right) = \cos\left(\pm |l_m - l_{m-1}| k_1 \pm |L_m - L_{m-1}| k_2\right) \qquad (5.2)$$

where

$$k_1 = \frac{2\pi n_1}{\lambda_v} \qquad (5.3)$$

and

$$k_2 = \frac{2\pi n_2}{\lambda_v} \qquad (5.4)$$

are the wave numbers of the two optical regions defined in Figures 5.7 through 5.10. Here, as we saw previously, $\lambda_1 = \lambda_v/n_1$ and $\lambda_2 = \lambda_v/n_2$, where λ_v is the vacuum wavelength and n_1 and n_2 are the corresponding indexes of refraction.

As previously explained in Chapter 4, the phase differences can be expressed exactly via the following geometrical expressions (Duarte, 1993):

$$|L_m - L_{m-1}| = \frac{2\xi_m d_m}{|L_m + L_{m-1}|} \qquad (5.5)$$

$$L_m^2 = D_{\langle x|j \rangle}^2 + \left(\xi_m + \frac{d_m}{2}\right)^2 \qquad (5.6)$$

$$L_{m-1}^2 = D_{\langle x|j \rangle}^2 + \left(\xi_m - \frac{d_m}{2}\right)^2 \qquad (5.7)$$

From the geometry of Figure 5.11 we can write

$$\sin \Phi_m = \frac{\xi_m + (d_m/2)}{L_m} \qquad (5.8)$$

and for the condition $D_{\langle x|j \rangle} \gg d_m$, we have $|L_m + L_{m-1}| \approx 2L_m$; then using Equations 5.5 and 5.6 we have

$$|L_m - L_{m-1}| \approx d_m \sin \Phi_m \qquad (5.9)$$

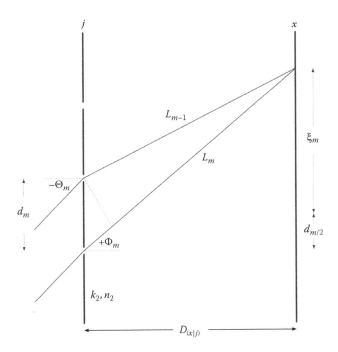

FIGURE 5.11
Close-up of the N-slit array, or transmission grating, plane (j) illustrating the path length difference and the angles of incidence $(-\Theta_m)$ and diffraction $(+\Phi_m)$ for the condition $D_{\langle x|j \rangle} \gg d_m$.

$$|l_m - l_{m-1}| \approx d_m \sin \Theta_m \tag{5.10}$$

where Θ_m and Φ_m are the angles of incidence and diffraction, respectively. Given that maxima occur at

$$\left(\pm |l_m - l_{m-1}| n_1 \pm |L_m - L_{m-1}| n_2 \right) \frac{2\pi}{\lambda_v} = M\pi \tag{5.11}$$

then using Equations 5.9 and 5.10

$$d_m \left(\pm n_1 \sin \Theta_m \pm n_2 \sin \Phi_m \right) \frac{2\pi}{\lambda_v} = M\pi \tag{5.12}$$

where $M = 0, 2, 4, 6, \ldots$ For $n_1 = n_2$, we have $\lambda = \lambda_v$, and this equation reduces to the *generalized diffraction grating equation*

$$d_m \left(\pm \sin \Theta_m \pm \sin \Phi_m \right) = m\lambda \tag{5.13}$$

where $m = 0, 1, 2, 3, \ldots$ are the various *diffraction orders*.

A most important observation is due here: in our discussion on the inter-ferometric equations, we have made explicit reference to the exact geomet-rical equations (Equations 5.5 through 5.7). However, in the derivation of Equations 5.12 and 5.13, we have used the approximation $D_{\langle x|j\rangle} \gg d_m$. Are we being consistent? The answer is yes! The exact Equations 5.5 through 5.7 are used in the generalized interferometric equation (Equation 5.1), while the approximation $D_{\langle x|j\rangle} \gg d_m$ has been applied in the derivation of the general-ized diffraction equation

$$d_m(\pm \sin \Theta_m \pm \sin \Phi_m) = m\lambda$$

that manifests itself in the *far field* as beautifully illustrated in Figures 5.2 through 5.5. From this equation it is clearly seen that beyond the zeroth order, m can take a series of $\pm$ values, that is, $m = \pm 1, \pm 2, \pm 3 \cdots$

5.2.2 Positive Diffraction

From the generalized diffraction equation (Equation 5.13) including both $\pm$ alternatives, the usual traditional equation can be stated as

$$d_m\left(\sin \Theta_m \pm \sin \Phi_m\right) = m\lambda \tag{5.14}$$

which was previously derived starting from (Duarte, 1997)

$$\cos\left((\theta_m - \theta_j) \pm (\phi_m - \phi_j)\right) = \cos\left(|\,l_m - l_{m-1}\,|\,k_1 \pm |\,L_m - L_{m-1}\,|\,k_2\right) \tag{5.15}$$

From Equation 5.14, setting $\Theta_m = \Phi_m = \Theta$, the diffraction grating equation for Littrow configuration emerges the well-known equation

$$m\lambda = 2d_m \sin \Theta \tag{5.16}$$

5.3 Positive and Negative Refraction

So far we have discussed interference and diffraction, and we have seen how diffraction manifests itself as the interferometric distribution propagates toward the far field. An additional fundamental phenomenon in optics is *refraction*.

Refraction is the change in the geometrical path, of a beam of light, due to transmission from the original medium of propagation to a second medium with a different refractive index. For example, refraction is the bending of a ray of light caused due to propagation in a glass, or crystalline, prism.

If in the diffraction grating equation d_m is made very small relative to a given λ, diffraction ceases to occur and the only solution can be found for $m = 0$ (Duarte, 1997).

That is, under these conditions, a grating made of grooves coated on a transparent substrate, such as optical glass, does not diffract and exhibits the refraction properties of the glass. For example, since the maximum value of $(\pm\sin\Theta_m \pm \sin\Phi_m)$ is 2, for a 5000-lines mm^{-1} transmission grating, let us say, no diffraction can be observed for the visible spectrum. Hence, for the condition $d_m \ll \lambda$ the diffraction grating equation can only be solved for

$$d_m \left(\pm n_1 \sin\Theta_m \pm n_2 \sin\Phi_m \right) \frac{2\pi}{\lambda_v} = 0 \qquad (5.17)$$

which leads to

$$(\pm n_1 \sin\Theta_m \pm n_2 \sin\Phi_m) = 0 \qquad (5.18)$$

For the case of incidence below the normal (–) and refraction above the normal (+) (Figure 5.7)

$$-n_1 \sin\Theta_m + n_2 \sin\Phi_m = 0 \qquad (5.19)$$

so that

$$n_1 \sin\Theta_m = n_2 \sin\Phi_m \qquad (5.20)$$

which is the well-known *equation of refraction*, also known as *Snell's law*. Under the present physical conditions, Θ_m is the angle of incidence, and Φ_m becomes the *angle of refraction*. The same outcome is obtained for incidence above the normal (+) and refraction below the normal (–) (Figure 5.10).

For the case of incidence above the normal (+) and refraction above the normal (+) (Figure 5.8),

$$+n_1 \sin\Theta_m + n_2 \sin\Phi_m = 0 \qquad (5.21)$$

so that

$$n_1 \sin\Theta_m = -n_2 \sin\Phi_m \qquad (5.22)$$

which is Snell's law for *negative refraction*. The same outcome is obtained for incidence below the normal (–) and diffraction below the normal (–) (Figure 5.9).

5.3.1 Focusing

Once the law of refraction is introduced, focusing is the next logical and natural step.

This is due to the fact that focusing naturally flows from the law of refraction, or Snell's law, acting on a curved surface. The relationships between surface radius of curvature, refractive index, and focal length for various lens types is discussed in detail in classical optics books such as *Fundamental of Optics* (Jenkins and White, 1957). For completeness, in Appendix C we provide an extensive table with focusing parameters for various lenses of interest using the *ABCD* propagation matrix formalism (Siegman, 1986; Duarte, 2003).

5.4 Reflection

The discussion on interference, up to now, has involved an *N*-slit array or a transmission grating. It should be indicated that the arguments and physics apply equally well to a reflection interferometer (Duarte, 2003), that is, to an interferometer incorporating a reflection, rather than a transmission, grating. Explicitly, if a mirror is placed at an infinitesimal distance immediately behind the *N*-slit array, as illustrated in Figure 5.12, then the transmission

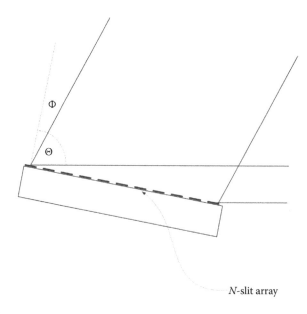

N-slit array

FIGURE 5.12

Approaching a mirror, at an infinitesimal distance, to an *N*-slit array is used to configure a reflection diffraction grating. The incidence angle is Θ and the diffraction angle is Φ.

interferometer becomes a reflection interferometer. Under those circumstances the equations

$$d_m(\pm n_1 \sin \Theta_m \pm n_2 \sin \Phi_m)\frac{2\pi}{\lambda_v} = M\pi$$

and

$$d_m(\pm \sin \Theta_m \pm \sin \Phi_m) = m\lambda$$

apply in the reflection domain, with Θ_m being the incidence angle and Φ_m the diffraction angle in the *reflection* domain. For the case of $d_m \ll \lambda$ and $n_1 = n_2$, we then have

$$(\pm \sin \Theta_m \pm \sin \Phi_m) = 0 \qquad (5.23)$$

For incidence above the normal (+) and reflection below the normal (−),

$$+\sin \Theta_m - \sin \Phi_m = 0 \qquad (5.24)$$

which means

$$\Theta_m = \Phi_m \qquad (5.25)$$

where
   Θ_m is the angle of incidence
   Φ_m is the angle of reflection. This is known as the *law of reflection*

5.5 Succinct Description of Optics

A summary of fundamental optical principles can now be given. Starting from the Dirac quantum principle (Dirac, 1939, 1978)

$$\langle x | s \rangle = \sum_{j=1}^{N} \langle x | j \rangle \langle j | s \rangle$$

the generalized 1D interferometric equation is derived (Duarte and Paine, 1989; Duarte, 1991)

$$|\langle x \mid s \rangle|^2 = \sum_{j=1}^{N} \Psi(r_j)^2 + 2 \sum_{j=1}^{N} \Psi(r_j) \left(\sum_{m=j+1}^{N} \Psi(r_m) \cos(\Omega_m - \Omega_j) \right)$$

From the phase term of this equation, the generalized diffraction equation

$$d_m(\pm n_1 \sin \Theta_m \pm n_2 \sin \Phi_m) \frac{2\pi}{\lambda_v} = M\pi$$

can be obtained, from which the generalized diffraction grating equation

$$d_m(\pm \sin \Theta_m \pm \sin \Phi_m) = m\lambda$$

can be arrived to. From the generalized diffraction equation and applying the condition $d_m \ll \lambda$, the generalized refraction equation

$$(\pm n_1 \sin \Theta_m \pm n_2 \sin \Phi_m) = 0$$

is obtained. And from this equation one can arrive at the law of reflection

$$\Theta_m = \Phi_m$$

Most other important optical phenomena, such as focusing and dispersion, can be explained from the principles outlined here. This hierarchical, orderly, and unified description of optics, *interference, diffraction, refraction,* and *reflection* (Duarte, 1997, 2003), illustrates that quantum principles are perfectly compatible with classical empirical phenomena. The phenomenon of multiple-prism dispersion originates from the derivatives of refraction and is treated in Chapter 6.

Going back to the Feynman statement, on interference and diffraction, we can confidently state that this whole phenomena is succinctly and beautifully described by the interferometric equation

$$|\langle x \mid s \rangle|^2 = \sum_{j=1}^{N} \Psi(r_j)^2 + 2 \sum_{j=1}^{N} \Psi(r_j) \left(\sum_{m=j+1}^{N} \Psi(r_m) \cos(\Omega_m - \Omega_j) \right)$$

Indeed, in reference to the interferometric progression illustrated in Figures 5.2 through 5.5, we see that the basic phenomenon is interference. Pure interference dominates in the near field. However, as the propagation distance increases toward the far field, diffraction orders ($m = \pm 1, \pm 2, \pm 3 \cdots$) do appear. This appearance of diffraction orders is analogous to a quantization of the interferometric distribution.

Problems

5.1 Show that the geometry depicted in Figure 5.11 leads to Equation 5.11.

5.2 Substitute Equations 5.9 and 5.10 into Equation 5.11 to obtain Equation 5.13.

5.3 Show that from the geometry outlined in Figure 5.7, Equation 5.20 follows.

5.4 Show that from the geometry outlined in Figure 5.8, Equation 5.22 follows.

References

Dirac, P. A. M. (1939). A new notation for quantum mechanics. *Math. Proc. Cam. Phil. Soc.* **35**, 416–418.

Dirac, P. A. M. (1978). *The Principles of Quantum Mechanics*, 4th edn. Oxford, London, U.K.

Duarte, F. J. (1991). Dispersive dye lasers. In *High Power Dye Lasers* (Duarte, F. J., ed.). Springer-Verlag, Berlin, Germany, Chapter 2.

Duarte, F. J. (1993). On a generalized interference equation and interferometric measurements. *Opt. Commun.* **103**, 8–14.

Duarte, F. J. (1997). Interference, diffraction, and refraction, via Dirac's notation. *Am. J. Phys.* **65**, 637–640.

Duarte, F. J. (2003). *Tunable Laser Optics*, Elsevier-Academic, New York.

Duarte, F. J. (2006). Multiple-prism dispersion equations for positive and negative refraction. *Appl. Phys. B* **82**, 35–38.

Duarte, F. J. and Paine, D. J. (1989). Quantum mechanical description of N-slit interference phenomena, in *Proceedings of the International Conference on Lasers'88* (Sze, R. C. and Duarte, F. J., eds.). STS Press, McLean, VA, pp. 42–47.

Feynman, R. P., Leighton, R. B., and Sands, M. (1965). *The Feynman Lectures on Physics*, Vol. I, Addison-Wesley, Reading, MA.

Jenkins, F. A. and White, H. E. (1957). *Fundamentals of Optics*, McGraw-Hill, New York.

Siegman, A. E. (1986). *Lasers*, University Science Books, Mill Valley, CA.

6

Generalized Multiple-Prism Dispersion

6.1 Introduction

Now that we have dealt with the fundamentals, we'll focus on the derived phenomenon of angular dispersion. Angular dispersion is an important quantity in optics that describes the ability for an optical element, such as a diffraction grating or prism, to geometrically spread a beam of light as a function of wavelength. Mathematically it is expressed by the differential $(\partial\theta/\partial\lambda)$. For spectrophotometers and wavelength meters based on dispersive elements, such as diffraction gratings and prism arrays, the dispersion should be as large as possible since that enables a higher wavelength spatial resolution.

Further, in the case of dispersive laser oscillators, a high dispersion leads to the achievement of narrow-linewidth emission since the dispersive cavity linewidth is given by

$$\Delta\lambda \approx \Delta\theta \left(\frac{\partial\theta}{\partial\lambda} \right)^{-1} \tag{6.1}$$

where $(\partial\theta/\partial\lambda)$ is the overall intracavity dispersion (Duarte, 1992). In Chapter 9, the cavity linewidth equation is derived from the Dirac quantum principles (Dirac, 1978) via the interferometric equation (Duarte, 1991, 1993)

$$|\langle x | s \rangle|^2 = \sum_{j=1}^{N} \Psi(r_j)^2 + 2\sum_{j=1}^{N} \Psi(r_j) \left(\sum_{m=j+1}^{N} \Psi(r_m) \cos(\Omega_m - \Omega_j) \right) \tag{6.2}$$

In this chapter, however, we concentrate on the dispersive term $(\partial\theta/\partial\lambda)$ of generalized multiple-prism arrays since these arrays are widely used in optics in a variety of optics and quantum optics applications such as the following:

1. Laser intracavity beam expanders, in narrow-linewidth tunable laser oscillators
2. Extracavity beam expanders

3. Laser pulse compressors, in femtosecond and ultrafast pulse lasers
4. Dispersive elements in a variety of optical instruments such as spectrometers

Albeit multiple-prism arrays were first introduced by Newton (1704), a mathematical description of their dispersion had to wait, a long time, until their application as intracavity beam expanders in narrow-linewidth tunable lasers (Duarte and Piper, 1982).

6.2 Generalized Multiple-Prism Dispersion

Generalized multiple-prism arrays are illustrated on Figure 6.1. The aim here is to obtain generalized angular dispersion equations based on the basic prismatic geometry and the generalized refraction equations obtained in Chapter 5 (Duarte and Piper, 1982; Duarte, 2006). Considering the mth prism, of the arrangements, the angular relations are given by

$$\phi_{1,m} + \phi_{2,m} = \varepsilon_m \pm \alpha_m \tag{6.3}$$

$$\psi_{1,m} + \psi_{2,m} = \alpha_m \tag{6.4}$$

$$\sin \phi_{1,m} = \pm n_m \sin \psi_{1,m} \tag{6.5}$$

$$\sin \phi_{2,m} = \pm n_m \sin \psi_{2,m} \tag{6.6}$$

As illustrated in Figure 6.1, $\phi_{1,m}$ and $\phi_{2,m}$ are the angles of incidence and emergence, and $\psi_{1,m}$ and $\psi_{2,m}$ are the corresponding angles of refraction, at the mth prism. The sign alternative $\pm$ allows for either positive refraction or negative refraction.

Differentiating Equations 6.5 and 6.6 and using

$$\frac{d\psi_{1,m}}{dn} = -\frac{d\psi_{2,m}}{dn} \tag{6.7}$$

the single-pass dispersion following the mth prism is given by (Duarte and Piper, 1982; Duarte, 2006)

$$\nabla_\lambda \phi_{2,m} = \pm \mathcal{H}_{2,m} \nabla_\lambda n_m \pm (k_{1,m} k_{2,m})^{-1} \left(\mathcal{H}_{1,m} \nabla_\lambda n_m (\pm) \nabla_\lambda \phi_{2,(m-1)} \right) \tag{6.8}$$

where $\nabla_\lambda = \partial/\partial\lambda$ and the following geometrical identities apply

$$k_{1,m} = \frac{\cos \psi_{1,m}}{\cos \phi_{1,m}} \tag{6.9}$$

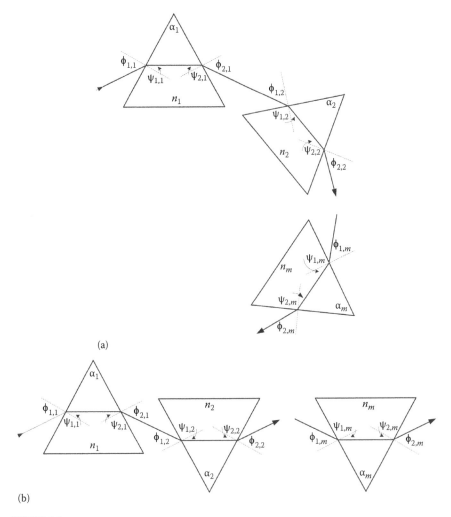

FIGURE 6.1
Generalized multiple-prism sequences: (a) positive configuration and (b) compensating configuration. Depiction of these generalized prismatic configurations was introduced by Duarte and Piper (1983).

$$k_{2,m} = \frac{\cos \phi_{2,m}}{\cos \psi_{2,m}} \tag{6.10}$$

$$\mathcal{H}_{1,m} = \frac{\tan \phi_{1,m}}{n_m} \tag{6.11}$$

$$\mathcal{H}_{2,m} = \frac{\tan \phi_{2,m}}{n_m} \tag{6.12}$$

The $k_{1,m}$ and $k_{2,m}$ factors represent the physical beam expansion experienced, at the mth prism, by the incidence and the emergence beams, respectively. In Equation 6.8 the sign $\pm$ alternative refers to either positive (+) or negative (−) refraction, while the same sign alternative in parenthesis (±) indicates whether the prismatic configuration is positive (+) or compensating (−). For positive refraction alone, Equation 6.8 becomes

$$\nabla_\lambda \phi_{2,m} = \mathcal{H}_{2,m} \nabla_\lambda n_m + (k_{1,m} k_{2,m})^{-1} \left(\mathcal{H}_{1,m} \nabla_\lambda n_m \pm \nabla_\lambda \phi_{2,(m-1)} \right) \tag{6.13}$$

The generalized *single-pass dispersion equation* indicates that the cumulative dispersion at the mth prism, namely, $\nabla_\lambda \phi_{2,m}$, is a function of the geometry of the mth prism, the position of the light beam relative to this prism, the material of this prism, and the cumulative dispersion up to the previous prism, $\nabla_\lambda \phi_{2,(m-1)}$ (Duarte and Piper, 1982, 1983).

For the special case of orthogonal beam exit, that is, $\phi_{2,m} = 0$ and $\psi_{2,m} = 0$, we have $\mathcal{H}_{2,m} = 0$, $k_{2,m} = 1$, and Equation 6.13 reduces to

$$\nabla_\lambda \phi_{2,m} = (k_{1,m})^{-1} \left(\mathcal{H}_{1,m} \nabla_\lambda n_m \pm \nabla_\lambda \phi_{2,(m-1)} \right) \tag{6.14}$$

For an array of r identical isosceles, or equilateral, prisms deployed symmetrically, in an additive configuration, for positive refraction, so that $\phi_{1,m} = \phi_{2,m}$, the cumulative dispersion reduces to (Duarte, 1990a)

$$\nabla_\lambda \phi_{2,r} = r \nabla_\lambda \phi_{2,1} \tag{6.15}$$

This is a simple dispersion equation that applies to the design of multiple-prism spectrometers incorporating identical, isosceles, or equilateral prisms arranged in symmetrical additive configurations.

The generalized single-pass dispersion equation for positive refraction (Equation 6.13) can be restated in a more practical and explicit notation (Duarte, 1989, 1990a)

$$\nabla_\lambda \phi_{2,r} = \sum_{m=1}^{r} (\pm 1) \mathcal{H}_{1,m} \left(\prod_{j=m}^{r} k_{1,j} \prod_{j=m}^{r} k_{2,j} \right)^{-1} \nabla_\lambda n_m$$

$$+ (M_1 M_2)^{-1} \sum_{m=1}^{r} (\pm 1) \mathcal{H}_{2,m} \left(\prod_{j=1}^{m} k_{1,j} \prod_{j=1}^{m} k_{2,j} \right) \nabla_\lambda n_m \tag{6.16}$$

where

$$M_1 = \prod_{j=1}^{r} k_{1,j} \tag{6.17}$$

$$M_2 = \prod_{j=1}^{r} k_{2,j} \qquad (6.18)$$

are the respective beam expansion factors. For the important practical case of r right angle prism, designed for orthogonal beam exit (i.e., $\phi_{2,m} = \psi_{2,m} = 0$), Equation 6.16 reduces to

$$\nabla_\lambda \phi_{2,r} = \sum_{m=1}^{r} (\pm 1)\mathcal{H}_{1,m}\left(\prod_{j=m}^{r} k_{1,j}\right)^{-1} \nabla_\lambda n_m \qquad (6.19)$$

If in addition the prism has identical apex angle ($\alpha_1 = \alpha_2 = \alpha_3 = \cdots = \alpha_m$) and is configured to have the same angle of incidence ($\phi_{1,1} = \phi_{1,2} = \phi_{1,3} = \cdots = \phi_{1,m}$), then Equation 6.19 can be written as (Duarte, 1985)

$$\nabla_\lambda \phi_{2,r} = \tan \psi_{1,1} \sum_{m=1}^{r} (\pm 1)\left(\frac{1}{k_{1,m}}\right)^{m-1} \nabla_\lambda n_m \qquad (6.20)$$

Further, if the angle of incidence for all prisms is Brewster's angle, then the single-pass dispersion reduces to the elegant expression

$$\nabla_\lambda \phi_{2,r} = \sum_{m=1}^{r} (\pm 1)\left(\frac{1}{n_m}\right)^m \nabla_\lambda n_m \qquad (6.21)$$

Alternative forms of expressing the generalized multiple-prism dispersion equation in series are given in Appendix D.

6.2.1 Example: Generalized Single-Prism Dispersion

For a single generalized prism, it is easy to show that the elegant generalized multiple-prism dispersion, Equation 6.13, reduces to

$$\nabla_\lambda \phi_{2,1} = \left(\frac{\sin \psi_{2,1}}{\cos \phi_{2,1}}\right)\nabla_\lambda n_1 + \left(\frac{\cos \psi_{2,1}}{\cos \phi_{2,1}}\right)\tan \psi_{1,1}\nabla_\lambda n_1 \qquad (6.22)$$

as given in well-known textbooks (Duarte, 1990a; Born and Wolf, 1999). Further, for the case of orthogonal beam exit ($\phi_{2,m} \approx \psi_{2,m} \approx 0$), Equation 6.22 reduces to (Wyatt, 1978)

$$\nabla_\lambda \phi_{2,1} \approx \tan \psi_{1,1}\nabla_\lambda n_1 \qquad (6.23)$$

The previously given examples are included to show that albeit general and elegant in its complete form, Equation 6.13 quickly leads to concrete results of practical interest to designers, laser practitioners, and optical engineers.

6.3 Double-Pass Generalized Multiple-Prism Dispersion

The evaluation of intracavity dispersion in tunable laser oscillators incorporating multiple-prism beam expanders requires the assessment of the double-pass, or return-pass, dispersion (Duarte and Piper, 1984; Duarte, 1990a). The double-pass dispersion of multiple-prism beam expanders was derived by thinking of the return pass as a mirror image of the first light passage as illustrated in Figure 6.2. The return-pass dispersion corresponds to the dispersion experienced by the return light beam at the first prism.

Thus, it is given by $\partial\phi'_{1,m}/\partial\lambda = \nabla_\lambda\phi'_{1,m}$ where the prime character indicates return pass (Duarte and Piper, 1982, 1984)

$$\nabla_\lambda\phi'_{1,m} = \mathcal{H}'_{1,m}\nabla_\lambda n_m + \left(k'_{1,m}k'_{2,m}\right)^{-1}\left(\mathcal{H}'_{2,m}\nabla_\lambda n_m \pm \nabla_\lambda\phi'_{1,(m+1)}\right) \tag{6.24}$$

where

$$k'_{1,m} = \frac{\cos\psi'_{1,m}}{\cos\phi'_{1,m}} \tag{6.25}$$

$$k'_{2,m} = \frac{\cos\phi'_{2,m}}{\cos\psi'_{2,m}} \tag{6.26}$$

$$\mathcal{H}'_{1,m} = \frac{\tan\phi'_{1,m}}{n_m} \tag{6.27}$$

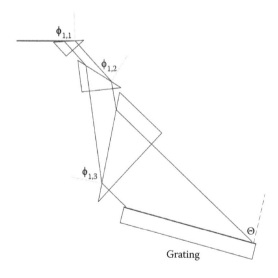

FIGURE 6.2
Multiple-prism grating assembly incorporating a three-prism beam expander designed for orthogonal beam exit.

$$\mathcal{H}'_{2,m} = \frac{\tan \phi'_{2,m}}{n_m} \tag{6.28}$$

Here, $\nabla_\lambda \phi'_{1,(m+1)}$ provides the cumulative single-pass multiple-prism dispersion plus the dispersion of the diffraction grating, that is,

$$\nabla_\lambda \phi'_{1,(m+1)} = \left(\nabla_\lambda \Theta_G \pm \nabla_\lambda \phi_{2,r} \right) \tag{6.29}$$

where $\nabla \lambda \Theta_G$ is the grating dispersion. If the grating is replaced by a mirror, then we simply have the prismatic contribution and

$$\nabla_\lambda \phi'_{1,(m+1)} = \nabla_\lambda \phi_{2,r} \tag{6.30}$$

Defining $\nabla_\lambda \phi'_{1,m} = \nabla \Phi_P$, where the capital ϕ stands for return pass and P for multiple prism, the explicit version of the generalized double-pass dispersion for a multiple-prism mirror system is given by (Duarte, 1985, 1989)

$$\nabla_\lambda \Phi_P = 2M_1 M_2 \sum_{m=1}^{r} (\pm 1) \mathcal{H}_{1,m} \left(\prod_{j=m}^{r} k_{1,j} \prod_{j=m}^{r} k_{2,j} \right)^{-1} \nabla_\lambda n_m$$

$$+ 2 \sum_{m=1}^{r} (\pm 1) \mathcal{H}_{2,m} \left(\prod_{j=1}^{m} k_{1,j} \prod_{j=1}^{m} k_{2,j} \right) \nabla_\lambda n_m \tag{6.31}$$

For the case of r right angle prism, designed for orthogonal beam exit (i.e., $\phi_{2,m} = \psi_{2,m} = 0$), Equation 6.31 reduces to

$$\nabla_\lambda \Phi_P = 2M_1 \sum_{m=1}^{r} (\pm 1) \mathcal{H}_{1,m} \left(\prod_{j=m}^{r} k_{1,j} \right)^{-1} \nabla_\lambda n_m \tag{6.32}$$

which can also be expressed as (Duarte, 1985)

$$\nabla_\lambda \Phi_P = 2 \sum_{m=1}^{r} (\pm 1) \left(\prod_{j=1}^{m} k_{1,j} \right) \tan \psi_{1,m} \nabla_\lambda n_m \tag{6.33}$$

If the angle of incidence for all prisms in the array is made equal to the Brewster angle, this equation simplifies further to (Duarte, 1990a)

$$\nabla_\lambda \Phi_P = 2 \sum_{m=1}^{r} (\pm 1) \left(n_m \right)^{m-1} \nabla_\lambda n_m \tag{6.34}$$

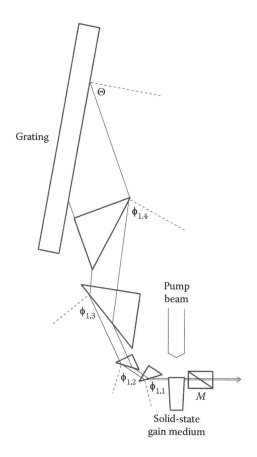

FIGURE 6.3
Dispersive long-pulse solid-state tunable laser oscillator incorporating a multiple-prism grating assembly. (Reproduced from Duarte, F.J. et al., *Appl. Opt.* 37, 3987, 1998, with permission from the Optical Society of America.)

The equations given in here are quite highly applicable to the design of multiple-prism beam expanders for narrow-linewidth tunable lasers (see Figure 6.3).

6.3.1 Design of Zero-Dispersion Multiple-Prism Beam Expanders

In practice, the dispersion of the grating, multiplied by the beam expansion, that is, $M(\nabla_\lambda \Theta_G)$, amply dominates the overall intracavity dispersion. Thus, it is sometimes advantageous to remove the dispersion component originating from the multiple-prism beam expander so that

$$\Delta\lambda \approx \Delta\theta_R \left(MR\nabla_\lambda\Theta_G \right)^{-1} \tag{6.35}$$

In such designs the tuning characteristics of the laser are those of the grating alone, around a specific wavelength.

The design of zero-dispersion, or quasi achromatic, multiple-prism beam expanders exhibiting orthogonal beam exit ($\phi_{2,m} \approx \psi_{2,m} \approx 0$), and made of identical material, involves the direct application of Equation 6.33 while setting $\nabla_\lambda \Phi_P = 0$. Thus, for a double-prism expander yielding zero dispersion, we obtain

$$(k_{1,1}) \tan \psi_{1,1} = (k_{1,1}) k_{1,2} \tan \psi_{1,2} \tag{6.36}$$

For a three-prism expander yielding zero dispersion, we obtain

$$(k_{1,1} + k_{1,1}k_{1,2}) \tan \psi_{1,1} = (k_{1,1}k_{1,2}) k_{1,3} \tan \psi_{1,3} \tag{6.37}$$

For a four-prism expander yielding zero dispersion, we obtain

$$(k_{1,1} + k_{1,1}k_{1,2} + k_{1,1}k_{1,2}k_{1,3}) \tan \psi_{1,1} = (k_{1,1}k_{1,2}k_{1,3}) k_{1,4} \tan \psi_{1,4} \tag{6.38}$$

For a five-prism expander yielding zero dispersion, we obtain

$$(k_{1,1} + k_{1,1}k_{1,2} + k_{1,1}k_{1,2}k_{1,3} + k_{1,1}k_{1,2}k_{1,3}k_{1,4}) \tan \psi_{1,1} = (k_{1,1}k_{1,2}k_{1,3}k_{1,4}) k_{1,5} \tan \psi_{1,5} \tag{6.39}$$

and so on. Here we should just reemphasize that all these configurations yield zero dispersion at the design wavelength, thus the use of the term quasi achromatic.

Optimized compact high-power solid-state multiple-prism laser oscillators have been demonstrated to yield single-longitudinal-mode oscillation at $\Delta\nu \approx 350$ MHz, at pulses $\Delta t \approx 3$ ns, near the limit allowed by the Heisenberg uncertainty principle (Duarte, 1999).

The oscillator, illustrated in Figure 6.4, requires the use of a small fused silica double-prism beam expander with $M \approx 42$, and $\phi_{2,m} \approx \psi_{2,m} \approx 0$, at $\lambda = 590$ nm. Thus we use Equation 6.33 to obtain Equation 6.36, which reduces to

$$\tan \psi_{1,1} = k_{1,2} \tan \psi_{1,2} \tag{6.40}$$

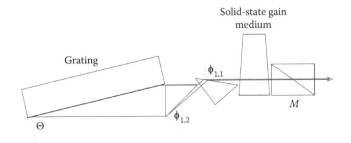

FIGURE 6.4
Optimized multiple-prism ($m = 2$) grating solid-state tunable laser oscillator delivering a linewidth ($\Delta\nu \approx 350$ MHz) near the limit allowed by the Heisenberg uncertainty principle. (Reproduced from Duarte, F.J., *Appl. Opt.* 38, 6347, 1999, with permission from the Optical Society of America.)

For $n = 1.4583$ and $\phi_{1,1} = 88.60°$, we get $\psi_{1,1} \approx 43.28°$, and $k_{1,1} \approx 29.80$. With these initial parameters, Equation 6.40 yields for the second prism $\phi_{1,2} \approx 53.93°$, $\psi_{1,2} \approx 33.66°$, and $k_{1,2} \approx 1.41$. Therefore, the overall intracavity beam expansion becomes

$$M = k_{1,1}k_{1,2} \approx 42.13$$

For a beam waist of $w = 100$ µm, this implies $2wM \approx 8.43$ mm. These dimensions require the first prism to have a hypotenuse of ~8 mm and the second prism a hypotenuse of ~10 mm. In this particular oscillator, this intracavity beam expansion is used to illuminate a 3300 lines/mm grating deployed at an angle of incidence ~77° in Littrow configuration (Duarte, 1999).

Duarte (2003) describes in detail the design of a zero-dispersion four-prism beam expander for $M = 103.48$, at $\lambda = 590$ nm. Shay and Duarte (2009) describe the design of a zero-dispersion five-prism beam expander for fused silica, at $\lambda = 1550$ nm ($n = 1.44402$), yielding an overall beam expansion of $M \approx 987$.

6.4 Multiple-Return-Pass Generalized Multiple-Prism Dispersion

Here we consider a multiple-prism grating or multiple-prism mirror assembly, for positive refraction, as illustrated in Figure 6.5. The light beam enters the first prism of the array, it is then expanded, and it is either diffracted back or reflected back into the multiple-prism array. In a dispersive laser oscillator, this process goes forth and back multiple times, thus giving rise to the

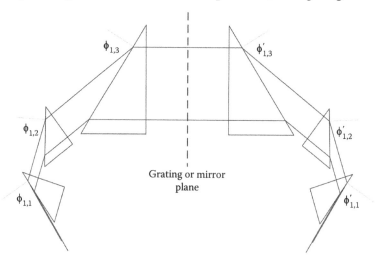

Grating or mirror plane

FIGURE 6.5

Multiple-prism grating assembly in its unfolded depiction. This type of description, for multiple-prism arrays, was first introduced by Duarte and Piper (1982).

concept of intracavity double pass or intracavity multiple-return pass. For the first return pass, toward the first prism in the array, the dispersion is given by

$$\nabla_\lambda \phi_{2,m} = \mathcal{H}_{2,m} \nabla_\lambda n_m + (k_{1,m} k_{2,m})^{-1} \left(\mathcal{H}_{1,m} \nabla_\lambda n_m \pm \nabla_\lambda \phi_{2,(m-1)} \right)$$

If N denotes the number of passes toward the grating, or reflecting element, and $2N$ the number of return passes, toward the first prism in the sequence, we have (Duarte and Piper, 1984)

$$(\nabla_\lambda \phi_{2,m})_N = \mathcal{H}_{2,m} \nabla_\lambda n_m + (k_{1,m} k_{2,m})^{-1} \left(\mathcal{H}_{1,m} \nabla_\lambda n_m \pm (\nabla_\lambda \phi_{2,(m-1)})_N \right) \tag{6.41}$$

and

$$(\nabla_\lambda \phi'_{1,m})_{2N} = \mathcal{H}'_{1,m} \nabla_\lambda n_m + (k'_{1,m} k'_{2,m})^{-1} \left(\mathcal{H}'_{2,m} \nabla_\lambda n_m \pm (\nabla_\lambda \phi'_{1,(m+1)})_{2N} \right) \tag{6.42}$$

For the first prism of the array (next to the gain medium), $(\nabla_\lambda \phi_{2,(m-1)})_N$ (with $N = 3, 5, 7...$) in Equation 6.41 is replaced by $(\nabla_\lambda \phi'_{1,1})_{2N}$ (with $N = 1, 2, 3 ...$).

Likewise, for the last prism of the assembly (next to the grating), $(\nabla_\lambda \phi'_{1,(m+1)})_{2N}$ (with $N = 1, 2, 3 ...$) in Equation 6.42 is replaced by $(\nabla_\lambda \Theta_G + (\nabla_\lambda \phi_{2,r})_N)$ (with $N = 1, 3, 5 ...$).

Thus, the multiple-return-pass dispersion for a multiple-prism grating assembly is given by (Duarte and Piper, 1984)

$$(\nabla_\lambda \theta)_R = \left(RM \nabla_\lambda \Theta_G + R \nabla_\lambda \Phi_P \right) \tag{6.43}$$

where $R = 2N$ is the number of return passes. This equation illustrates the very important fact that in the return-pass dispersion of a multiple-prism grating assembly, the dispersion of a grating is multiplied by the factor RM, where M is the overall beam magnification of the multiple-prism beam expander. Subsequently, the multiple-return-pass linewidth equation becomes (Duarte, 2001)

$$\Delta\lambda = \Delta\theta_R \left(RM \nabla_\lambda \Theta_G + R \nabla_\lambda \Phi_P \right)^{-1} \tag{6.44}$$

where $\Delta\theta_R$ is the multiple-return-pass beam divergence (Duarte, 2001, 2003) described in Chapter 9. Once again, if the grating is replaced by a mirror, that is, $\nabla_\lambda \Theta_G = 0$, the dispersion reduces to

$$(\nabla_\lambda \theta)_R = R \nabla_\lambda \Phi_P \tag{6.45}$$

which implies that the multiple-prism intracavity dispersion increases linearly as a function of R. The finite number R can be determined experimentally from the time delay observed between the leading edge of the excitation pulse and the leading edge of the narrow-linewidth emission pulse (Duarte and Piper, 1984). For narrow-linewidth high-power dispersive dye-laser oscillators, this number is typically $R \approx 3$ (Duarte, 2001).

6.4.1 Multiple-Prism Beam Compressors

Observing Figure 6.5, it becomes immediately apparent that, in a multiple-prism array, propagation from left to right leads to beam expansion, while propagation from right to left leads to *beam compression* as described by Duarte (2006). This is a geometrical beam compression effect different from temporal pulse compression as described in the next section.

Beam expansion and beam compression, in a given multiple-prism configuration, are symmetric phenomena. If beam expansion occurs in one direction, then beam compression occurs in the opposite direction as illustrated in Figure 6.5. Certainly, the equations of multiple-prism dispersion given here equally apply to both beam compressors and beam expanders. An explicit beam compressor is depicted in Figure 6.6 showing propagation from right to left. Numerous further examples of geometrical beam compressors are given by Duarte (2006). The beauty of multiple-prism beam compressors is that, ideally, they reduce the cross section of the propagating beam without inducing traditional focusing that leads to divergence of the beam beyond the focal point.

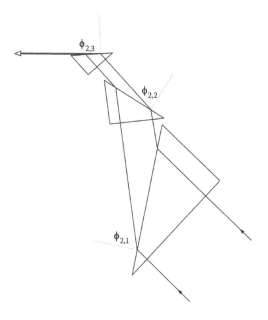

FIGURE 6.6
Multiple-prism geometrical beam compressor. In this configuration the incident beam enters orthogonally to the first larger prism. The arrows indicate the direction of the propagation. If the direction of the beam is reversed, then the compressor becomes an expander.

6.5 Multiple-Prism Dispersion and Laser Pulse Compression

The generation of femtosecond, attosecond, or ultrashort laser pulses is of fundamental interest to optics, quantum optics, and laser science in general (Diels and Rudolph, 2006).

From the uncertainty relation

$$\Delta v \Delta t \approx 1 \qquad (6.46)$$

it is immediately apparent that the generation of ultrashort time pulses (Δt) requires the simultaneous generation of a very wide spectral distribution (Δv). From the cavity linewidth equation

$$\Delta \lambda \approx \Delta \theta \left(\frac{\partial \theta}{\partial \lambda} \right)^{-1}$$

it also clear that the generation of very wide spectral emission requires the least amount of intracavity dispersion. Thus, it is necessary to understand and control all aspects of intracavity dispersion.

Pulse compression in ultrashort pulse, or femtosecond, lasers requires control of the first, second, and third derivatives of the intracavity dispersion. Using the identity

$$\nabla_n \phi_{2,m} = \nabla_\lambda \phi_{2,m} (\nabla_\lambda n_m)^{-1} \qquad (6.47)$$

Equation 6.13 can be restated as (Duarte, 2009)

$$\nabla_n \phi_{2,m} = \mathcal{H}_{2,m} + (\mathcal{M})^{-1} \left(\mathcal{H}_{1,m} \pm \nabla_n \phi_{2,(m-1)} \right) \qquad (6.48)$$

where the identity

$$k_{1,m}^{-1} k_{2,m}^{-1} = (\mathcal{M})^{-1} \qquad (6.49)$$

applies. Hence, the complete second derivative of the refraction angle, or first derivative of the dispersion $\nabla_\lambda \phi_{2,m}$, is given by (Duarte, 1987)

$$\nabla_n^2 \phi_{2,m} = \nabla_n \mathcal{H}_{2,m}$$
$$+ \left(\nabla_n \mathcal{M}^{-1} \right) \left(\mathcal{H}_{1,m} \pm \nabla_n \phi_{2,(m-1)} \right)$$
$$+ (\mathcal{M}^{-1}) \left(\nabla_n \mathcal{H}_{1,m} \pm \nabla_n^2 \phi_{2,(m-1)} \right) \qquad (6.50)$$

The second derivative of the dispersion $\nabla_\lambda \phi_{2,m}$ is given by (Duarte, 2009)

$$\nabla_n^3 \phi_{2,m} = \nabla_n^2 \mathcal{H}_{2,m} + \left(\nabla_n^2 \mathcal{M}^{-1}\right)\left(\mathcal{H}_{1,m} \pm \nabla_n \phi_{2,(m-1)}\right)$$

$$+ 2\left(\nabla_n \mathcal{M}^{-1}\right)\left(\nabla_n \mathcal{H}_{1,m} \pm \nabla_n^2 \phi_{2,(m-1)}\right) + \left(\mathcal{M}^{-1}\right)\left(\nabla_n^2 \mathcal{H}_{1,m} \pm \nabla_n^3 \phi_{2,(m-1)}\right)$$

$$(6.51)$$

the third derivative of the dispersion $\nabla_\lambda \phi_{2,m}$ is given by (Duarte, 2009)

$$\nabla_n^4 \phi_{2,m} = \nabla_n^3 \mathcal{H}_{2,m} + \left(\nabla_n^3 \mathcal{M}^{-1}\right)\left(\mathcal{H}_{1,m} \pm \nabla_n \phi_{2,(m-1)}\right) + 3\left(\nabla_n^2 \mathcal{M}^{-1}\right)\left(\nabla_n \mathcal{H}_{1,m} \pm \nabla_n^2 \phi_{2,(m-1)}\right)$$

$$+ 3\left(\nabla_n \mathcal{M}^{-1}\right)\left(\nabla_n^2 \mathcal{H}_{1,m} \pm \nabla_n^3 \phi_{2,(m-1)}\right) + \left(\mathcal{M}^{-1}\right)\left(\nabla_n^3 \mathcal{H}_{1,m} \pm \nabla_n^4 \phi_{2,(m-1)}\right) \quad (6.52)$$

and the fourth derivative of the dispersion $\nabla_\lambda \phi_{2,m}$ is given by (Duarte, 2009)

$$\nabla_n^5 \phi_{2,m} = \nabla_n^4 \mathcal{H}_{2,m} + \left(\nabla_n^4 \mathcal{M}^{-1}\right)\left(\mathcal{H}_{1,m} \pm \nabla_n \phi_{2,(m-1)}\right) + 4\left(\nabla_n^3 \mathcal{M}^{-1}\right)\left(\nabla_n \mathcal{H}_{1,m} \pm \nabla_n^2 \phi_{2,(m-1)}\right)$$

$$+ 6\left(\nabla_n^2 \mathcal{M}^{-1}\right)\left(\nabla_n^2 \mathcal{H}_{1,m} \pm \nabla_n^3 \phi_{2,(m-1)}\right) + 4\left(\nabla_n \mathcal{M}^{-1}\right)\left(\nabla_n^3 \mathcal{H}_{1,m} \pm \nabla_n^4 \phi_{2,(m-1)}\right)$$

$$+ \left(\mathcal{M}^{-1}\right)\left(\nabla_n^4 \mathcal{H}_{1,m} \pm \nabla_n^5 \phi_{2,(m-1)}\right)$$

$$(6.53)$$

and so on, for higher derivatives. By inspection, as stated by Duarte (2009), it can be seen that from the second term on the numerical factors can be pre-determined from Pascal's triangle relative to N where $(N + 1)$ is the order of the derivative.

Albeit the preceding exposition might appear a little bit abstract, these equations lead to specific numerical results (Duarte, 1987, 1990b). Osvay et al. (2004, 2005) have used the lower-order derivatives, given here, in practical femtosecond lasers to determine dispersions and laser pulse durations, for double-prism compressors, with excellent agreement between theory and experiments. The equations described here represent the complete description of the generalized multiple-prism dispersion theory applicable to pulse compression prismatic arrays in femtosecond, or ultrashort, pulse lasers and nonlinear optics.

Exact numerical calculations to determine $\nabla_n \phi_{2,m}$, and $\nabla_n^2 \phi_{2,m}$, for $m = 1, 2, 3, 4$, were performed by Duarte (1990). In these calculations the angle of incidence was deviated by minute amounts from the Brewster angle of incidence. Duarte (2009) provides exact values, as a function of the refractive index n, for $\nabla_n \phi_{2,m}$, $\nabla_n^2 \phi_{2,m}$, and $\nabla_n^3 \phi_{2,m}$. Simplifying assumptions include incidence at the Brewster angle of incidence, prisms of identical isosceles geometry, and made of the same material with refractive index $n_m = n$.

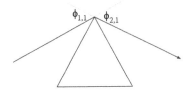

FIGURE 6.7
Single-prism pulse compressor.

6.5.1 Example: Single-Prism Pulse Compressor

For the single-prism laser pulse compressor (Dietel et al., 1983) ($m = 1$) of isosceles geometry, made of material with refractive index $n_m = n$ and deployed at the Brewster angle of incidence (see Figure 6.7), we find

$$\nabla_n \phi_{2,1} = 2 \tag{6.54}$$

$$\nabla_n^2 \phi_{2,1} = \left(4n - 2n^{-3} \right) \tag{6.55}$$

$$\nabla_n^3 \phi_{2,1} = \left(24n^2 + 8n^0 - 12n^{-2} + 6n^{-4} + 6n^{-6} \right) \tag{6.56}$$

6.5.2 Example: Double-Prism Pulse Compressor

A prism pulse compressor integrated by $m = 2$ prisms of identical isosceles geometry, made of the same material with refractive index $n_m = n$ and deployed at the Brewster angle of incidence in *compensating configuration*, is depicted in Figure 6.8 (Diels et al., 1985). The compensating configuration requires the subtraction of the previous dispersive derivatives and the use of $\nabla_n \phi_{1,2} = \nabla_n \phi_{2,1}$ as a geometrical nexus between stages. Careful evaluation of all relevant identities and their correct substitution yield

$$\nabla_n \phi_{2,2} = 0 \tag{6.57}$$

$$\nabla_n^2 \phi_{2,2} = 0 \tag{6.58}$$

$$\nabla_n^3 \phi_{2,2} = 0 \tag{6.59}$$

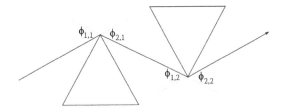

FIGURE 6.8
Double-prism pulse compressor.

as would be expected from geometrical and symmetrical arguments. However, the correct mathematical evaluation of $(\nabla_n^3\phi_{2,2})$ requires considerable attention to detail given the numerous identities involved.

6.5.3 Example: Four-Prism Pulse Compressor

A four-prism compressor (Fork et al., 1984) is formed by unfolding the double-prism configuration about a symmetry axis perpendicular to the exit beam depicted in Figure 6.7, as illustrated in Figure 6.9. For exactly balanced compensating prism arrays composed of two pairs of compensating prisms, it can be shown that

$$\nabla_n\phi_{2,1} = \nabla_n\phi_{2,3} = 2 \tag{6.60}$$

$$\nabla_n^2\phi_{2,1} = \nabla_n^2\phi_{2,3} = \left(4n - 2n^{-3}\right) \tag{6.61}$$

$$\nabla_n^3\phi_{2,1} = \nabla_n^3\phi_{2,3} = \left(24n^2 + 8n^0 - 12n^{-2} + 6n^{-4} + 6n^{-6}\right) \tag{6.62}$$

$$\nabla_n\phi_{2,2} = \nabla_n\phi_{2,4} = 0 \tag{6.63}$$

$$\nabla_n^2\phi_{2,2} = \nabla_n^2\phi_{2,4} = 0 \tag{6.64}$$

$$\nabla_n^3\phi_{2,2} = \nabla_n^3\phi_{2,4} = 0 \tag{6.65}$$

A six-prism pulse compressor has been used in semiconductor laser pulse compression by Pang et al. (1992).

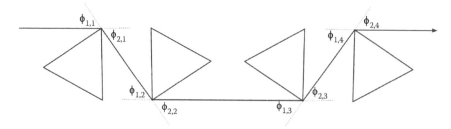

FIGURE 6.9
Four-prism pulse compressor.

Problems

6.1 Use Equation 6.4 to obtain the identity given in Equation 6.7.

6.2 Derive Equation 6.14 from 6.13 using for $\phi_{2,m} \approx \psi_{2,m} \approx 0$.

6.3 Derive Equation 6.15.

6.4 Derive Equation 6.20 from 6.19.

6.5 Use Equation 6.13 to obtain the single-prism Equation 6.22.

6.6 Use Equation 6.33, set $\nabla\lambda\Phi_p = 0$, to derive Equation 6.39, for a five-prism zero-dispersion beam expander.

6.7 For a single-prism laser pulse compressor, derive Equations 6.54 through 6.56 (Hint: see Duarte, 2009).

6.8 Using the methodology described in the three-prism pulse compressor examples given, develop a set of equations applicable to a symmetrical six-prism pulse compressor (Hint: the dispersions of the first three prisms, deployed in additive configuration, are compensated by the second set of three prisms).

References

Born, M. and Wolf, E. (1999). *Principles of Optics*, 7th edn. Cambridge, New York.

Diels, J. -C., Dietel, W., Fontaine, J. J., Rudolph, W., and Wilhelmi, B. (1985). Analysis of a mode-locked ring laser: Chirped-solitary-pulse solutions. *J. Opt. Soc. Am. B*, **2**, 680–686.

Diels, J. -C. and Rudolph, W. (2006). *Ultrafast Laser Pulse Phenomena*, 2nd edn. Academic, New York.

Dietel, W., Fontaine, J. J., and Diels, J. -C. (1983). Intracavity pulse compression with glass: A new method of generating pulses shorter than 60 fs. *Opt. Lett.* **8**, 4–6.

Dirac, P. A. M. (1978). *The Principles of Quantum Mechanics*, 4th edn. Oxford, London, U.K.

Duarte, F. J. (1985). Note on achromatic multiple-prism beam expanders. *Opt. Commun.* **53**, 259–262.

Duarte, F. J. (1987). Generalized multiple-prism dispersion theory for pulse compression in ultrafast dye lasers. *Opt. Quantum Electron.* **19**, 223–229.

Duarte, F. J. (1989). Transmission efficiency in achromatic nonorthogonal multiple-prism laser beam expanders. *Opt. Commun.* **71**, 1–5.

Duarte, F. J. (1990a). Narrow-linewidth pulsed dye laser oscillators. In *Dye Laser Principles* (Duarte, F. J. and Hillman, L. W., eds.). Academic, New York, Chapter 4.

Duarte, F. J. (1990b). Prismatic pulse compression: Beam deviations and geometrical perturbations. *Opt. Quantum Electron.* **22**, 467–471.

Duarte, F. J. (1991). Dispersive dye lasers. In *High Power Dye Lasers* (Duarte, F. J., ed.). Springer-Verlag, Berlin, Germany, Chapter 2.

Duarte, F. J. (1992). Cavity dispersion equation $\Delta\lambda \approx \Delta\theta(\partial\theta/\partial\lambda)^{-1}$: A note on its origin. *Appl. Opt.* **31**, 6979–6982.

Duarte, F. J. (1993). On a generalized interference equation and interferometric measurements. *Opt. Commun.* **103**, 8–14.

Duarte, F. J. (1999). Multiple-prism grating solid-state dye laser oscillator: Optimized architecture. *Appl. Opt.* **38**, 6347–6349.

Duarte, F. J. (2001). Multiple-return-pass beam divergence and the linewidth equation. *Appl. Opt.* **40**, 3038–3041.

Duarte, F. J. (2003). *Tunable Laser Optics*, Elsevier-Academic, New York.

Duarte, F. J. (2006). Multiple-prism dispersion equations for positive and negative refraction. *Appl. Phys. B* **82**, 35–38.

Duarte, F. J. (2009). Generalized multiple-prism dispersion theory for laser pulse compression: Higher order phase derivatives. *Appl. Phys. B* **96**, 809–814.

Duarte, F. J. and Piper, J. A. (1982). Dispersion theory of multiple-prism beam expander for pulsed dye lasers. *Opt. Commun.* **43**, 303–307.

Duarte, F. J. and Piper, J. A. (1983). Generalized prism dispersion theory. *Am. J. Phys.* **51**, 1132–1134.

Duarte, F. J. and Piper, J. A. (1984). Multi-pass dispersion theory of prismatic pulsed dye lasers. *Optica. Acta* **31**, 331–335.

Duarte, F. J., Taylor, T. S., Costela, A., Garcia-Moreno, I., and Sastre, R. (1998). Longpulse narrow-linewidth dispersive solid-state dye-laser oscillator. *Appl. Opt.* **37**, 3987–3989.

Fork, R. L., Martinez, O. M., and Gordon, J. P. (1984). Negative dispersion using pairs of prisms. *Opt. Lett.* **9**, 150–152.

Newton, I. (1704). *Opticks*, Royal Society, London, U.K.

Osvay, K., Kovács, A. P., Heiner, Z., Kurdi, G., Klebniczki, J., and Csatári, M. (2004). Angular dispersion and temporal change of femtosecond pulses from misaligned pulse compressors. *IEEE J. Selec. Top. Quantum Electron.* **10**, 213–220.

Osvay, K., Kovács, A. P., Kurdi, G., Heiner, Z., Divall, M., Klebniczki, J., and Ferincz, I. E. (2005). Measurement of non-compensated angular dispersion and the subsequent temporal lengthening of femtosecond pulses in a CPA laser. *Opt. Commun.* **248**, 201–209.

Pang, L. Y., Fujimoto, J. G., and Kintzer, E. S. (1992). Ultrashort-pulse generation from high-power diode arrays by using intracavity optical nonlinearities. *Opt. Lett.* **17**, 1599–1601.

Shay, T. M. and Duarte, F. J. (2009). Tunable fiber lasers. In *Tunable Laser Applications*, 2nd edn. (Duarte, F. J., ed.). CRC, New York, Chapter 6.

Wyatt, R. (1978). Comment on "On the dispersion of a prism used as a beam expander in a nitrogen laser." *Opt. Commun.* **26**, 9–11.

7

Dirac Notation Identities

7.1 Useful Identities

As described by Dirac (1978) and Feynman et al. (1965), the Dirac notation includes various mathematical properties and allows for various abstractions and permutations. Here, a few useful set of identities and properties of the notation are described.

First, the complex conjugate of $\langle \phi | \psi \rangle$ is defined as

$$\langle \phi | \psi \rangle = \langle \psi | \phi \rangle^* \tag{7.1}$$

Also, the probability amplitude

$$\langle \phi | \psi \rangle = \langle \phi | j \rangle \langle j | \psi \rangle \tag{7.2}$$

can be expressed in abstract form as

$$| \psi \rangle = | j \rangle \langle j | \psi \rangle \tag{7.3}$$

An additional form of abstract notation is

$$\langle \chi | A | \phi \rangle = \langle \chi | i \rangle \langle i | A | j \rangle \langle j | \phi \rangle \tag{7.4}$$

where A is

$$A = | i \rangle \langle i | A | j \rangle \langle j | \tag{7.5}$$

Another abstraction is illustrated by

$$A | \phi \rangle = | i \rangle \langle i | A | j \rangle \langle j | \phi \rangle \tag{7.6}$$

Further, A can be multiplied by B so that

$$\langle \chi | BA | \phi \rangle = \langle \chi | i \rangle \langle i | B | j \rangle \langle j | A | k \rangle \langle k | \phi \rangle \tag{7.7}$$

To express

$$\langle \chi | A | \phi \rangle = \langle \chi | i \rangle \langle i | A | j \rangle \langle j | \phi \rangle \tag{7.8}$$

in the abstract form

$$\langle \chi | \psi \rangle = \langle \chi | i \rangle \langle i | \psi \rangle \tag{7.9}$$

it is necessary that

$$\langle i | \psi \rangle = \langle i | A | j \rangle \langle j | \phi \rangle = \langle i | A | \phi \rangle \tag{7.10}$$

which means that

$$\langle \chi | \psi \rangle = \langle \chi | A | \phi \rangle \tag{7.11}$$

Further abstracting leads to

$$| \psi \rangle = | A | \phi \rangle \tag{7.12}$$

Other examples of abstractions include

$$\langle i | \phi \rangle = C_i \tag{7.13}$$

$$\langle i | \chi \rangle = D_i \tag{7.14}$$

$$| \phi \rangle = \sum_i | i \rangle C_i \tag{7.15}$$

$$| \chi \rangle = \sum_i | i \rangle D_i \tag{7.16}$$

$$\langle \chi | = \sum_i D_i^* \langle i | \tag{7.17}$$

which is the abstracted version of

$$\langle \chi \,|\, \phi \rangle = \sum_{ij} D_j^* \langle j \,|\, i \rangle C_i \tag{7.18}$$

and since $\langle j|i \rangle = \delta_{ij}$

$$\langle \chi \,|\, \phi \rangle = \sum_i D_i^* C_i \tag{7.19}$$

Finally, using the ultimate abstraction suggested by Dirac (Feynman et al., 1965), Equation 7.19 can be expressed as

$$|= \sum_i |i\rangle\langle i| \tag{7.20}$$

7.1.1 Example

The probability amplitude describing interference in a Mach–Zehnder interferometer can be described as (Duarte, 2003)

$$\langle x \,|\, s \rangle = \sum_{kj} \langle x \,|\, k \rangle\langle k \,|\, j \rangle\langle j \,|\, s \rangle \tag{7.21}$$

Defining

$$\langle j \,|\, s \rangle = C_j \tag{7.22}$$

$$\langle k \,|\, x \rangle = D_k \tag{7.23}$$

$$\langle x \,|\, s \rangle = \sum_{kj} D_k^* \langle k \,|\, j \rangle C_j \tag{7.24}$$

and using $\langle k|j \rangle = \delta_{kj}$, we get

$$\langle x \,|\, s \rangle = \sum_j D_j^* C_j \tag{7.25}$$

7.2 Linear Operations

Dirac (1978) describes various mathematical properties related to his *ket* vectors. First, if c_1 and c_2 are complex numbers, *ket* vectors can be multiplied by these complex numbers and added to produce a new *ket* vector

$$c_1 |\phi\rangle + c_2 |\psi\rangle = |\theta\rangle \qquad (7.26)$$

Superposition of a state, with itself, yields the original state

$$c_1 |\phi\rangle + c_2 |\phi\rangle = (c_1 + c_2) |\phi\rangle \qquad (7.27)$$

Additional sum and product conditions are illustrated by

$$\langle\phi|(|\psi\rangle + |\chi\rangle) = \langle\phi|\psi\rangle + \langle\phi|\chi\rangle \qquad (7.28)$$

$$\langle\phi|(c|\psi\rangle) = c\langle\phi|\psi\rangle \qquad (7.29)$$

and

$$((\langle\phi| + \langle\chi|)|\psi\rangle = \langle\phi|\psi\rangle + \langle\chi|\psi\rangle \qquad (7.30)$$

$$(c\langle\phi|)|\psi\rangle = c\langle\phi|\psi\rangle \qquad (7.31)$$

If α is a linear operator, then

$$|\vartheta\rangle = \alpha|\psi\rangle \qquad (7.32)$$

and

$$\alpha(|\psi\rangle + |\chi\rangle) = \alpha|\psi\rangle + \alpha|\chi\rangle \qquad (7.33)$$

$$\alpha(c|\psi\rangle) = c\alpha|\psi\rangle \qquad (7.34)$$

$$(\alpha + \beta)|\psi\rangle = \alpha|\psi\rangle + \beta|\psi\rangle \qquad (7.35)$$

$$(\alpha\beta)|\psi\rangle = \alpha(\beta|\psi\rangle) \qquad (7.36)$$

$$((\langle\phi|\alpha)|\psi\rangle = \langle\phi|(\alpha|\psi\rangle) \qquad (7.37)$$

Further useful identities introduced by Dirac (1978) are

$$|\phi\rangle |\psi\rangle = |\psi\rangle |\phi\rangle \qquad (7.38)$$

$$|\phi\rangle |\psi\rangle = |\psi\rangle |\phi\rangle = |\phi\psi\rangle \qquad (7.39)$$

$$|\phi\rangle |\psi\rangle |\chi\rangle ... = |\phi\psi\chi...\rangle \qquad (7.40)$$

Also, for more than one particle, Dirac (1978) gives the *ket* for the assembly as

$$|X\rangle = |a_1\rangle |b_2\rangle |c_3\rangle ... |g_n\rangle \qquad (7.41)$$

7.2.1 Example

The Pryce–Ward probability amplitude, prior to normalization, for entangled photons, with polarizations x and y, traveling in opposite directions 1 and 2, is given by

$$|\psi\rangle = \left(|x\rangle_1 |y\rangle_2 - |y\rangle_1 |x\rangle_2\right) \qquad (7.42)$$

which can also be expressed as

$$|\psi\rangle = \left(|x,y\rangle - |y,x\rangle\right) \qquad (7.43)$$

Problems

7.1 Write in abstract form the probability amplitude corresponding to a Sagnac interferometer given by (Duarte, 2003)

$$\langle x|s\rangle = \sum_{kj} \langle x|k\rangle\langle k|j\rangle\langle j|s\rangle$$

assuming that $\langle k|j\rangle = 1$ (see Chapter 10).

7.2 The probability amplitude for a multiple-beam interferometer (see Chapter 10) can be expressed as (Duarte, 2003)

$$\langle x|s\rangle = \langle x|m\rangle\langle m|l\rangle\langle l|k\rangle\langle k|j\rangle\langle j|s\rangle$$

Use the various abstract identities, given in this chapter, to efficiently abstract this probability amplitude.

References

Dirac, P. A. M. (1978). *The Principles of Quantum Mechanics*, 4th edn. Oxford, London, U.K.

Duarte, F. J. (2003). *Tunable Laser Optics*, Elsevier-Academic, New York.

Feynman, R. P., Leighton, R. B., and Sands, M. (1965). *The Feynman Lectures on Physics*, Vol. III, Addison-Wesley, Reading, MA.

8

Laser Excitation

8.1 Introduction

Lasers are essentially quantum devices. The process of stimulated emission is a quantum phenomenon. Stimulated emission is essential to the generation of spatially and spectrally coherent radiation, which is also quantum phenomena. Thus, albeit initially a macroscopic device, the laser emits radiation that is intrinsically quantum in character. In this chapter, we consider a first and essential step in the creation of a laser: laser excitation.

Here, we consider the process of laser excitation and emission, in a gain medium, from a semi classical and quantum perspective. In Chapter 9 we examine the optics phenomena applicable to generate spatially and spectrally coherent radiation.

8.2 Brief Laser Overview

The word *laser* has its origin in an acronym of the words *light amplification* by *stimulated emission* of *radiation*. However, the laser is readily associated with the spatial and spectral coherence characteristics of its emission.

A laser is a device that transforms electrical energy, chemical energy, or incoherent optical energy into coherent optical emission. This coherence is both spatial and spectral. Spatial coherence means a highly directional light beam, with little divergence, and spectral coherence means an extremely pure color of emission. These concepts of spatial and spectral coherence are intimately related to the Heisenberg uncertainty principle:

$$\Delta p \Delta x \approx h$$

An alternative way to cast this idea is to think of the laser as a device that transforms ordinary incoherent energy into an extremely well-defined form of energy, in both the spatial and the spectral domain.

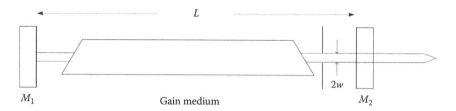

FIGURE 8.1
Basic laser resonator. It is comprised of an atomic, or molecular, gain medium and two mirrors aligned along the optical axis. The length of the cavity is L and the diameter of the beam is $2w$. The gain medium can be excited either optically or electrically.

Physically, the laser consists of an atomic or molecular gain medium optically aligned within an optical resonator, or optical cavity, as depicted in Figure 8.1. When excited by electrical energy, or optical energy, the atoms or molecules in the gain medium oscillate at optical frequencies. This oscillation is maintained and sustained by the optical resonator, or optical cavity. In this regard, the laser is analogous to a mechanical or radio oscillator but oscillating at extremely high frequencies. For the green color of $\lambda \approx 500$ nm, the equivalent frequency is $v \approx 5.99 \times 10^{14}$ Hz. A direct comparison between a *laser oscillator* and an old radio oscillator makes the atomic, or molecular gain medium, equivalent to the vacuum tube and the elements of the optical cavity equivalent to the resistance capacitances and inductances.

The spectral purity of the emission of a laser is related to how narrow its linewidth (Δv) is. High-power broadband tunable lasers can exhibit a linewidth in the $4.5 \le \Delta\lambda \le 10$ nm range (see, e.g., Schäfer et al., 1966). High-power pulsed narrow-linewidth lasers can have single-longitudinal-mode linewidths of $\Delta v \approx 350$ MHz (i.e., $\Delta\lambda \approx 0.0004$ nm, at 590 nm) near the limit allowed by the Heisenberg uncertainty principle (Duarte, 1999). Single longitudinal mode means that all the emission radiation is contained in a single electromagnetic mode.

Low-power continuous-wave (CW) narrow-linewidth lasers can offer much narrower linewidths approaching the kHz regime. Cooled-stabilized CW lasers can yield $\Delta v \approx 1$ Hz or even less (Kessler et al., 2012).

In the language of the laser literature, a laser emitting narrow-linewidth radiation is referred to as a *laser oscillator*, or *master oscillator* (MO). High-power narrow-linewidth emission is attained when an MO is used to inject a *laser amplifier*, or *power amplifiers* (PAs). Large high-power systems include several MOPA chains with each chain including several amplifiers. The difference between an oscillator and an amplifier is that the amplifier simply stores energy to be released up on the arrival of the narrow-linewidth oscillator signal. In some cases the amplifiers are configured within unstable resonator cavities in what is referred to as a *forced oscillator* (FO). When that is the case, the amplifier is called an FO and the integrated configuration is referred to as a MOFO system.

Appendix A includes a number of tables summarizing the spectral emission from various gas-, liquid-, and solid-state lasers.

8.2.1 Laser Optics

Laser optics, as defined in Duarte (2003), refers to the individual optics elements that comprise laser cavities, to the optics ensembles that comprise laser cavities, and to the physics that results from the propagation of laser radiation. In addition, the subject of laser optics includes instrumentation employed to characterize laser radiation and instrumentation that incorporates lasers. The physics and architecture of tunable narrow-linewidth laser oscillators is presented in Chapter 9, while a broad survey of laser cavities is given in Appendix B.

8.3 Laser Excitation

As already mentioned, lasers can be excited via several forms of energy including electrical, optical, chemical, and even nuclear. Electrical excitation can be for lasers in either the gaseous or the solid state. Electrical excitation in the gaseous state gives origin to gas lasers, while in the solid state it gives origin mainly to semiconductor lasers.

Optical excitation can be used in gaseous-, liquid-, or solid-state, gain media. Here we provide a brief survey of examples of laser gain media in the gaseous, liquid, and solid state.

8.3.1 Electrically Excited Gas Lasers

Electrically excited gas lasers, form a very broad class of lasers that includes high-power excimer lasers, metal-vapor lasers, and CO_2 lasers. It also includes an array of CW metal ion lasers. A list of these lasers, including their respective transitions, is included in Appendix A.

Here, to illustrate how some of these lasers are excited, we'll refer to metal ion lasers such as the He–Zn laser and in particular to a subclass of lasers known as the He–Zn halogen lasers, that is, He–$ZnBr_2$, He–$ZnCl_2$, and He–ZnI_2.

These lasers need a buffer gas that is a rare gas. In this case that rare gas is helium. Very briefly, the rare-gas metal hollow-cathode discharge is excited electrically as the impedance of an electrical circuit as depicted in Figure 8.2.

The metal, in this case zinc, is evaporated into the discharge, thus creating a He–Zn discharge. The metal species are excited via Duffendack reactions, that is,

$$He^+ + M \rightarrow He + (M^+)^* + \Delta E \tag{8.1}$$

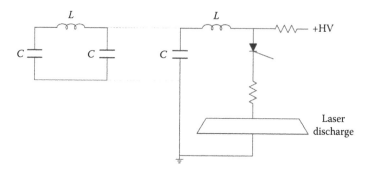

FIGURE 8.2
Transmission line excitation circuit of rare-gas metal-vapor laser discharge. An LC circuit of up to 10 sections serves as a pulsed forming network and the pulse is switched via a high-voltage thyristor. (From Duarte, F.J., *Excitation Processes in Continuous Wave Rare Gas-Metal Halide Vapour Lasers*, Macquarie University, Sydney, 1977.)

This means that electrons ionize helium atoms, thus creating ions He^+ that in turn collide with metal atoms, thus yielding excited metal ions $(M^+)^*$. The energy defect of the reaction is ΔE. These lasers need very energetic electrons to ionize helium and thus utilize hollow-cathode discharges (Piper and Gill, 1975).

An additional excitation mechanism is known as Penning ionization in which electrons excite helium atoms to a metastable state $He^*(2^3S_1)$, so that

$$He^*(2^3S_1) + M \rightarrow He + (M^+)^* + \Delta E \qquad (8.2)$$

An energy level diagram illustrating the zinc transitions due to both Duffen-dack reactions and Penning ionization is illustrated in Figure 8.3. Notice that since the laser transitions occur between specific atomic levels, these transitions are specific in wavelength and intrinsically narrow linewidth.

In lasers such as He–CdI$_2$, in addition to the metal transitions, iodine transitions are added to the emission, thus giving rise to white-light lasing (Piper, 1976). An advantage of the metal-halide vapor lasers over the pure metal-vapor lasers is that they need lower operational temperatures. However, this complicates the excitation cycle since the metal-halide molecule needs to be dissociated prior to excitation and needs to recombine following laser emission. Chemical recombination is critical to the successful continuation of the excitation cycle. A time-resolved study in the He–ZnX$_2$ systems, where X$_2$ refers to the halogen, led to the conclusion that second-order reactions of the form

$$M + X_m \xrightarrow{k_1} MX_2 \qquad (8.3)$$

are the likely process of recombination. The solution is

$$M(t) = \left([X_m]_0 - [M]_0\right)\left(\frac{[X_m]_0}{[M]_0} e^{\tau_1 t} - 1\right)^{-1} \qquad (8.4)$$

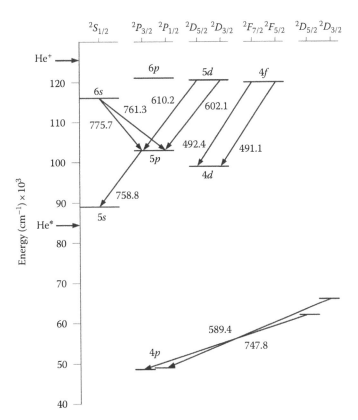

FIGURE 8.3
Partial energy level diagram of the He–Zn laser. Upper transitions are excited via Duffendack
reactions in hollow-cathode discharge lasers, while the lower transitions originate from
Penning ionization. The energies corresponding to helium ionization and the relevant helium
metastable are indicated. Transition wavelengths are given in nm. See Appendix A for a listing
of visible ionic laser transitions from various elements.

where the initial concentrations are $[X_m]_0 > [M]_0$. Here, the decay rate (in s^{-1})
is given by

$$\tau_1 = ([X_m]_0 - [M]_0)k_1 \tag{8.5}$$

For hollow-cathode rare-gas metal-vapor lasers, the measured neutral metal
decay rates are $(1.6 \le \tau_1 \le 2.0) \times 10^4 \, s^{-1}$ for $ZnBr_2$ and $(0.8 \le \tau_1 \le 1.0) \times 10^4 \, s^{-1}$
for ZnI_2 (Duarte, 1977; Duarte and Piper, 1985).

  In summary, for rare-gas metal-halide vapor lasers, the excitation cycle
begins with the dissociation of the MX_2 molecules, followed by either
Duffendack or Penning excitation of the metal atom. Following excitation
and emission, the metal atom decays to its neutral state and undergoes
chemical recombination with the halogen, and the cycle continues.

The population dynamics of this type of laser can be analyzed using rate equations, while the transition cross sections can be either measured or estimated using quantum methods (Willett, 1974).

8.3.2 Optically Pumped Gas and Liquid Lasers

Optical excitation of lasers can be accomplished either via the use of incoherent means, such as flashlamps, or by using direct laser excitation (see, e.g., Duarte, 2003). Here, we consider two examples of laser excitation of two types of distinct molecular lasers.

First, we briefly consider the laser-pumped molecular iodine dimer laser, I_2. This molecule has sufficient vapor pressure at room temperature and can be excited longitudinally using a laser compatible with its absorption characteristics. Excitation lasers include pulsed copper vapor lasers (Kaslin et al., 1980), frequency-doubled Nd:YAG lasers (Byer et al., 1972), and narrow-linewidth tunable dye lasers (Duarte and Piper, 1986). Laser excitation of molecular dimer lasers has also been accomplished in the CW regime (see, e.g., Wellegehausen, 1979).

The transitions for molecular iodine belong to the $B^3\Pi_{ou}^+ - X^1\Sigma_g^+$ electronic system. Specific narrow-linewidth excitation of lower-lying vibrational–rotational levels in the lower electronic state $X^1\Sigma_g^+$ results in the population of higher-lying vibrational–rotational levels in the $B^3\Pi_{ou}^+$ state. Subsequently, transitions from those higher-lying vibrational–rotational levels, in the $B^3\Pi_{ou}^+$ state, are observed toward higher-lying vibrational–rotational levels, in the $X^1\Sigma_g^+$ state. This type of selective narrow-linewidth excitation leads to a series of specific, and discrete, narrow-linewidth vibrational–rotational transitions. For instance, narrow-linewidth excitation at $\lambda_p \approx 510.55$ yields a series of *discrete* lines in the red to near infrared (Duarte and Piper, 1986).

A different class of optically pumped molecular laser is the liquid organic dye laser. Organic laser dyes are enormous molecules with very large molecular weights (in the 175–1000 m_u range; see Duarte, 2003). A consequence of this extraordinary characteristic is that each electronic state of a laser-dye molecule includes multitudes of closely lying, and overlapping, vibrational–rotational levels. This is the feature that provides the continuous tunability of the dye laser. Figure 8.4 shows the molecular structure of the coumarin 545 tetramethyl laser dye that exhibits an approximate tuning range of 500 $\leq \lambda \leq 570$ nm (Duarte et al., 2006). The tuning curve of a simple grating resonator using this green laser dye is shown in Figure 8.5. The emission available from laser dyes spans the spectrum continuously from ~330 to ~900 nm (Duarte, 2003).

The excitation dynamics of dye lasers is described later in this chapter. Liquid organic dye lasers are enormously versatile and lase either in the CW regime (see, e.g., Hollberg, 1990) or in the pulsed regime (Duarte, 2003). Their liquid gain media are particularly apt to the removal of excess heat. Hence, dye lasers are very suitable to the generation of high average

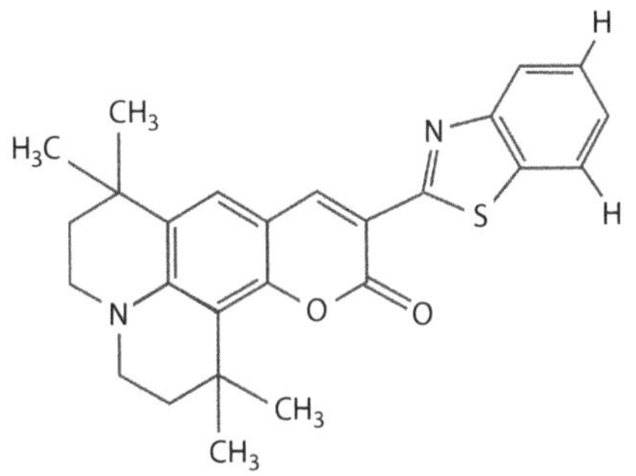

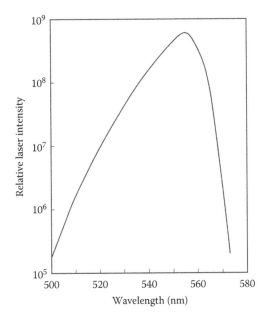

FIGURE 8.4
Molecular structure of Coumarin 545 T laser dye. A laser-dye molecule, such as this, in an ethanol solution becomes the gain medium for a dye laser. (Reproduced from Duarte, F.J. et al., *J. Opt. A: Pure Appl. Opt.* 8, 172, 2006, with permission from the Institute of Physics.)

FIGURE 8.5
Tuning curve of the emission from the Coumarin 545 T laser dye. (Reproduced from Duarte, F.J. et al., *J. Opt. A: Pure Appl. Opt.* 8, 172, 2006, with permission from the Institute of Physics.)

powers and very large pulsed energy. Output characteristics of dye lasers are summarized in Appendix A.

Transverse laser excitation of a narrow-linewidth tunable dye-laser oscillator is illustrated in Figure 8.6. As it will be discussed later, the population dynamics of dye lasers can be analyzed using rate equations and the transition cross section is mainly obtained from measurements (Duarte, 2003).

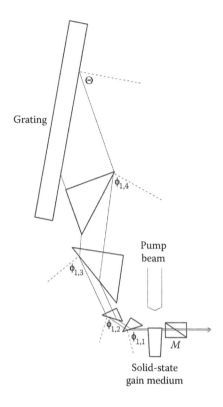

FIGURE 8.6

Transverse excitation of narrow-linewidth dye-laser oscillator. The gain medium here is a laser-dye-doped polymer. (Reproduced from Duarte, F.J. et al., *Appl. Opt.* 37, 3987, 1998, with permission from the Optical Society of America.)

8.3.3 Optically Pumped Solid-State Lasers

The first visible laser was a flashlamp-pumped crystalline laser: the ruby laser (Maiman, 1960). Cr^{3+}–Al_3O_2 lases at $\lambda = 694.3$ nm via the $^2E(\bar{E}) - {}^4A_2$ transition in a three-level energy excitation scheme as illustrated in Figure 8.7. On the other hand, the widely tunable Ti–sapphire laser emits in the $660 \leq \lambda \leq 986$ nm range in a two-level energy system that operates like a four-level laser, thus allowing its wide tuning range (Barnes, 1995a).

There is a large variety of optically pumped solid-state lasers that include transition metal solid-state lasers (Barnes, 1995a) and optical parametric oscillators (Barnes, 1995b; Orr et al., 2009). More recently, fiber lasers have become highly developed and widely used in many applications (Popov, 2009).

Diode-laser excitation has become widely applied in the excitation of solid-state lasers. Figure 8.8 shows the simplified energy level diagram for diode-laser excitation of a Nd:YAG laser and a schematic of a longitudinal excitation scheme. Again, a brief survey of widely used optically pumped solid-state lasers is given in Appendix A.

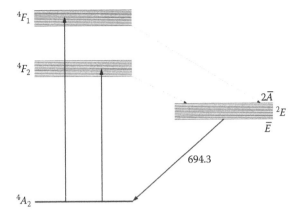

FIGURE 8.7
Three-level energy diagram for the ruby laser. Optical pumping to either $^4F_1$ or $^4F_2$ results in rapid nonradiative decay to ^2E from where laser action takes place to the ground level $^4A_2$. The wavelength corresponding to the transition $^2E(\bar{E}) - {}^4A_2$ is $\lambda = 694.3$ nm.

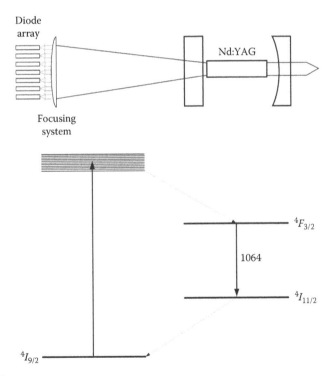

FIGURE 8.8
Diode-pumped Nd:YAG laser, using longitudinal pumping with a near IR diode-laser array and corresponding four-level energy diagram. Optical pumping leads to rapid nonradiative decay to $^4F_{3/2}$ from where laser action takes place to the $^4I_{11/2}$ level. The wavelength corresponding to the transition $^4F_{3/2} - {}^4I_{11/2}$ is $\lambda = 1064$ nm.

The population dynamics of optically excited solid-state lasers can be analyzed using rate equations, while the transition cross sections can be derived from spectral measurements (Barnes, 1995a).

8.3.4 Electrically Excited Semiconductor Lasers

The excitation and emission process in semiconductor lasers can be described via Schrödinger's equation as discussed in Chapter 12. The beauty of semiconductor lasers is that they can be directly excited using basic electric circuitry as illustrated in Figure 8.9.

In semiconductor laser materials, emission occurs between a conduction band and a lower valence band as illustrated in Figure 8.10. This is intrinsically a quantum effect.

The emission wavelength in these lasers depends on the energy difference between the conduction and the valence band, the *band gap* ($E_G = h\nu$), and is inherently tunable. Emission spectral characteristics of representative semiconductor lasers, such as GaAlAs, are given in Appendix A.

Most semiconductor lasers emit in the visible and near infrared. An interesting semiconductor laser is the quantum cascade laser (QCL) that covers an impressive segment of the infrared spectrum $3 \leq \lambda \leq 24$ μm (Silfvast, 2008). These lasers operate on transitions between quantized conduction band states of multiple-quantum well structures. The only carries are electrons. A single stage consists of an injector and an active region. An electron is injected at $n = 3$ of the quantum well, and as a photon is emitted, the electron transitions to $n = 2$. This is a multiple process so that

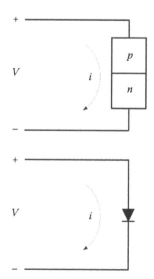

FIGURE 8.9
Simple excitation circuit of generic semiconductor laser.

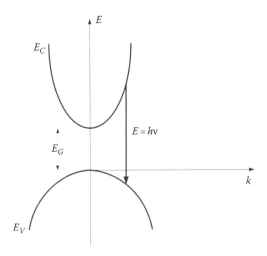

FIGURE 8.10
Conduction and valence bands in a semiconductor emitter.

one electron can emit a large number of photons. The emission wavelength is given by (see Chapter 12)

$$\lambda = (3^2 - 2^2)^{-1} \frac{8mcL_x^2}{h} \tag{8.6}$$

where L_x is the thickness of the quantum well. Further emission details on QCLs are given in Appendix A.

8.4 Excitation and Emission Dynamics

There are various methods and approaches to describe the dynamics of excitation in the gain media of lasers. Approaches range from complete quantum mechanical treatments to rate equation descriptions (Haken, 1970). A complete survey of energy level diagrams corresponding to gain media in the gaseous, the liquid, and solid states is given by Silfvast (2008). Here, a basic description of laser excitation mechanisms is given using energy levels and classical rate equations applicable to tunable organic molecular gain media. This description is based on the standard approach to the subject (see, e.g., Peterson, 1979) and follows a review given by Duarte (2003).

8.4.1 Rate Equations for a Two-Level System

A simplified two-level molecular system is depicted in Figure 8.11. The pump laser intensity $I_p(t)$ populates the upper energy level N_1 from the ground

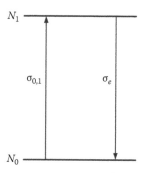

FIGURE 8.11
Simplified two-level energy diagram used to describe a basic rate equation system.

state N_0. The emission from the upper state is designated as $I_l(x, t, \lambda)$. Thus, the time evolution of the upper- or excited-state population is written as

$$\frac{\partial N_1}{\partial t} = N_0 \sigma_{0,1} I_p(t) - N_1 \sigma_e I_l(x,t,\lambda) \tag{8.7}$$

where
$\sigma_{0,1}$ is the absorption cross section
σ_e is the emission cross section

Cross sections have units of cm², time has units of seconds (s), the populations have units of molecules cm⁻³, and the intensities have units of photons cm⁻² s⁻¹. The transition cross sections are quantum mechanical in origin and are described later in this chapter.

The dynamics of the pump intensity $I_p(t)$ is described by

$$c^{-1}\frac{\partial I_p(t)}{\partial t} = -N_0\sigma_{0,1}I_p(t) \tag{8.8}$$

where c is the speed of light. In reference to Figure 8.11, the dynamics of the emission intensity $I_l(x, t, \lambda)$ depends on the difference between the upper-level population and lower-level population so that

$$c^{-1}\frac{\partial I_l(x,t,\lambda)}{\partial t} + \frac{\partial I_l(x,t,\lambda)}{\partial x} = \left(N_1\sigma_e - N_0\sigma'_{0,1}\right)I_l(x,t,\lambda) \tag{8.9}$$

In the steady state this equation reduces to

$$\frac{\partial I_l(x,\lambda)}{\partial x} \approx \left(N_1\sigma_e - N_0\sigma'_{0,1}\right)I_l(x,\lambda) \tag{8.10}$$

and integration yields

$$I_l(x,\lambda) = I_l(0,\lambda) e^{(N_1 \sigma_e - N_0 \sigma_{0,1}^l)L} \qquad (8.11)$$

The exponential terms in Equation 8.11 are referred to as the *gain*. It can be easily seen that laser threshold is reached for $N_1 \sigma_e \geq N_0 \sigma_{0,1}^l$ and that for strong laser action we need

$$N_1 \sigma_e \gg N_0 \sigma_{0,1}^l$$

8.4.2 Dynamics of a Multiple-Level System

The literature on rate equations includes the works of Ganiel et al. (1975), Teschke et al. (1976), Penzkofer and Falkenstein (1978), Dujardin and Flamant (1978), Peterson (1979), Munz and Haag (1980), Haag et al. (1983), Nair and Dasgupta (1985), and Jensen (1991). The rate equation approach given here incorporates several of the elements common in the published literature and emphasizes the frequency-selective aspects of the dynamics as outlined by Duarte (2003). This approach applies to laser-dye gain media in either the *liquid* or the *solid state*.

An energy level diagram for a laser-dye molecule (see, Figure 8.4) is included in Figure 8.12. S_0, S_1, and S_2 are the *electronic states* of the molecule, while T_1 and T_2 represent the *triplet states*, which are detrimental to laser emission. Laser emission occurs via $S_1 \to S_0$ transitions.

An important feature of laser dyes is that each electronic state contains a large number of overlapping vibrational–rotational levels. This multitude of closely lying vibrational–rotational levels is the origin of the broadband gain and tunability in dye lasers.

In reference to the energy level diagram of Figure 8.12 and considering only vibrational manifolds at each electronic state, a set of rate equations for transverse excitations can be written as (Duarte, 2003)

$$N = \sum_{S=0}^{m} \sum_{v=0}^{m} N_{S,v} + \sum_{T=1}^{m} \sum_{v=0}^{m} N_{T,v} \qquad (8.12)$$

$$\frac{\partial N_{1,0}}{\partial t} \approx \sum_{v=0}^{m} N_{0,v} \sigma_{0,10,v} I_p(t) + \sum_{v=0}^{m} N_{0,v} \sigma_{0,10,v}^l I_l(x,t,\lambda_v) + \frac{N_{2,0}}{\tau_{2,1}}$$

$$- N_{1,0} \left(\sum_{v=0}^{m} \sigma_{1,20,v} I_p(t) + \sum_{v=0}^{m} \sigma_{e0,v} I_l(x,t,\lambda_v) \right.$$

$$\left. + \sum_{v=0}^{m} \sigma_{1,20,v}^l I_l(x,t,\lambda_v) + \left(k_{S,T} + \tau_{1,0}^{-1} \right) \right) \qquad (8.13)$$

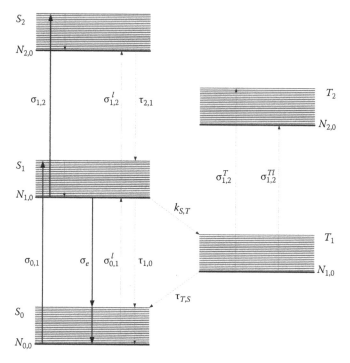

FIGURE 8.12

Energy level diagram corresponding to a laser-dye molecule. The electronic singlet states are S_0, S_1, and S_2 plus the triplet levels T_1 and T_2. Notice that each electronic level includes a multitude of closely lying vibrational–rotational levels. Laser emission takes place from the lowest vibro-rotational level in S_1 to S_0. The presence of a manifold of closely lying vibro-rotational levels at S_0, allowing a range of energies, is what gives rise to wavelength tunability. Using cavity design techniques described in Chapter 9, and Appendix B, tunable narrow-linewidth oscillation can be achieved.

$$\frac{\partial N_{T_{1,0}}}{\partial t} \approx N_{1,0} k_{S,T} - \frac{N_{T_{1,0}}}{\tau_{T,S}} - N_{T_{1,0}} \left(\sum_{v=0}^{m} \sigma_{1,20,v}^{T} I_p(t) + \sum_{v=0}^{m} \sigma_{1,20,v}^{Tl} I_l(x,t,\lambda_v) \right) \tag{8.14}$$

$$c^{-1} \frac{\partial I_p(t)}{\partial t} \approx - \left(N_{0,0} \sum_{v=0}^{m} \sigma_{0,10,v} + N_{1,0} \sum_{v=0}^{m} \sigma_{1,20,v} + N_{T_{1,0}} \sum_{v=0}^{m} \sigma_{1,20,v}^{T} \right) I_p(t) \tag{8.15}$$

$$c^{-1} \frac{\partial I_l(x,t,\lambda)}{\partial t} + \frac{\partial I_l(x,t,\lambda)}{\partial x} \approx N_{1,0} \sum_{v=0}^{m} \sigma_{e0,v} I_l(x,t,\lambda_v) - \sum_{v=0}^{m} N_{0,v} \sigma_{0,10,v}^{l} I_l(x,t,\lambda_v)$$

$$- N_{1,0} \sum_{v=0}^{m} \sigma_{1,20,v}^{l} I_l(x,t,\lambda_v) - N_{T_{1,0}} \sum_{v=0}^{m} \sigma_{1,20,v}^{Tl} I_l(x,t,\lambda_v) \tag{8.16}$$

$$I_l(x,t,\lambda) = \sum_{v=0}^{m} I_l(x,t,\lambda_v) \tag{8.17}$$

$$I_l(x,t,\lambda) = I_l^+(x,t,\lambda) + I_l^-(x,t,\lambda) \tag{8.18}$$

In this set of equations, frequency dependence is incorporated via the summation terms, and variables, depending on the vibrational assignment v. The equation parameters are as follows (see Figure 8.12):

1. $I_p(t)$ is the intensity of the pump laser beam. Units are photons $cm^{-2}\,s^{-1}$.
2. $I_l(x, t, \lambda)$ is the laser emission from the gain medium. Units are photons $cm^{-2}\,s^{-1}$.
3. $N_{S,v}$ refers to the population of the S electronic state at the v vibrational level. It is given as a number per unit volume (cm^{-3}).
4. $N_{T,v}$ refers to the population of the T triplet state at the v vibrational level. It is given as a number per unit volume (cm^{-3}).
5. The absorption cross sections, such as $\sigma_{0,10,v}$, are identified by a subscript S'', $S'_{v'',v'}$ that refers the electronic $S'' \rightarrow S'$ transition and the vibrational transition $v'' \rightarrow v'$. The same convention applies to the triplet levels. Units are cm^2.
6. The *emission* cross sections, $\sigma_{e0,v}$, are identified by the subscript $e_{v',v''}$. Units are cm^2.
7. Radiationless decay times, such as $\tau_{1,0}$, are identified by subscripts that denote the corresponding $S' \rightarrow S''$ transition. Units are s.
8. $k_{S,T}$ is a radiationless decay rate from the singlet to the triplet. Units are s^{-1}.

Ignoring the vibrational manifolds and other finer details, Equations 8.12 through 8.16 can be expressed in reduced form as

$$N = N_0 + N_1 + N_T \tag{8.19}$$

$$\frac{\partial N_1}{\partial t} \approx N_0\sigma_{0,1}I_p(t) + \left(N_0\sigma_{0,1}^l - N_1\sigma_e - N_1\sigma_{1,2}^l\right)I_l(x,t,\lambda) - N_1\left(k_{S,T} + \tau_{1,0}^{-1}\right) \tag{8.20}$$

$$\frac{\partial N_T}{\partial t} = N_1 k_{S,T} - N_T\tau_{T,S}^{-1} - N_T\sigma_{1,2}^{Tl}I_l(x,t,\lambda) \tag{8.21}$$

$$c^{-1}\frac{\partial I_p(t)}{\partial t} = -\left(N_0\sigma_{0,1} + N_1\sigma_{1,2}\right)I_p(t) \tag{8.22}$$

$$c^{-1}\frac{\partial I_l(x,t,\lambda)}{\partial t} + \frac{\partial I_l(x,t,\lambda)}{\partial x} = \left(N_1\sigma_e - N_0\sigma_{0,1}^l - N_1\sigma_{1,2}^l - N_T\sigma_{1,2}^{Tl}\right)I_l(x,t,\lambda) \tag{8.23}$$

TABLE 8.1

Laser Excitation Parameters for the Rhodamine 6G Molecule

Symbol	Measured Value	λ (nm)	Reference
$\sigma_{0,1}$	1.66×10^{-16} cm²	510	Hargrove and Kan (1980)
$\sigma_{0,1}$	4.50×10^{-16} cm²	530	Everett (1991)
$\sigma_{1,2}$	0.40×10^{-16} cm²	510	Hammond (1979)
σ_e	1.86×10^{-16} cm²	572	Hargrove and Kan (1980)
σ_e	1.30×10^{-16} cm²	600	Everett (1991)
$\sigma_{0,1}^l$	1.0×10^{-19} cm²	600	Everett (1991)
$\sigma_{1,2}^l$	1.0×10^{-17} cm²	600	Everett (1991)
$\sigma_{1,2}^T$	1.0×10^{-17} cm²	530	Everett (1991)
$\sigma_{1,2}^{Tl}$	4.0×10^{-17} cm²	600	Everett (1991)
$\tau_{1,0}$	4.8×10^{-9} s		Tuccio and Strome (1972)
$\tau_{2,1}$	1.0×10^{-12} s		Hargrove and Kan (1980)
$\tau_{T,S}$	1.1×10^{-7} s		Tuccio and Strome (1972)
$k_{S,T}$	8.2×10^{6} s⁻¹		Tuccio and Strome (1972)

This simplified set of equations is similar to the equations disclosed by Teschke et al. (1976). This type of rate equations can be effectively applied to simulate numerically the dynamics of dye-laser intensity as a function of the laser-pump intensity and dye molecular concentration. Relevant cross sections and excitation rates are given in Table 8.1.

It should be noted that since organic dye gain media exhibit homogeneous broadening, the introduction of intracavity frequency-selective optics (see Chapter 9) enables all the molecules to contribute efficiently to tunable narrow-linewidth emission.

8.4.3 Long-Pulse Approximation

For long-pulse or CW emission, a simplified set of equations is possible, thus opening the alternative to closed form solutions. Assuming that the time derivatives vanish, Equations 8.20 through 8.23 reduce to

$$N_0\sigma_{0,1}I_p + \left(N_0\sigma_{0,1}^l - N_1\sigma_e - N_1\sigma_{1,2}^l\right)I_l(x,\lambda) = N_1\left(k_{S,T} + \tau_{1,0}^{-1}\right) \tag{8.24}$$

$$N_1 k_{S,T} = N_T\tau_{T,S}^{-1} + N_T\sigma_{1,2}^{Tl}I_l(x,\lambda) \tag{8.25}$$

$$N_0\sigma_{0,1} = -N_1\sigma_{1,2} \tag{8.26}$$

$$\frac{\partial I_l(x,\lambda)}{\partial x} = \left(N_1\sigma_e - N_0\sigma_{0,1}^l - N_1\sigma_{1,2}^l - N_T\sigma_{1,2}^{Tl}\right)I_l(x,\lambda) \tag{8.27}$$

Using *triplet-level quenchers* such as O_2 and C_8H_8 (see, e.g., Duarte, 1990), it is possible to neutralize the effect of triplets so that the intensity given in Equation 8.27 simplifies to

$$\frac{\partial I_l(x,\lambda)}{\partial x} = \left(N_1\sigma_e - N_0\sigma_{0,1}^l - N_1\sigma_{1,2}^l\right)I_l(x,\lambda) \tag{8.28}$$

from which it follows that

$$I_l(x,\lambda) = I(0,\lambda)e^{\left(N_1(\sigma_e - \sigma_{1,2}^l) - N_0\sigma_{0,1}^l\right)L} \tag{8.29}$$

Thus, the gain can be expressed as

$$g = \left(N_1(\sigma_e - \sigma_{1,2}^l) - N_0\sigma_{0,1}^l\right)L \tag{8.30}$$

From this relation it can be deduced that in the absence of triplet losses, gain can be achieved for

$$N_1(\sigma_e - \sigma_{1,2}^l) > N_0\sigma_{0,1}^l \tag{8.31}$$

8.4.4 Example

Equation 8.24, and the excitation parameters given in Table 8.1, can be used to determine the laser pump intensity necessary to overcome threshold. In this regard, just below threshold, that is, $I_l(x, \lambda) \approx 0$, Equation 8.24 reduces to

$$I_p \approx \left(\sigma_{0,1}\tau_{1,0}\right)^{-1}\frac{N_1}{N_0} \tag{8.32}$$

in the absence of triplets. This means that to approach population inversion, using rhodamine 6G under visible laser excitation, pump intensities exceeding $\sim 5 \times 10^{23}$ photons cm^{-2} s^{-1} are necessary (see also Dienes and Yankelevich, 1998).

8.5 Quantum Transition Probabilities and Cross Sections

Albeit the dynamics of laser excitation can be described using classical rate equations, an examination of transition probabilities, and transition cross sections, requires a quantum treatment. Here, this is done via the Dirac notation (Dirac, 1978) while adopting the Feynman approach (Feynman et al., 1965). An introduction to the Dirac notation is given in Chapter 4 and a summary of useful identities is available in Chapter 7. The following description is based on a review given by Duarte (2003).

Here, we begin from the basic Dirac principles:

$$\langle \phi | \psi \rangle = \sum_j \langle \phi | j \rangle \langle j | \psi \rangle \tag{8.33}$$

$$\langle \phi | \psi \rangle = \langle \psi | \phi \rangle^* \tag{8.34}$$

$$\langle i | j \rangle = \delta_{ij} \tag{8.35}$$

For $j = 1, 2$, Equation 8.33 can be expanded into

$$\langle \phi | \psi \rangle = \langle \phi | 2 \rangle \langle 2 | \psi \rangle + \langle \phi | 1 \rangle \langle 1 | \psi \rangle \tag{8.36}$$

or

$$\langle \phi | \psi \rangle = \langle \phi | 2 \rangle C_2 + \langle \phi | 1 \rangle C_1 \tag{8.37}$$

where

$$C_1 = \langle 1 | \psi \rangle \tag{8.38}$$

and

$$C_2 = \langle 2 | \psi \rangle \tag{8.39}$$

Following Feynman, we express the derivative of the C_j amplitudes, with respect to time, as (Feynman et al., 1965; Dirac, 1978)

$$i\hbar \frac{dC_j}{dt} = \sum_k^2 H_{jk} C_k \tag{8.40}$$

where H_{jk} is the *Hamiltonian*.

Next, using the Feynman notation, new amplitudes C_I and C_{II} are defined as linear combinations of C_1 and C_2. Furthermore, since

$$\langle II | II \rangle = \langle II | 1 \rangle \langle 1 | II \rangle + \langle II | 2 \rangle \langle 2 | II \rangle = 1 \tag{8.41}$$

the normalization factor $2^{-1/2}$ is introduced, so that

$$C_{II} = \frac{1}{\sqrt{2}} (C_1 + C_2) \tag{8.42}$$

$$C_I = \frac{1}{\sqrt{2}} (C_1 - C_2) \tag{8.43}$$

Next, the Hamiltonian of the molecule under the effect of an electric field is allowed to be perturbed so that the matrix elements of the Hamiltonian become

$$H_{11} = E_0 + \mu \mathcal{E} \tag{8.44}$$

$$H_{12} = -A \tag{8.45}$$

$$H_{21} = -A \tag{8.46}$$

$$H_{22} = E_0 - \mu \mathcal{E} \tag{8.47}$$

where

$$\mathcal{E} = \mathcal{E}_0 (e^{i\omega t} + e^{-i\omega t}) \tag{8.48}$$

and μ corresponds to the electric dipole moment. The term $\mathcal{E}\mu$ is known as the *perturbation term*. Expanding Equation 8.40, followed by subtraction and addition, leads to

$$i\hbar \frac{dC_I}{dt} = (E_0 + A)C_I + \mu \mathcal{E} C_{II} \tag{8.49}$$

$$i\hbar \frac{dC_{II}}{dt} = (E_0 - A)C_{II} + \mu \mathcal{E} C_I \tag{8.50}$$

Assuming a small electric field, solutions are of the form

$$C_I = D_I e^{-iE_I / \hbar t} \tag{8.51}$$

$$C_{II} = D_{II} e^{-iE_{II} / \hbar t} \tag{8.52}$$

where

$$E_I = E_0 + A \tag{8.53}$$

and

$$E_{II} = E_0 - A \tag{8.54}$$

Assuming that $(\omega + \omega_0)$ oscillates too rapidly to contribute to the rate of change of D_I and D_{II}, we can write

$$i\hbar \frac{dD_I}{dt} = \mu \mathscr{E}_0 D_{II} e^{-i(\omega-\omega_0)t} \tag{8.55}$$

$$i\hbar \frac{dD_{II}}{dt} = \mu \mathscr{E}_0 D_I e^{i(\omega-\omega_0)t} \tag{8.56}$$

If at $t = 0$, $D_I \approx 1$, then integration of Equation 8.56 leads to

$$D_{II} = \frac{\mu \mathscr{E}_0}{\hbar} \left(\frac{1 - e^{i(\omega-\omega_0)T}}{(\omega - \omega_0)} \right) \tag{8.57}$$

and following multiplication with its complex conjugate,

$$|D_{II}|^2 = \left(\frac{\mu \mathscr{E}_0}{\hbar} \right)^2 \left(\frac{2 - 2\cos(\omega - \omega_0)T}{(\omega - \omega_0)^2} \right) \tag{8.58}$$

which can be written as (Feynman et al., 1965)

$$|D_{II}|^2 = \left(\frac{\mu \mathscr{E}_0 T}{\hbar} \right)^2 \frac{\sin^2\left(\frac{1}{2}(\omega - \omega_0)T\right)}{\left(\frac{1}{2}(\omega - \omega_0)T\right)^2} \tag{8.59}$$

This is the probability for the transition $I \rightarrow II$ during the time segment T. This result is central to the theory of absorption and radiation of light by atoms and molecules. It can be further shown that

$$|D_I|^2 = |D_{II}|^2 \tag{8.60}$$

which means that the physics for the stimulated emission probability is the same as the physics for the absorption probability.

Using $\mathscr{S} = 2\varepsilon_0 c \mathscr{E}_0^2$ and replacing μ by $3^{-1/2}\mu$ (Sargent et al., 1974), the transition probability can be written as

$$|D_{II}|^2 = \frac{2\pi}{3} \left(\frac{\mu^2 T^2}{4\pi\varepsilon_0 c\hbar^2} \right) \mathscr{S}(\omega_0) \frac{\sin^2\left(\frac{1}{2}(\omega - \omega_0)T\right)}{\left(\frac{1}{2}(\omega - \omega_0)T\right)^2} \tag{8.61}$$

where
  μ is the dipole moment in units of Cm
  $(1/4\pi\varepsilon_0)$ is in units of Nm^2 C^{-2}
  $\mathscr{S}(\omega_0)$ is the intensity in units of J s^{-1} m^{-2}

Integrating Equation 8.61 with respect to the frequency ω, the dimensionless transition probability becomes

$$|D_{II}|^2 = \frac{4\pi^2}{3}\left(\frac{\mu^2}{4\pi\varepsilon_0 c\hbar^2}\right)\left(\frac{\mathcal{S}(\omega_0)}{\Delta\omega}\right)T \qquad (8.62)$$

It then follows that the cross section for the transition can be written as

$$\sigma = \frac{4\pi^2}{3}\left(\frac{\mu^2}{4\pi\varepsilon_0 c\hbar}\right)\left(\frac{\omega}{\Delta\omega}\right) \qquad (8.63)$$

in units of m^2 (although the more widely used unit is cm^2; see Table 8.1) For a simple atomic, or molecular, system, the dipole moment can be calculated from the definition (Feynman et al., 1965)

$$\mu_{mn}\xi = \langle m\,|\,H\,|\,n\rangle = H_{mn} \qquad (8.64)$$

where H_{mn} is the matrix element of the Hamiltonian. For a simple diatomic molecule, the dependence of this matrix element on the Franck–Condon factor ($q_{v',v''}$) and the square of the transition moment ($|R_e|^2$) are described by Chutjian and James (1969).

Byer et al. (1972) wrote an expression for the gain, of vibrational–rotational transitions, of the form

$$g = \sigma NL \qquad (8.65)$$

or more specifically

$$g = \frac{4\pi^2}{3}\left(\frac{1}{4\pi\varepsilon_0 c\hbar}\right)\left(\frac{\omega}{\Delta\omega}\right)|R_e|^2\,q_{v',v''}\left(\frac{S_{J''}}{(2J''+1)}\right)NL \qquad (8.66)$$

where
$S_{J''}$ is known as the line strength
J'' identifies a specific rotational level

In practice, however, cross sections are mostly determined experimentally as in the case of those listed in Table 8.1.

Going back to Equation 8.62 and rearranging its terms, it follows that the intensity can be expressed as a function of the transition probability

$$\mathcal{S}(\omega_0) = \frac{3}{\pi}\left(\frac{\varepsilon_0 c\hbar^2}{\mu^2}\right)\left(\frac{\Delta\omega}{T}\right)|D_{II}|^2 \qquad (8.67)$$

or

$$\mathcal{S}(\omega_0) = \frac{3}{\pi} \kappa \left(\frac{\Delta\omega}{T} \right) |D_{II}|^2 \tag{8.68}$$

with

$$\kappa = \left(\frac{\varepsilon_0 c \hbar^2}{\mu^2} \right) \tag{8.69}$$

where the units for the constant κ are J s m^{-2}. Subsequently, the intensity $\mathcal{S}(\omega_0)$ has units of J s^{-1} m^{-2} or W m^{-2}.

8.5.1 Long-Pulse Approximation

For a very long pulse, Equation 8.61 can be approximated as

$$|D_{II}|^2 \approx \frac{2}{3} \left(\frac{\mu^2}{\varepsilon_0 c \hbar^2} \right) (\omega - \omega_0)^{-2} \mathcal{S}(\omega_0) \tag{8.70}$$

which can be written as

$$\mathcal{S}(\omega) \approx \frac{3}{2} \kappa \left(\Delta\omega \right)^2 |D_{II}|^2 \tag{8.71}$$

This approximation indicates that the intensity is proportional to the square of the frequency difference multiplied by the probability of the transition, or $(\Delta\omega)^2 |D_{II}|^2$, and is given in units of J s^{-1} m^{-2} or W m^{-2}.

Problems

8.1 Show that in the steady state Equation 8.20 becomes Equation 8.24.

8.2 Show that in the steady state Equation 8.23 becomes Equation 8.27.

8.3 Show that by neglecting the triplet state, Equation 8.27 can be expressed as Equation 8.29.

8.4 Starting from Equations 8.49 to 8.50, derive an expression for $|D_I|^2$ and show that it is equal to $|D_{II}|^2$.

8.5 Use Equation 8.62 to arrive at the expression for the transition cross section given in Equation 8.63.

8.6 Show that the dimensions of the intensity given in Equation 8.71 are in W m^{-2}.

References

Barnes, N. P. (1995a). Transition metal solid-state lasers. In *Tunable Lasers Handbook* (Duarte, F. J., ed.). Academic, New York, Chapter 6.

Barnes, N. P. (1995b). Optical parametric oscillators. In *Tunable Lasers Handbook* (Duarte, F. J., ed.). Academic, New York, Chapter 7.

Byer, R. L., Herbst, R. L., Kildal, H., and Levenson, M. D. (1972). Optically pumped molecular iodine vapor-phase laser. *Appl. Phys. Lett.* **20**, 463–466.

Chutjian, A. and James, T. C. (1969). Intensity measurements in the $B^3\Pi_u^+ - X^1\Sigma_g^+$ system of I_2. *J. Chem. Phys.* **51**, 1242–1249.

Dienes A. and Yankelevich, D. R. (1998). Tunable dye lasers. In *Encyclopedia of Applied Physics*, Vol. 22 (Trigg, G. L., ed.). Wiley-VCH, New York, pp. 299–334.

Dirac, P. A. M. (1978). *The Principles of Quantum Mechanics*, 4th edn. Oxford University, London, U.K.

Duarte, F. J. (1977). *Excitation Processes in Continuous Wave Rare Gas-Metal Halide Vapour Lasers*, Macquarie University, Sydney.

Duarte, F. J. (1990). Technology of pulsed dye lasers. In *Dye Laser Principles* (Duarte, F. J. and Hillman, L. W., eds.). Academic, New York, Chapter 6.

Duarte, F. J. (1999). Solid-state multiple-prism grating dye laser oscillator: Optimized architecture. *Appl. Opt.* **38**, 6347–6349.

Duarte, F. J. (2003). *Tunable Laser Optics*, Elsevier-Academic, New York.

Duarte, F. J., Liao, L. S., Vaeth, K. M., and Miller, A. M. (2006). Widely tunable green laser emission using the coumarin 545 tetramethyl dye as the gain medium. *J. Opt. A: Pure Appl. Opt.* **8**, 172–174.

Duarte, F. J. and Piper, J. A. (1985). Measurements of neutral metal decay rates in CW rare-gas metal-halide vapor lasers. In *Proceedings of the International Conference on Lasers '84* (Corcoran, K. M. ed.). STS, McLean, VA, pp. 261–267.

Duarte, F. J. and Piper, J. A. (1986). Pulsed dye laser study of the B-X system in the I_2 laser medium. In *Proceedings of the International Conference on Lasers '85* (Wang, C. P. ed.). STS, McLean, VA, pp. 789–796.

Duarte, F. J., Taylor, T. S., Costela, A., Garcia-Moreno, I., and Sastre, R. (1998). Longpulse narrow-linewidth dispersive solid-state dye-laser oscillator. *Appl. Opt.* **37**, 3987–3989.

Dujardin, G. and Flamant, P. (1978). Amplified spontaneous emission and spatial dependence of gain in dye amplifiers. *Opt. Commun.* **24**, 243–247.

Everett, P. N. (1991). Flashlamp-excited dye lasers. In *High Power Dye Lasers* (Duarte, F. J., ed.). Springer-Verlag, Berlin, Germany, pp. 183–245.

Feynman, R. P., Leighton, R. B., and Sands, M. (1965). *The Feynman Lectures on Physics*, Vol. III, Addison-Wesley, Reading, MA.

Ganiel, U., Hardy, A., Neumann, G., and Treves, D. (1975). Amplified spontaneous emission and signal amplification in dye-laser systems. *IEEE J. Quantum Electron.* **QE-11**, 881–892.

Haag, G., Munz, M., and Marowski, G. (1983). Amplified spontaneous emission (ASE) in laser oscillators and amplifiers. *IEEE J. Quantum Electron.* **QE-19**, 1149–1160.

Haken, H. (1970). *Light and Matter,* Springer-Verlag, Berlin, Germany.

Hammond, P. (1979). Spectra of the lowest excited singlet states of rhodamine 6G and rhodamine B. *IEEE J. Quantum Electron.* **QE-15**, 624–632.

Hargrove, R. S. and Kan, T. K. (1980). High power efficient dye amplifier pumped by copper vapor lasers. *IEEE J. Quantum Electron.* **QE-16**, 1108–1113.

Hollberg, L. (1990). CW dye lasers. In *Dye Laser Principles* (Duarte, F. J. and Hillman, L. W., eds.). Academic, New York, Chapter 5.

Jensen, C. (1991). Pulsed dye laser gain analysis and amplifier design. In *High Power Dye Lasers* (Duarte, F. J., ed.). Springer-Verlag, Berlin, Germany, Chapter 3.

Kaslin, V. M., Petrash, G. G., and Yakushev, O. F. (1980). Laser action at the B-X transition of the I_2 molecule using a copper-vapor laser for optical pumping. *Sov. Phys. JETP* **51**, 679–686.

Kessler, T., Hagemann, C., Grebing, C., Legero, T., Sterr, U., Riehle, F., Martin, M. J., Chen, L., and Ye, J. (2012). A sub-40-mHz-linewidth laser based on a silicon single-crystal optical cavity. *Nat. Photon.* **6**, 687–692.

Maiman, T. H. (1960). Stimulated optical radiation in ruby. *Nature* **187**, 493–494.

Munz, M. and Haag, G. (1980). Optimization of dye-laser output coupling by consideration of the spatial gain distribution. *Appl. Phys.* **22**, 175–184.

Nair, L. G. and Dasgupta, K. (1985). Amplified spontaneous emission in narrow-band pulsed dye laser oscillators-theory and experiment. *IEEE J. Quantum Electron.* **21**, 1782–1794.

Orr, B. J., He, Y., and White, R. T. (2009). Spectroscopic applications of pulsed tunable optical parametric oscillators. In *Tunable Laser Applications*, 2nd edn. (Duarte, F. J., ed.). CRC, New York, Chapter 2.

Penzkofer, A. and Falkenstein, W. (1978). Theoretical investigation of amplified spontaneous emission with picosecond light pulses in dye solutions. *Opt. Quantum Electron.* **10**, 399–423.

Peterson, O. G. (1979). Dye lasers. In *Methods of Experimental Physics*, Vol. 15 (Tang, C. L. ed.). Academic, New York, Chapter 5.

Piper, J. A. (1976). Simultaneous CW laser oscillation on transitions of Cd^+ and I^+ in a hollow-cathode He-CdI$_2$ discharge. *Opt. Commun.* **19**, 189–192.

Piper, J. A. and Gill, P. (1975). Output characteristics of the He-Zn laser. *J. Phys. D: Appl. Phys.* **8**, 127–134.

Popov, S. (2009). Fiber laser overview and medical applications. In *Tunable Laser Applications*, 2nd edn. (Duarte, F. J., ed.). CRC, New York, Chapter 7.

Sargent, M., Scully, M. O., and Lamb, W. E. (1974). *Laser Physics*, Addison Wesley, Reading, MA.

Schäfer, F. P., Schmidt, W., and Volze, J. (1966). Organic dye solution laser. *Appl. Phys. Lett.* **9**, 306–309.

Silfvast, W. T. (2008). *Laser Fundamentals*, 2nd edn. Cambridge University, Cambridge.

Teschke, O., Dienes, A., and Whinnery, J. R. (1976). Theory and operation of high-power CW and long-pulse dye lasers. *IEEE J. Quantum Electron.* **QE-12**, 383–395.

Tuccio, S. A. and Strome, F. C. (1972). Design and operation of a tunable continuous dye laser. *Appl. Opt.* **11**, 64–73.

Wellegehausen, B. (1979). Optically pumped CW dimer lasers. *IEEE J. Quantum Electron.* **QE-15**, 1108–1130.

Willett, C. S. (1974). *An Introduction to Gas Lasers: Population Inversion Mechanisms*, Pergamon, New York.

9

Laser Oscillators Described via the Dirac Notation

9.1 Introduction

Here we derive the classical linewidth cavity equation

$$\Delta\lambda \approx \Delta\theta \left(\frac{\partial\theta}{\partial\lambda}\right)^{-1} \tag{9.1}$$

using the Dirac notation approach. First, we notice that in this equation $\Delta\theta$ is the beam divergence previously related to the uncertainty principle (see Chapter 3):

$$\Delta p\,\Delta x \approx h \tag{9.2}$$

and $(\partial\theta/\partial\lambda)^{-1}$ is the overall cavity angular dispersion (Duarte, 2003).

We should also mention that Equation 9.1 is the single-pass version of the *multiple-pass* linewidth cavity equation (Duarte and Piper, 1984; Duarte, 1990, 2001):

$$\Delta\lambda = \Delta\theta_R \left(MR\nabla_\lambda \Theta_G + R\nabla_\lambda \Phi_P \right)^{-1} \tag{9.3}$$

where the multiple-return-pass beam divergence is given by (Duarte, 1989, 1990)

$$\Delta\theta_R = \frac{\lambda}{\pi w}\left(1+\left(\frac{L_\mathfrak{R}}{B_R}\right)^2+\left(\frac{A_R L_\mathfrak{R}}{B_R}\right)^2\right)^{1/2} \tag{9.4}$$

R is the number of return-cavity passes necessary to reach laser threshold
$L_\mathfrak{R} = (\pi w^2/\lambda)$ is the Rayleigh length (Duarte, 1990)
w is the beam waist
A_R and B_R are the corresponding multi-return-pass matrix elements (Duarte, 2003) as indicated in Appendix C

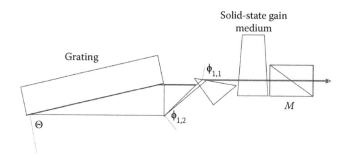

FIGURE 9.1
Optimized multiple-prism grating solid-state dye-laser oscillator using a 3000 lines mm^{-1} diffraction grating deployed in the Littrow configuration. The measured laser linewidth is $\Delta \nu \approx 350$ MHz (Duarte, 1999). This is a *closed cavity* configuration; see Appendix B.

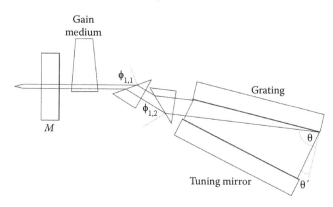

FIGURE 9.2
Hybrid multiple-prism near-grazing incidence (HMPGI) grating oscillator (Duarte and Piper, 1981; Duarte, 1990). For the organic solid-state HMPGIG oscillator, the measured laser linewidth is $\Delta \nu \approx 375$ MHz (Duarte, 1997a). This is a *closed cavity* configuration; see Appendix B.

For high-power, high-gain, tunable narrow-linewidth oscillators (as depicted in Figures 9.1 and 9.2), the factor R has been measured to be $R \approx 3$ (Duarte, 2001). Equation 9.3 has been found to be fairly successful to predict, and account for, measured laser linewidths in high-gain, pulsed tunable lasers (Duarte, 2001).

In this chapter we shall see that equations and concepts previously developed in a classical context can also be outlined and derived from a quantum perspective.

9.2 Transverse and Longitudinal Modes

The unrestricted initial emission from laser gain media is both spatially and spectrally broad; in other words, it lacks spectral purity. This is the essence of broadband emission.

Spatially, this broad emission contains many spatial modes and each of those spatial modes includes a multitude of longitudinal modes. Albeit exhibiting a directional beam, this type of broadband emission is essentially chaotic and exhibits a high degree of entropy.

In order to achieve highly selective, controllable, narrow-linewidth, spectrally pure emission, it is imperative to

1. Restrict the spatial emission to a single transverse electromagnetic mode, that is, TEM_{00}
2. Restrict the longitudinal modes, *within* that single transverse mode (TEM_{00}), to a single-longitudinal mode (SLM)

In other words, highly coherent, spectrally pure, low-entropy laser emission requires the selection of a single transverse electromagnetic mode, followed by the selection of an SLM within that single transverse mode. The discussion that follows next includes concepts and elements from a review given by Duarte (2003).

9.2.1 Transverse-Mode Structure

A fundamental laser cavity is comprised of a gain medium and two mirrors, as illustrated in Figure 9.3. The physical dimension of the intracavity aperture $(2w)$ relative to the separation of mirrors, or cavity length (L), determines the number of transverse electromagnetic modes. A typical broadband laser cavity, under optical excitation, has an aperture in the few mm range and a cavity length of about 10 cm. For the same cavity length, the aperture size in a narrow-linewidth cavity is reduced to the $100 \leq 2w \leq 200$ μm range.

The narrower the width of the intracavity aperture and the longer the cavity length, the lower the number of transverse modes (Duarte, 2003). The single-pass transverse-mode structure in one dimension can be characterized using the generalized interferometric equation introduced in Chapter 4 (Duarte, 1991, 1993):

$$|\langle x|s\rangle|^2 = \sum_{j=1}^{N}\Psi(r_j)^2 + 2\sum_{j=1}^{N}\Psi(r_j)\left(\sum_{m=j+1}^{N}\Psi(r_m)\cos(\Omega_m - \Omega_j)\right) \qquad (9.5)$$

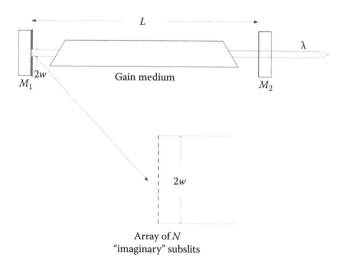

FIGURE 9.3
Mirror–mirror laser cavity. The physical dimensions of the intracavity aperture relative to the cavity length determine the number of transverse modes. Parameters that enter in the calculation are laser wavelength (λ), cavity length (L), and number of subslits (N).

and in two dimensions by (Duarte, 1995a)

$$|\langle x|s\rangle|^2 = \sum_{z=1}^{N}\sum_{y=1}^{N}\Psi(r_{zy})\sum_{q=1}^{N}\sum_{p=1}^{N}\Psi(r_{pq})e^{i(\Omega_{qp}-\Omega_{zy})} \tag{9.6}$$

The single-pass approximation to estimate the transverse-mode structure assumes that in a laser with a given cavity length, most of the emission generated next to the output-coupler mirror is in the form of spontaneous emission and thus highly divergent. Thus, only the emission generated at the opposite end of the cavity and that propagates via an intracavity length L contributes to the initial transverse-mode structure.

In order to illustrate the use of these equations, let us consider a hypothetical laser with a 10 cm cavity emitting at $\lambda = 590$ nm incorporating a 1D aperture ($2w$) = 2 mm wide. Using Equation 9.5 the intensity distribution of the emission is calculated as shown in Figure 9.4. Each ripple represents a transverse mode. An estimate of this number can be obtained by counting the ripples in Figure 9.4 that yield an approximate number of 17. The Fresnel number (Siegman, 1986) for the given dimensions is

$$N_F = \left(\frac{w^2}{L\lambda}\right) = 16.95 \tag{9.7}$$

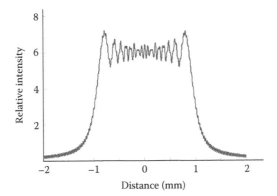

FIGURE 9.4
Cross section of diffraction distribution corresponding to a large number of transverse modes. Here, $\lambda = 590$ nm, $2w = 2$ mm, $L = 10$ cm, and $N_F \approx 17$. The wide aperture is assumed to be composed of $N = 1000$ "imaginary" subslits.

For the same wavelength at $\lambda = 590$ nm and cavity length ($L = 10$ cm), if the aperture is reduced to $2w = 250$ µm, the calculated intensity distribution, using Equation 9.5, is given in Figure 9.5. In this case the Fresnel number becomes $N_F \approx 0.26$. The distribution in Figure 9.5 indicates that most of the emission intensity is contained in a central near-Gaussian distribution. A measured single-transverse-mode beam, with an ovaloid profile, from a narrow-linewidth tunable solid-state dye laser (Duarte, 1995b), is displayed in Figure 9.6

In summary, reducing the transverse-mode distribution to TEM$_{00}$ emission is the first step in the design of narrow-linewidth tunable lasers.

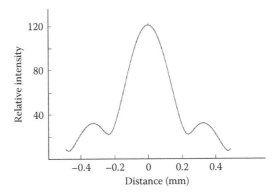

FIGURE 9.5
Cross section of diffraction distribution corresponding to a near single transverse mode corresponding to $\lambda = 590$ nm, $2w = 250$ µm, $L = 10$ cm, and $N_F \approx 0.26$. In practice, the lower-intensity higher-order maxima are not observed due to cavity losses and often only the central mode remains.

FIGURE 9.6
Single-transverse-mode beam originating from an SLM ($\Delta v \approx 420$ MHz) multiple-prism grating solid-state dye laser. (Reproduced from Duarte, F.J., *Opt. Commun.* 117, 480, 1995b, with permission from Elsevier.)

The task of the designer consists in achieving TEM_{00} emission within desirable geometrical parameters that include the shortest possible cavity length.

9.2.2 Double- and Single-Longitudinal-Mode Emission

Successful discrimination toward a single transverse mode is the first step toward the attainment of tunable narrow-linewidth emission. The next task consists in controlling the number of longitudinal modes in the cavity. In a laser resonator with cavity length L, the longitudinal-mode spacing (δv), in the frequency domain, is given by an alternative form of the uncertainty principle (see Chapter 3):

$$\delta v = \frac{c}{2L} \tag{9.8}$$

and the number of longitudinal modes N_{LM} is given

$$N_{LM} = \frac{\Delta v}{\delta v} \tag{9.9}$$

where Δv is the measured laser linewidth (Duarte, 2003). From Equations 9.8 and 9.9, it is clear that the number of allowed longitudinal modes (N_{LM}) decreases as the cavity length decreases. Thus, the importance of cavity compactness is highlighted.

An additional, and complementary, approach to achieve SLM emission is to optimize the beam divergence and to increase the intracavity dispersion to yield a narrower cavity linewidth that would restrict oscillation to the SLM regime. In this context the linewidth

$$\Delta\lambda \approx \Delta\theta \left(\frac{\partial\theta}{\partial\lambda}\right)^{-1} \tag{9.10}$$

is converted to $\Delta\nu$ units (Hz) using the identity

$$\Delta\nu = \Delta\lambda \left(\frac{c}{\lambda^2}\right) \tag{9.11}$$

and applying the criterion

$$\Delta\nu \le \delta\nu \tag{9.12}$$

to guide the design of the dispersive oscillator.

Multiple-longitudinal-mode emission is complex and chaotic, both in the frequency and temporal domains. Double-longitudinal-mode (DLM) and SLM emission can be characterized in the frequency domain using Fabry–Perot interferometry or in the temporal domain by observing the shape of the temporal pulsed. In the case of DLM emission, the interferometric rings appear to be double. In the temporal domain, *mode beating* is still observed when the intensity ratio of the primary to the secondary mode is 100:1 or even higher. Mode beating of two longitudinal modes, as illustrated in Figure 9.7, is characterized using a wave representation where each mode of amplitude

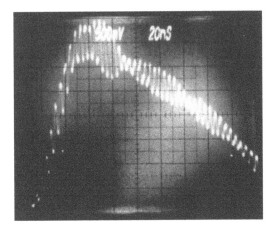

FIGURE 9.7
Measured mode beating resulting from DLM oscillation. Temporal scale is 20 ns div⁻¹.
(Reproduced from Duarte, F.J. et al., *Appl. Opt.* 27, 843, 1988, with permission from the Optical Society of America.)

E_1 and E_2, with frequencies ω_1 and ω_2, combines to produce a resulting field (Pacala et al., 1984):

$$E = E_1 \cos(\omega_1 t - k_1 z) + E_2 \cos(\omega_2 t - k_2 z) \qquad (9.13)$$

For incidence at $z = 0$ on a square-law temporal detector, the intensity can be expressed as

$$E^2 = \frac{1}{2}(E_1^2 + E_2^2) + \frac{1}{2}\left(E_1^2 \cos 2\omega_1 t + E_2^2 \cos \omega_2 t\right)$$

$$+ E_1 E_2 \cos(\omega_1 + \omega_2)t + E_1 E_2 \cos(\omega_1 - \omega_2)t \qquad (9.14)$$

For detection in the nanosecond regime, we can neglect the second and third terms, so that

$$E^2 \approx \frac{1}{2}(E_1^2 + E_1^2) + E_1 E_2 \cos(\omega_1 - \omega_2)t \qquad (9.15)$$

Using this approximation and a non-Gaussian temporal representation, derived from experimental data, for the amplitudes of the form

$$E_1(t) = (a_2 t^2 + a_1 t + a_0)(b_1 t + b_0)^{-1} \qquad (9.16)$$

a calculated version of the experimental waveform exhibiting mode beating can be obtained as shown in Figure 9.8 (Duarte et al., 1988). For the oscillator under consideration, which was lasing in the DLM regime, the ratio of frequency jitter $\delta\omega$ to cavity mode spacing $\Delta\omega \approx (\omega_1 - \omega_2)$ was represented by a sinusoidal function at 20 MHz. The initial mode intensity ratio is 200:1 (Duarte et al., 1988; Duarte, 1990).

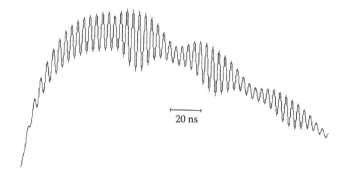

20 ns

FIGURE 9.8
Calculated temporal pulse assuming interference between the two longitudinal modes (Reproduced from Duarte, F.J. et al., *Appl. Opt.* 27, 843, 1988, with permission from the Optical Society of America.)

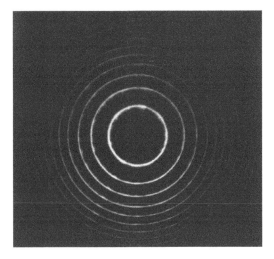

FIGURE 9.9
Fabry–Perot interferogram corresponding to SLM emission at $\Delta\nu \approx 350$ MHz. (Reproduced from Duarte, F.J., *Appl. Opt.* 38, 6347, 1999, with permission from the Optical Society of America.)

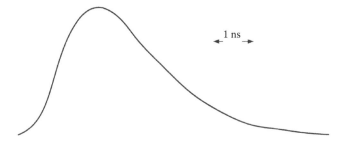

FIGURE 9.10
Near-Gaussian temporal pulse corresponding to SLM emission. The temporal scale is 1 ns div^{-1}. (Reproduced from Duarte, F.J., *Appl. Opt.* 38, 6347, 1999, with permission from the Optical Society of America.)

In the case of SLM emission, the Fabry–Perot interferometric rings appear singular and well defined (see Figure 9.9). Mode beating in the temporal domain is absent and the pulses assume a near-Gaussian distribution (see Figure 9.10). These results were obtained in an optimized solid-state multiple-prism grating dye-laser oscillator for which $\Delta\nu\,\Delta t \approx 1$, which is near the limit allowed by the Heisenberg uncertainty principle (Duarte, 1999).

9.2.2.1 Example

For a laser with a 15 cm cavity length ($\delta\nu \approx 1$ GHz, using Equation 9.8), and a measured linewidth of $\Delta\nu = 3$ GHz, the number of longitudinal modes becomes $N_{LM} \approx 3$ (using Equation 9.9). If the cavity length is reduced to 10 cm ($\delta\nu \approx 1.5$ GHz), then the number of longitudinal modes is reduced to $N_{LM} \approx 2$

and the emission would be called DLM emission. Furthermore, if the cavity length is reduced to 5 cm, then $N_{LM} \approx 1$ and the laser is said to be undergoing SLM oscillation. This example highlights the advantages of compact cavity designs.

9.3 Laser Cavity Equation: An Intuitive Approach

We now describe an intuitive approach to the laser cavity equation using the Dirac notation as applied to a multiple-prism grating cavity as illustrated in Figures 9.1 and 9.2 (Duarte, 1992). A close-up view of the frequency-selective assembly, in an unfolded configuration, is shown in Figure 9.11.

In reference to this figure, the probability amplitude describing the propagation from the active region emitter s to the entrance of the multiple-prism array at an incidence angle $\phi_{1,m}$ can be expressed as

$$\langle \phi_{1,m} | s \rangle$$

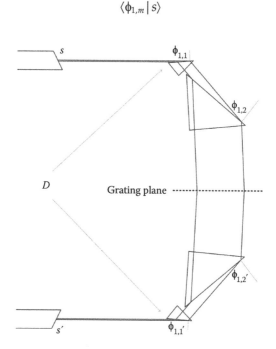

FIGURE 9.11
Unfolded optical path of a dispersive multiple-prism grating configuration showing the emission source s followed by the entrance of the multiple-prism grating assembly ($\phi_{1,m}$ with $m = 1$), the dispersive assembly D, and the corresponding quantities to the return path back to the gain region. (Adapted from Duarte, F.J., *Appl. Opt.* 31, 6979, 1992, with permission from the Optical Society of America.)

while the probability amplitude to propagate, through the dispersive multiple-prism grating assembly D, from an incidence angle $\phi_{1,m}$ to a return angle $\phi'_{1,m}$ can be written as

$$\langle \phi'_{1,m} | D | \phi_{1,m} \rangle$$

Similarly, the probability amplitude to exit the multiple-prism array at an angle $\phi'_{1,m}$ back to the gain medium is

$$\langle s' | \phi'_{1,m} \rangle$$

Thus, the overall probability amplitude for a photon to propagate from the active medium s, through the dispersive multiple-prism grating assembly, and back to the gain medium, is

$$\langle s' | D | s \rangle = \sum_{\phi'_{1,m}} \sum_{\phi_{1,m}} \langle s' | \phi'_{1,m} \rangle \langle \phi'_{1,m} | D | \phi_{1,m} \rangle \langle \phi_{1,m} | s \rangle \tag{9.17}$$

Since $\phi_{1,m}$ is a *unique angle* of incidence on the gain axis, at the multiple-prism expander, that is necessary to induce diffraction at the grating followed by an exact return passage to the gain medium, the probability amplitude can be reduced to (Duarte, 1992)

$$\langle s' | D | s \rangle = \langle s' | \phi'_{1,m} \rangle \langle \phi'_{1,m} | D | \phi_{1,m} \rangle \langle \phi_{1,m} | s \rangle \tag{9.18}$$

so that the probability for the intracavity photon propagation just described becomes

$$| \langle s' | D | s \rangle |^2 = | \langle s' | \phi'_{1,m} \rangle |^2 | \langle \phi'_{1,m} | D | \phi_{1,m} \rangle |^2 | \langle \phi_{1,m} | s \rangle |^2 \tag{9.19}$$

The different components of this probability can be identified by describing the propagation at each segment. Immediately, at the exit of the gain region, in the first segment, the probability of narrow-linewidth emission propagation is inversely proportional to the beam divergence $\Delta\theta$ of the emission, so that

$$| \langle \phi_{1,m} | s \rangle |^2 = \kappa_1 \left(\frac{1}{\Delta\theta} \right) \tag{9.20}$$

Once the photon flux arrives at the dispersive multiple-prism grating assembly, the return of resonant narrow-linewidth emission is proportional to the dispersion of the configuration. The higher the dispersion, the narrower the linewidth:

$$| \langle \phi'_{1,m} | D | \phi_{1,m} \rangle |^2 = \kappa_2 \left(\frac{\partial\theta}{\partial\lambda} \right) \tag{9.21}$$

For highly selective resonant emission that returns precisely at $\phi'_{1,m}$, the probability to return to the gain region for further amplification is high so that

$$|\langle s'|\phi'_{1,m}\rangle|^2 \approx 1 \tag{9.22}$$

Now, since the overall probability for resonant narrow-linewidth amplification is inversely proportional to the wavelength spread of the emission,

$$|\langle s'|D|s\rangle|^2 = \kappa_3\left(\frac{1}{\Delta\lambda}\right) \tag{9.23}$$

Combining Equations 9.20 through 9.23 into 9.19, we get

$$\Delta\lambda \approx \Delta\theta\left(\frac{\kappa_1\kappa_2}{\kappa_3}\right)\left(\frac{\partial\theta}{\partial\lambda}\right)^{-1} \tag{9.24}$$

which, for $\kappa_3 \approx (\kappa_1\kappa_2)$, takes the form

$$\Delta\lambda \approx \Delta\theta\left(\frac{\partial\theta}{\partial\lambda}\right)^{-1} \tag{9.25}$$

Albeit fairly intuitive, this approach lends itself to illustrate the refinement process that occurs with multiple intracavity passes.

9.4 Laser Cavity Equation via the Interferometric Equation

By now, the reader should be getting the message that interference is at the heart of many phenomena in laser optics. We just saw how the interferometric equation applied to a single aperture can be used to describe the transverse-mode structure of a laser cavity. In other words, we saw how the geometrical ratio of aperture width to cavity length affects the transverse-mode distribution of the emission. Now, we apply the generalized, 1D, interferometric equation

$$|\langle x|s\rangle|^2 = \sum_{j=1}^{N}\Psi(r_j)^2 + 2\sum_{j=1}^{N}\Psi(r_j)\left(\sum_{m=j+1}^{N}\Psi(r_m)\cos(\Omega_m - \Omega_j)\right)$$

to describe the origin of the cavity linewidth equation. We do this by focusing attention into the phase term of the N-slit interferometric equation (Duarte, 1997b)

$$\cos\left((\theta_m - \theta_j) \pm (\phi_m - \phi_j)\right) = \cos\left(|\, l_m - l_{m-1}\,|\, k_1 \pm |\, L_m - L_{m-1}\,|\, k_2\right) \quad (9.26)$$

from which the well-known *grating equation* can be derived (see Chapter 5)

$$d_m(\sin\Theta_m \pm \sin\Phi_m) = m\lambda \quad (9.27)$$

where $m = 0, \pm 1, \pm 2, \pm 3, \dots$. For a grating deployed in the reflection domain, and at the *Littrow* configuration, $\Theta_m = \Phi_m = \Theta$ (i.e., the diffracted light goes back at the same angle of the incident beam) so that the grating equation reduces to

$$m\lambda = 2d_m \sin\Theta \quad (9.28)$$

where $m = 0, 1, 2, 3, \dots$ are the various *diffraction orders*.

Considering two slightly different wavelengths, an expression for the wavelength difference can be written as

$$\Delta\lambda = \frac{2d_m}{m}(\sin\Theta_1 - \sin\Theta_2) \quad (9.29)$$

for $\Theta_1 \approx \Theta_2 (= \Theta)$, this equation can be restated as

$$\Delta\lambda \approx \frac{2d_m}{m}\Delta\theta\left(1 - \frac{3\Theta^2}{3!} + \frac{5\Theta^4}{5!} - \cdots\right) \quad (9.30)$$

Differentiation of the grating equation leads to

$$\left(\frac{\partial\theta}{\partial\lambda}\right)\cos\Theta = \frac{m}{2d_m} \quad (9.31)$$

and substitution into Equation 9.30 yields

$$\Delta\lambda \approx \Delta\theta\left(\frac{\partial\theta}{\partial\lambda}\right)^{-1}\left(1 - \frac{\Theta^2}{2!} + \frac{\Theta^4}{4!} - \cdots\right)(\cos\Theta)^{-1} \quad (9.32)$$

which reduces to the well-known cavity linewidth equation (Duarte, 1992)

$$\Delta\lambda \approx \Delta\theta\left(\frac{\partial\theta}{\partial\lambda}\right)^{-1} \quad (9.33)$$

or

$$\Delta\lambda \approx \Delta\theta(\nabla_\lambda\theta)^{-1} \qquad (9.34)$$

where $\nabla_\lambda\theta = (\partial\theta/\partial\lambda)$. This equation has been used extensively to determine the emission linewidth in pulsed narrow-linewidth dispersive laser oscillators (Duarte, 1990). It originates in the generalized N-slit interference equation and incorporates $\Delta\theta$ whose value can be determined either from the Heisenberg uncertainty principle or from the interferometric equation itself. This equation is also well known in the field of classical spectrometers where it has been introduced using geometrical arguments (Robertson, 1955). In addition to its technical and computational usefulness, Equation 9.33 and/or 9.34 illustrates the inherent interdependence between spectral and spatial coherence.

Problems

9.1 Show that for $R = 1$, and in the absence of a grating, Equation 9.3 reduces to an equation of the form of Equation 9.1 where the dispersion is provided by the multiple-prism assembly.

9.2 For an optimized multiple-prism grating oscillator, as shown in Figure 9.1, the measured laser linewidth is $\Delta\nu \approx 350$ MHz. Given that the free spectral range of the cavity is $FSR \approx 1.6$ GHz, determine (a) the approximate length of the cavity and (b) the value of the overall intracavity dispersion given that the measured beam divergence is $\Delta\theta \approx 2.2$ m rad.

9.3 For the case of laser radiation in the visible spectrum, show that Equation 9.14 reduces to Equation 9.15 for detectors with a time response in the nanosecond regime.

9.4 Show in detail how Equation 9.30 reduces to the cavity linewidth equation: Equation 9.33.

References

Duarte, F. J. (1989). Ray transfer matrix analysis of multiple-prism dye laser oscillators. *Opt. Quant. Electron.* **21**, 47–54.
Duarte, F. J. (1990). Narrow linewidth pulsed dye laser oscillators. In *Dye Laser Principles* (Duarte, F. J. and Hillman, L. W. eds.). Academic Press, New York, Chapter 4.
Duarte, F. J. (1991). Dispersive dye lasers. In *High Power Dye Lasers* (Duarte, F. J. ed.). Springer-Verlag, Berlin, Germany, Chapter 2.

Duarte, F. J. (1992). Cavity dispersion equation $\Delta\lambda \approx \Delta\theta(\partial\theta/\partial\lambda)^{-1}$: A note on its origin. *Appl. Opt.* **31**, 6979–6982.

Duarte, F. J. (1993). On a generalized interference equation and interferometric measurements. *Opt. Commun.* **103**, 8–14.

Duarte, F. J. (1995a). Interferometric imaging. In *Tunable Laser Applications* (Duarte, F. J., ed.). Marcel Dekker, New York, Chapter 5.

Duarte, F. J. (1995b). Solid-state dispersive dye laser oscillator: Very compact cavity. *Opt. Commun.* **117**, 480–484.

Duarte, F. J. (1997a). Multiple-prism near-grazing-incidence grating solid-state dye laser oscillator. *Opt. Laser Technol.* **29**, 513–516.

Duarte, F. J. (1997b). Interference, diffraction, and refraction, via Dirac's notation. *Am. J. Phys.* **65**, 637–640.

Duarte, F. J. (1999). Multiple-prism grating solid-state dye laser oscillator: Optimized architecture. *Appl. Opt.* **38**, 6347–6349.

Duarte, F. J. (2001). Multiple-return-pass beam divergence and the linewidth equation. *Appl. Opt.* **40**, 3038–3041.

Duarte, F. J. (2003). *Tunable Laser Optics.* Elsevier-Academic, New York.

Duarte, F. J., Ehrlich, J. J., Patterson, S. P., Russell, S. D., and Adams, J. E. (1988). Linewidth instabilities in narrow-linewidth flashlamp-pumped dye laser oscillators. *Appl. Opt.* **27**, 843–846.

Duarte, F. J. and Piper, J. A. (1981). A prism preexpanded grazing incidence pulsed dye laser. *Appl. Opt.* **20**, 2113–2116.

Duarte, F. J. and Piper, J. A. (1984). Multi-pass dispersion theory of prismatic pulsed dye lasers. *Opt. Acta* **31**, 331–335.

Pacala, T. J., McDermid, I. S., and Laudenslager, J. B. (1984). Single-longitudinal-mode operation of an XeCl laser. *Appl. Phys. Lett.* **45**, 507–509.

Robertson, J. K. (1955). *Introduction to Optics: Geometrical and Physical.* Van Nostrand, New York.

Siegman, A. E. (1986). *Lasers.* University Science Books. Mill Valley, CA.

10

Interferometry *via* the Dirac Notation

10.1 Interference *à la* Dirac

The genesis of quantum optics can be found in the Dirac discussion (from the 1930s) on interference as disclosed in *The Principles of Quantum Mechanics* (Dirac, 1978). This is a masterful and prophetic discussion that begins by considering a *roughly monochromatic light* source. The discussion continues by considering *a beam of light consisting of a large number of photons*. In other words, Dirac is considering a beam of light with a definite spectral linewidth and high power, a beam of light as available from narrow-linewidth high-power lasers (Duarte, 1998, 2003). For the discussion at hand, the term *monochromatic* is reserved for single-photon emission, while quasi-monochromatic, semi-monochromatic, or nearly monochromatic relates to spectrally narrow emission as available from narrow-linewidth high-power lasers. These narrow-linewidth lasers provide populations of indistinguishable photons. The narrower the laser linewidth, the more indistinguishable the photons are. In the case of optimized pulsed laser oscillators, the emission linewidth can be as narrow as allowed by the Heisenberg uncertainty principle, that is, $\Delta\nu \approx 350$ MHz (or $\Delta\lambda \approx 0.0004$ nm at $\lambda \approx 590$ nm) for a $\Delta t \approx 3$ ns pulse (Duarte, 1999).

Thus, in the Dirac discussion on interference, we are dealing with a population, or ensemble, of indistinguishable photons. He then goes on to associate *the translational state of a photon with one of the wave functions of ordinary wave optics* (Dirac, 1978).

He argues that the association is only statistical and that the wave function provides information *about the probability of our finding the photon in any particular place*. This idea is reinforced with a similar sentence stating that the wave function gives us information about *one photon being in a particular place* (Dirac, 1978). Going back to his thought experiment, he then considers a beam of light with a large number of nearly monochromatic, that is, indistinguishable photons, and divides it into two beams of equal intensity and the two sub-beams are made to interfere. In this regard, Dirac's thought experiment applies directly to a high-power laser beam (of linewidth $\Delta\nu$) made to interfere in a Mach–Zehnder interferometer, for instance. Next, Dirac explains that each photon then goes partly into the two interfering sub-beams: *each photon then interferes with itself* (Dirac, 1978). This statement is directly applicable to the experiment at hand in which a large number of indistinguishable photons are divided into two sub-beams.

Under this circumstance each individual photon in that ensemble of indistinguishable photons goes partly into each beam (given that they have a large coherent length defined by $\Delta x \approx c/\Delta \nu$) and interferes with itself. Now, mathematically, what interfere are the probability amplitudes associated with each photon.

At this stage three explanations are necessary. First, as Dirac himself explains, when the beam is divided into two subcomponents, it does not mean that a particular photon goes into one sub-beam and a different photon goes into the other sub-beam. Secondly, under the Dirac description, the interference between two beams from different lasers emitting at the same frequency is perfectly allowed since the photons are indistinguishable and therefore *the same*. In other words, the interference from two laser sources, of the same frequency and linewidth, appears the same as if the beam of one of the lasers were divided and then allowed to interfere.

Thirdly, as explained elsewhere in this book, the equation for N-slit interference is derived for a single photon. However, it applies equally well to a population of indistinguishable photons. If light from a narrow-linewidth laser is used (that would be a nearly perfect population of indistinguishable photons), the interferograms are perfectly sharp and exhibit a visibility (Michelson, 1927)

$$\mathcal{V} = \frac{I_1 - I_2}{I_1 + I_2} \tag{10.1}$$

of near unity ($\mathcal{V} \approx 1$). If, on the other hand, broadband emission is used, the interference equation becomes part of an interferometric distribution including the interferograms corresponding to all the different wavelengths used. Thus the interferograms become broad, with decreased spatial definition and decreased visibility (Duarte, 2008). This effect is explained in detail in Chapter 4.

Besides Dirac (1978), useful references on interferometry include Steel (1967), Meaburn (1976), and Born and Wolf (1999). Note that as mentioned previously, the reference cited here for Dirac's book corresponds to the 9th printing of the 4th edition.

The first edition was published in 1930, the second in 1935, the third in 1947, and the fourth in 1958.

10.2 Hanbury Brown–Twiss Interferometer

The Hanbury Brown–Twiss effect originates in interferometric measurements performed by an "intensity interferometer" used for astronomical observations (Hanbury Brown and Twiss, 1956). A diagram of the stellar

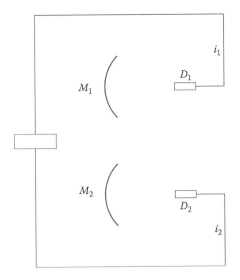

FIGURE 10.1
The Hanbury Brown and Twiss interferometer: the light, from an astronomical source, is collected at mirrors M_1 and M_2 and focused onto detectors D_1 and D_2. The current generated at
these detectors interferes to produce an interference signal characterized by an equation in the
form of Equation 10.2.

intensity interferometer used to determine the diameter of stars is depicted
in Figure 10.1. Feynman in one of his exercises to the *Feynman Lectures in
Physics* (Feynman et al., 1965) explains that the electric currents from the
two detectors are mixed in a coincidence circuit where the currents become
indistinguishable. Feynman then asks to show that the coincidence counting rate, in the Hanbury Brown–Twiss configuration, is proportional to an
expression of the form

$$2 + 2\cos k(R_2 - R_1) \tag{10.2}$$

where R_1 and R_2 are the distances from detector 1 and detector 2 to the
source. Using the N-slit interferometric equation (Duarte, 1991, 1993)

$$|\langle x \,|\, s \rangle|^2 = \sum_{j=1}^{N} \Psi\left(r_j\right)^2 + 2\sum_{j=1}^{N} \Psi\left(r_j\right)\left(\sum_{m=j+1}^{N} \Psi\left(r_m\right)\cos\left(\Omega_m - \Omega_j\right)\right) \tag{10.3}$$

with $N = 2$, one immediately arrives at

$$|\langle x \,|\, s \rangle|^2 = \Psi(r_1)^2 + \Psi(r_2)^2 + 2\Psi(r_1)\Psi(r_2)\cos(\Omega_2 - \Omega_1) \tag{10.4}$$

and setting $\Psi(r_1) = \Psi(r_2) = 1$

$$|\langle x|s\rangle|^2 = 2 + 2\cos(\Omega_2 - \Omega_1) \tag{10.5}$$

Now, using (as suggested by Feynman) $\Omega_1 = kR_1$ and $\Omega_2 = kR_2$

$$|\langle x|s\rangle|^2 = 2 + 2\cos k(R_2 - R_1) \tag{10.6}$$

From the measured signal distribution and these equations, the angular spread of the emission can be determined, and knowing the distance from the source to the detector, it becomes possible to estimate the diameter of the aperture at the emission, in other words, the diameter of the star under observation.

The Hanbury Brown–Twiss interferometric argument was not easily accepted by the physics community at the time given that many physicists erroneously thought that since visible photons either arrived at detector 1 or detector 2, these were not correlated. In other words, many had not accepted the Dirac description of interference and did not understand the concept of coherence length. Further, the Hanbury Brown–Twiss observations could also be explained classically using conventional Fourier techniques, thus adding to the confusion.

10.3 Two-Beam Interferometers

Two-beam interferometers are optical devices that divide and then recombine a light beam. It is on recombination of the beams that interference occurs. The most well-known two-beam interferometers are the Sagnac interferometer, the Mach–Zehnder interferometer, and the Michelson interferometer.

For a highly coherent light beam, such as the beam from a narrow-linewidth laser, the coherence length

$$\Delta x \approx \frac{c}{\Delta v} \tag{10.7}$$

can be rather large thus allowing a relatively large optical path length in the two-beam interferometer of choice. Alternatively, this relation provides an avenue to accurately determine the linewidth of a laser by increasing the optical path length until interference ceases to be observed.

Parts of the following discussion on interferometry are based on a review by Duarte (2003).

10.3.1 Sagnac Interferometer

The Sagnac, or cyclic, interferometer is illustrated in Figure 10.2. In this interferometer, the incident light beam is divided into two sub-beams by a

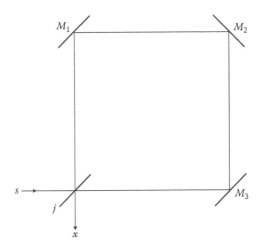

FIGURE 10.2
Sagnac interferometer. All three mirrors M_1, M_2, and M_3 are assumed to be identical. For the description given in the text, the beam splitter is assumed to be lossless and to divide the incident beam exactly into two components half the intensity of the original beam.

beam splitter. The reflected beam, on the incidence beam splitter, is then sent into a path defined by the reflections on M_1, M_2, and M_3 mirrors. The transmitted beam, on the incidence beam splitter, is sent into a path defined by the reflections on M_3, M_2, and M_1 mirrors. Both counterpropagating beams are recombined at the beam splitter. The interference mechanics of the counterpropagating round trips can be described using the Dirac notation via the probability amplitude

$$\langle x \,|\, s \rangle = \langle x \,|\, j \rangle\langle j \,|\, M_3 \rangle\langle M_3 \,|\, M_2 \rangle\langle M_2 \,|\, M_1 \rangle\langle M_1 \,|\, j \rangle\langle j \,|\, s \rangle$$

$$+ \langle x \,|\, j' \rangle\langle j' \,|\, M_1 \rangle\langle M_1 \,|\, M_2 \rangle\langle M_2 \,|\, M_3 \rangle\langle M_3 \,|\, j' \rangle\langle j' \,|\, s \rangle \qquad (10.8)$$

where
  j refers to the reflection mode of the beam splitter (BS)
  j' to the transmission mode of the beam splitter

Assuming that

$$\langle j \,|\, M_3 \rangle\langle M_3 \,|\, M_2 \rangle\langle M_2 \,|\, M_1 \rangle\langle M_1 \,|\, j \rangle = 1 \qquad (10.9)$$

and

$$\langle j' \,|\, M_1 \rangle\langle M_1 \,|\, M_2 \rangle\langle M_2 \,|\, M_3 \rangle\langle M_3 \,|\, j' \rangle = 1 \qquad (10.10)$$

Then, Equation 10.8 reduces to

$$\langle x \,|\, s \rangle = \langle x \,|\, j \rangle\langle j \,|\, s \rangle + \langle x \,|\, j' \rangle\langle j' \,|\, s \rangle \qquad (10.11)$$

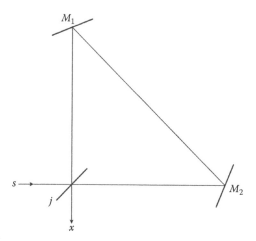

FIGURE 10.3
Sagnac interferometer in triangular configuration.

If $j' = 1$ represents the beam splitter in a transmission mode and $j = 2$ the reflection mode, then Equation 10.11 can be written as (Duarte, 2003)

$$\langle x \mid s \rangle = \langle x \mid 2 \rangle \langle 2 \mid s \rangle + \langle x \mid 1 \rangle \langle 1 \mid s \rangle \qquad (10.12)$$

which, for $N = 2$, can be expressed as

$$\langle x \mid s \rangle = \sum_{j=1}^{N=2} \langle x \mid j \rangle \langle j \mid s \rangle \qquad (10.13)$$

An alternative triangular Sagnac interferometer, with only two mirrors (M_1 and M_2), is shown in Figure 10.3.

10.3.2 Mach–Zehnder Interferometer

The Mach–Zehnder interferometer is illustrated in Figure 10.4. In this interferometer the incident light beam is divided into two sub-beams by a beam splitter. The reflected beam, on the incidence beam splitter, is then sent into a path defined by the reflection on M_1 toward the exit beam splitter. The transmitted beam, on the incidence beam splitter, is sent into a path defined by the reflection on M_2 toward the exit beam splitter.

Both counterpropagating beams are recombined at the exit beam splitter. The interference mechanics of the counterpropagating beams can be described using the Dirac notation via the probability amplitude

$$\langle x \mid s \rangle = \langle x \mid k' \rangle \langle k' \mid M_1 \rangle \langle M_1 \mid j \rangle \langle j \mid s \rangle + \langle x \mid k \rangle \langle k \mid M_2 \rangle \langle M_2 \mid j' \rangle \langle j' \mid s \rangle \quad (10.14)$$

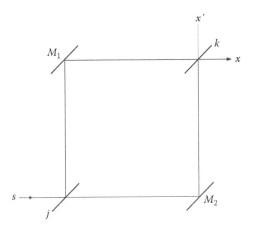

FIGURE 10.4
Mach–Zehnder interferometer configured with entrance (j) and exit (k) beam splitters and internal mirrors M_1 and M_2. In the schematics x' represents a weak secondary output.

which can be abstracted to

$$\langle x \,|\, s \rangle = \langle x \,|\, k' \rangle \langle k' \,|\, j \rangle \langle j \,|\, s \rangle + \langle x \,|\, k \rangle \langle k \,|\, j' \rangle \langle j' \,|\, s \rangle \qquad (10.15)$$

If $j' = k' = 1$ represent the beam splitters in a transmission mode and $j = k = 2$ in a reflection mode, then Equation 10.15 can be written as (Duarte, 2003)

$$\langle x \,|\, s \rangle = \langle x \,|\, 1 \rangle \langle 1 \,|\, 2 \rangle \langle 2 \,|\, s \rangle + \langle x \,|\, 2 \rangle \langle 2 \,|\, 1 \rangle \langle 1 \,|\, s \rangle \qquad (10.16)$$

The same result can be obtained from

$$\langle x \,|\, s \rangle = \sum_{k=1}^{N} \sum_{j=1}^{N} \langle x \,|\, k \rangle \langle k \,|\, j \rangle \langle j \,|\, s \rangle \qquad (10.17)$$

which leads to

$$\langle x \,|\, s \rangle = \langle x \,|\, 1 \rangle \langle 1 \,|\, 2 \rangle \langle 2 \,|\, s \rangle + \langle x \,|\, 1 \rangle \langle 1 \,|\, 1 \rangle \langle 1 \,|\, s \rangle$$
$$+ \langle x \,|\, 2 \rangle \langle 2 \,|\, 2 \rangle \langle 2 \,|\, s \rangle + \langle x \,|\, 2 \rangle \langle 2 \,|\, 1 \rangle \langle 1 \,|\, s \rangle \qquad (10.18)$$

However, since $\langle 1|1 \rangle$ and $\langle 2|2 \rangle$ illuminate x' rather than x, then the probability amplitude, for this geometry, reduces to that given in Equation 10.16.

A prismatic Mach–Zehnder interferometer is illustrated in Figure 10.5. In this prismatic version of the Mach–Zehnder, there is asymmetry in regard to the intra-interferometric beam dimensions. The P_1–M_2–P_2 beam is expanded relatively to the P_1–M_1–P_2 beam. Also, in this particular example

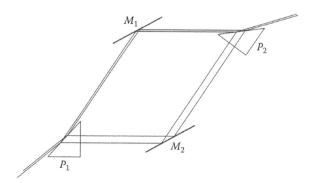

FIGURE 10.5
Prismatic Mach–Zehnder interferometer configured with $P_1 \rightarrow j$, M_1, M_2, and exit prism $P_2 \rightarrow k$.

(based on a prism with a magnification of $k_{1,1} \approx 5$, see Chapter 15), there is also a power asymmetry since the unexpanded beam propagating in the P_1–M_1–P_2 arm has about 30% of the incident power while the expanded beam P_1–M_2–P_2 carries the remaining 70% of the incident power for light polarized parallel to the plane of incidence. In this regard, it should be possible to design a prismatic Mach–Zehnder where the power density (Wm^{-2}) in each arm is balanced. Applications for this type of interferometer include imaging and microscopy. Additional Mach–Zehnder interferometric configurations include transmission gratings as beam splitters (Steel, 1967).

10.3.3 Michelson Interferometer

The Michelson interferometer (Michelson, 1927) is illustrated in Figure 10.6. In this interferometer the incident light beam is divided into two sub-beams by a beam splitter that serves as both input and output element. The reflected

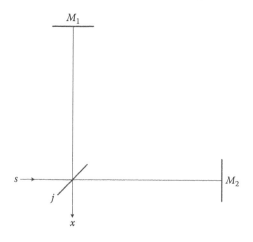

FIGURE 10.6
Michelson interferometer includes a single beam splitters and mirrors M_1 and M_2.

beam, on the incidence beam splitter, is then sent into a path defined by the reflection on M_1 and back toward the exit beam splitter. The transmitted beam, on the incidence beam splitter, is sent into a path defined by the reflection on M_2 and back toward the exit beam splitter. Both beams are recombined interferometrically at the beam splitter. For the Michelson interferometer, the interference can be characterized using a probability amplitude of the form

$$\langle x \,|\, s \rangle = \langle x \,|\, j \rangle \langle j \,|\, M_2 \rangle \langle M_2 \,|\, j' \rangle \langle j' \,|\, s \rangle + \langle x \,|\, j' \rangle \langle j' \,|\, M_1 \rangle \langle M_1 \,|\, j \rangle \langle j \,|\, s \rangle \quad (10.19)$$

which can be abstracted to

$$\langle x \,|\, s \rangle = \langle x \,|\, j \rangle \langle j \,|\, j' \rangle \langle j' \,|\, s \rangle + \langle x \,|\, j' \rangle \langle j' \,|\, j \rangle \langle j \,|\, s \rangle \quad (10.20)$$

If $j' = 1$ represents the function of the beam splitter in the transmission mode and $j = 2$ represents the function of the beam splitter in the reflection mode,

$$\langle x \,|\, s \rangle = \langle x \,|\, 2 \rangle \langle 2 \,|\, 1 \rangle \langle 1 \,|\, s \rangle + \langle x \,|\, 1 \rangle \langle 1 \,|\, 2 \rangle \langle 2 \,|\, s \rangle \quad (10.21)$$

It is clear that the substitution of the appropriate wave functions for the various terms in Equations 10.11, 10.15, and 10.20 and the multiplication of these equations with their respective complex conjugates yield probability equations of an interferometric character. A variant of the Michelson interferometer uses retroreflectors (Steel, 1967).

10.4 Multiple-Beam Interferometers

An N-slit interferometer, which can be considered as a multiple-beam interferometer, was introduced in Chapter 4 and is depicted in Figure 10.7. In this configuration, an expanded beam of light illuminates simultaneously an array of N slits. Following propagation the N sub-beams interfere at a plane perpendicular to the plane of propagation. The probability amplitude is given by

$$\langle x \,|\, s \rangle = \sum_{j=1}^{N} \langle x \,|\, j \rangle \langle j \,|\, s \rangle \quad (10.22)$$

and the probability is

$$|\langle x \,|\, s \rangle|^2 = \sum_{j=1}^{N} \Psi(r_j) \sum_{m=1}^{N} \Psi(r_m) e^{i(\Omega_m - \Omega_j)} \quad (10.23)$$

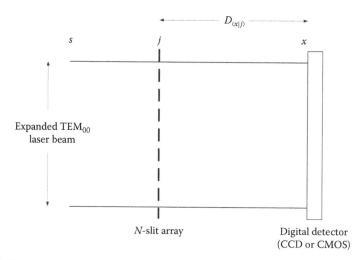

FIGURE 10.7
N-Slit interferometer.

which can also be expressed as Equation 10.3 (Duarte, 1991, 1993):

$$|\langle x \mid s \rangle|^2 = \sum_{j=1}^{N} \Psi(r_j)^2 + 2\sum_{j=1}^{N} \Psi(r_j)\left(\sum_{m=j+1}^{N} \Psi(r_m)\cos(\Omega_m - \Omega_j)\right)$$

Expressions for 2D and 3D cases are given in Chapter 4. As seen previously, this approach is also applicable to the two-beam interferometer introduced by Hanbury Brown and Twiss (1956) (shown in Figure 10.1) and to other multiple-beam, or multiple-slit, interferometers used in astronomical applications.

The second multiple-beam interferometer is the Fabry–Perot interferometer depicted in Figure 10.8. This interferometer is also considered in Appendix B as an *intracavity etalon*. Generally, intracavity etalons are a solid slab of optical glass, or fused silica, with highly parallel surfaces coated to increase reflectivity (Figure 10.8b).

These are also known as Fabry–Perot etalons. Fabry–Perot interferometers, on the other hand, are constituted by two separate slabs of optical flats with their inner surfaces coated as shown in Figure 10.8a. The space between the two coated surfaces is filled with air or other inert gas. The optical flats in a Fabry–Perot interferometer are mounted on rigid metal bars, with a low thermal expansion coefficient, such as invar. The plates can be moved, with micrometer precision or better, to vary the free spectral range (*FSR*).

These interferometers are widely used to characterize and quantify the laser linewidth. The mechanics of multiple-beam interferometry can be described in some detail considering the multiple reflection, and refraction, of a beam incident on two parallel surfaces separated by a region of refractive index n

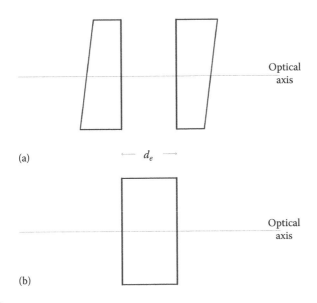

(a)   — d_e —

(b)

FIGURE 10.8
(a) Fabry–Perot interferometer and (b) Fabry–Perot etalon. Darker lines represent coated surfaces.

as illustrated in Figure 10.9. In this configuration, at each point of reflection and refraction, a fraction of the beam, or a sub-beam, is transmitted toward the boundary region. Following propagation, these sub-beams interfere. In this regard, the physics is similar to that of the N-slit interferometer with the exception that each parallel beam has less intensity due to the increasing

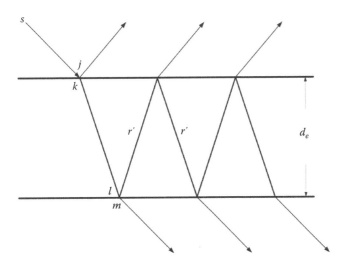

FIGURE 10.9
Multiple-beam interferometer diagram illustrating the multiple internal reflection geometry.

number of reflections. Here, for transmission, interference can be described using a series of probability amplitudes representing the events depicted in Figure 10.9 (Duarte, 2003):

$$\langle x \mid s \rangle = \sum_{m=1}^{N} \sum_{l=1}^{N} \sum_{k=1}^{N} \sum_{j=1}^{N} \langle x \mid m \rangle \langle m \mid l \rangle \langle l \mid k \rangle \langle k \mid j \rangle \langle j \mid s \rangle \qquad (10.24)$$

where
   j is at the reflection surface of incidence
   k is immediately next to the surface of reflection
   l is at the second surface of reflection
   m is immediately next to the second surface of reflection as illustrated in Figure 10.9

If the incident beam is assumed to be a narrow beam incident at a single point j, then the propagation of this single beam proceeds to l and is represented by the incidence amplitude A_j, which is a complex number, attenuated by a transmission factor t, so that the first three probability amplitudes can be unified by an expression of the form (Duarte, 2003)

$$\langle l \mid k \rangle \langle k \mid j \rangle \langle j \mid s \rangle = A_i t \qquad (10.25)$$

and Equation 10.24 can be written as

$$\langle x \mid s \rangle = A_i t \sum_{m=1}^{N} \langle x \mid m \rangle \langle m \mid l \rangle \qquad (10.26)$$

which, using the established notation, can be expressed as (Steel, 1967; Born and Wolf, 1999)

$$A_t(p) = A_i t (t' + t' r'^2 e^{i\delta} + t' r'^4 e^{i2\delta} + \cdots + t' r'^{2(p-1)} e^{i(p-1)\delta}) \qquad (10.27)$$

Defining

$$\mathcal{T} = tt' \qquad (10.28)$$

and

$$\mathcal{R} = r'^2 \qquad (10.29)$$

and taking the limit as $p \to \infty$, Equation 10.27 reduces to (Born and Wolf, 1999)

$$A_t = \mathcal{T}(1 - \mathcal{R} e^{i\delta})^{-1} A_i \qquad (10.30)$$

and multiplication with its complex conjugate yields an expression for the intensity

$$I_t = \mathcal{T}^2(1 + \mathcal{R}^2 - 2\mathcal{R}\cos\delta)^{-1}I_i \tag{10.31}$$

which is known as the *Airy formula* or *Airy function*.

In Chapter 3 we saw that alternative forms of the uncertainty principle

$$\Delta x \, \Delta p \approx h \tag{10.32}$$

are

$$\Delta x \, \Delta\lambda \approx \lambda^2 \tag{10.33}$$

and

$$\Delta x \, \Delta\nu \approx c \tag{10.34}$$

For *solid* Fabry–Perot etalons made of optical glass with refractive index n, $\Delta x = 2nd_e$, and $\Delta\lambda$ becomes the *FSR* so that

$$FSR \approx \frac{\lambda^2}{2nd_e} \tag{10.35}$$

and in the frequency domain

$$FSR \approx \frac{c}{2nd_e} \tag{10.36}$$

For an air-spaced Fabry–Perot interferometer, $n = 1$, and $\Delta x = 2d_e$.

The *FSR* corresponds to the separation of the rings in Figure 10.10, and a measure of the width of the rings determines the linewidth of the emission being observed. The minimum resolvable linewidth $\Delta\nu_{FRS}$ is given by

$$\Delta\nu_{FRS} = \frac{FRS}{\mathcal{F}} \tag{10.37}$$

where $\mathcal{F}$ is the *effective finesse*. Thus, a Fabry–Perot etalon with an $FSR = 7.00$ GHz and $\mathcal{F} = 50$ provides discrimination down to ~140 MHz. The finesse is a function of the flatness of the surfaces (often in the $\lambda/100$–$\lambda/50$ range), the dimensions of the aperture, and the reflectivity of the surfaces. The *FSR* concept also applies to laser cavities as discussed in Chapter 9.

The effective finesse of a Fabry–Perot interferometer is given by (Meaburn, 1976)

$$\mathcal{F}^{-2} = \mathcal{F}_R^{-2} + \mathcal{F}_F^{-2} + \mathcal{F}_A^{-2} \tag{10.38}$$

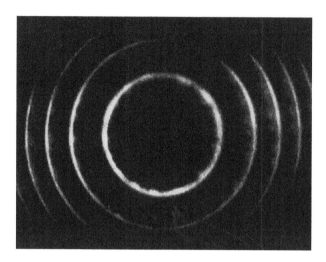

FIGURE 10.10
Fabry–Perot interferogram depicting single-longitudinal-mode oscillation, at $\Delta \nu \approx 500$ MHz, from a tunable multiple-prism grating solid-state oscillator. (Reproduced from Duarte, F.J., *Opt. Commun.* 117, 480, 1995, with permission from Elsevier.)

Here $\mathcal{F}_R, \mathcal{F}_F$, and $\mathcal{F}_A$ are the reflective, flatness, and aperture finesses, respectively. The reflective finesse is given by (Steel, 1967; Born and Wolf, 1999)

$$\mathcal{F}_R = \frac{\pi \sqrt{\mathcal{R}}}{(1-\mathcal{R})} \tag{10.39}$$

where $\mathcal{R}$ is the reflectivity.

10.5 N-Slit Interferometer as a Wavelength Meter

Interferometry signals and profiles are a function of the wavelength of the radiation that produces them. Thus, interferometers are well suited to be applied as wavelength meters specially when a digital detector array is used to record the resulting interferogram. As such, a variety of interferometric configurations have been used in the measurement of tunable laser wavelengths. For a review in this subject, the reader should refer to Demtröder (2003).

The wavelength sensitivity of multiple-beam interferometry has its origin in the phase information of the equations describing the behavior of the interferometric signal. In the case of the N-slit interferometer, the interferometric profile is characterized by the interferometric equation

$$|\langle x|s\rangle|^2 = \sum_{j=1}^{N} \Psi(r_j)^2 + 2\sum_{j=1}^{N} \Psi(r_j)\left(\sum_{m=j+1}^{N} \Psi(r_m)\cos(\Omega_m - \Omega_j)\right)$$

which includes a phase difference term that, as explained in Chapter 4, can be expressed as

$$\cos\left((\theta_m - \theta_j) \pm (\phi_m - \phi_j)\right) = \cos\left(|\,l_m - l_{m-1}\,|\,k_1 \pm |\,L_m - L_{m-1}\,|\,k_2\right) \qquad (10.40)$$

where

$$k_1 = \frac{2\pi n_1}{\lambda_v} \qquad (10.41)$$

and

$$k_2 = \frac{2\pi n_2}{\lambda_v} \qquad (10.42)$$

Here, $\lambda_1 = \lambda_v / n_1$ and $\lambda_1 = \lambda_v / n_2$ where λ_v is the vacuum wavelength and n_1 and n_2 are the corresponding indexes of refraction (Wallenstein and Hänsch, 1974; Born and Wolf, 1999). Hence, it is easy to see that different wavelengths will produce different interferograms. To illustrate this point in Figures 10.11 and 10.12, two calculated interferograms, using Equation 10.3, for the N-slit interferometer, with $N = 50$ and $D_{\langle x|j\rangle} = 25$ cm, are shown for $\lambda_1 = 589$ nm and $\lambda_2 = 590$ nm, respectively. For a given set of geometrical parameters, measured interferograms can be matched, in an iterative process, with theoretical interference patterns to determine the wavelength of the radiation. Again, the resolution depends on the optical path length between the slit array and the digital detector and on the size of the pixels and the linearity of the detector. In this regard, increased resolution in CCD, and CMOS

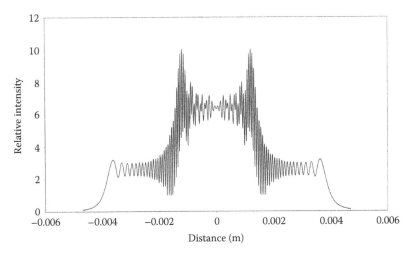

FIGURE 10.11
Interferogram at $\lambda_1 = 589$ nm. These calculations are for slits 30 µm wide, separated by 30 µm, and $N = 50$. The j–x distance is $D_{\langle x|j\rangle} = 25$ cm.

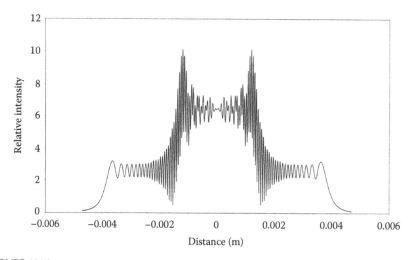

FIGURE 10.12
Interferogram at $\lambda_2 = 590$ nm. These calculations are for slits 30 μm wide, separated by 30 μm, and $N = 50$. The j–x distance is $D_{\langle x|j \rangle} = 25$ cm.

detectors, should improve significantly the wavelength resolution achievable with the N-slit interferometer. This, coupled with the simplicity of the optics, should enhance considerably the application perspectives of this interferometer as a wavelength meter.

Further applications of the NSLI are described by Duarte (2009) and include interferometric imaging, microdensitometry, microscopy, and secure optical communications (see Chapter 11).

10.6 Ramsey Interferometer

Finally, we very briefly touch on a different type of interferometer: the Ramsey interferometer, which was discovered by Norm Ramsey around 1950 (Ramsey, 1950, 1990). This is a very different type of interferometer to those described previously since it uses atoms rather than photons as the interfering entities. In the Ramsey interferometer the laser, or photon source, is replaced by an atom source as described in Figure 10.13. In its path toward the detector, the atom beam is allowed to interact with microwave radiation at two places. The first place is near the source and the second place near the detector. At each of those spatial sections, the microwave field is allowed to modulate the state of the atoms. Defining P_g as the probability to be in the ground state and P_e the probability to be in the excited state (Haroche et al., 2013),

$$P_g = (1 - P_e) = \frac{(1 - \cos\phi)}{2}$$

(10.43)

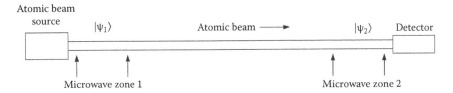

FIGURE 10.13
Simplified depiction of the Ramsey interferometer showing the source of atomic beam and microwave field regions 1 and 2. Sources of atomic beams include laser-cooled cesium and rubidium.

where the angle ϕ represents the phase difference between the ground state and the excited state. If $\phi = 2\pi m$ (where m is an integer), the states are in phase and thus there is constructive interference. In other words, when $\phi = 2\pi m$ there is atomic coherence.

Using the relevant Dirac notation, introduced in Chapters 16 and 17, one can define the probability amplitude of the initial unmodulated state as

$$|\psi_1\rangle = \frac{1}{\sqrt{2}}(|a\rangle + |b\rangle) \tag{10.44}$$

and the probability amplitude of the modulated state as

$$|\psi_2\rangle = \frac{1}{\sqrt{2}}(|a\rangle + e^{i\phi}|b\rangle) \tag{10.45}$$

Multiplying Equation 10.45 with its complex conjugate can lead to

$$P_e = \frac{(1+\cos\phi)}{2} \tag{10.46}$$

and using $P_g = (1 - P_e)$ enables us to write

$$P_g = \frac{(1-\cos\phi)}{2}$$

as given in Equation 10.43.

The phase angle itself is a function of the frequency difference between the frequency of the transition (between the ground state and the excited state, i.e., ν_{eg}) and the microwave frequency ν (Haroche, 2013): $\phi = 2\pi(\nu_{eg} - \nu)\Delta t$. Thus, the interference pattern produced is a function of $(\nu_{eg} - \nu)$. Locking the microwave frequency to the transition frequency, that is, $\nu = \nu_{eg}$, yields a time standard anchored to the atomic transition (Haroche, 2013). This principle, coupled to laser-cooled atomic beams, is central to the technology of atomic clocks.

Problems

10.1 Interferometric measurements with a Hanbury Brown–Twiss type interferometer yield a measured angular spread of $\Delta\theta \approx 5.94$ marcsec from the emission of the Sirius star. Determine the radius of this star (hint: 1 marcsec is 4.8481368×10^{-9} rad and the distance to Sirius is approximately 8.6 light-years).

10.2 A laser beam fails to provide interference fringes when the distance from the beam splitter to the mirrors, in a Michelson interferometer, is 1 m. Estimate the linewidth of the laser in Hz units.

10.3 Using the usual complex-wave representation for probability amplitudes, use Equation 10.15 to arrive at an equation for the probability of transmission in a Mach–Zehnder interferometer.

10.4 List the simplifying assumptions that lead from Equations 10.24 to 10.27.

10.5 Starting from Equation 10.45, arrive at Equation 10.46 [Hint: Consult Chapter 17].

References

Born, M. and Wolf, E. (1999). *Principles of Optics*, 7th edn. Cambridge University, Cambridge, U.K.

Demtröder, W. (2003). *Laser Spectroscopy*, 3rd edn. Springer-Verlag, Berlin, Germany.

Dirac, P. A. M. (1978). *The Principles of Quantum Mechanics*, 4th edn. Oxford University, Oxford, U.K.

Duarte, F. J. (1991). Dispersive dye lasers. In *High Power Dye Lasers* (Duarte, F. J., ed.). Springer-Verlag, Berlin, Germany, Chapter 2.

Duarte, F. J. (1993). On a generalized interference equation and interferometric measurements. *Opt. Commun.* **103**, 8–14.

Duarte, F. J. (1995). Solid-state dispersive dye laser oscillator: Very compact cavity. *Opt. Commun.* **117**, 480–484.

Duarte, F. J. (1998). Interference of two independent sources. *Am. J. Phys.* **66**, 662–663.

Duarte, F. J. (1999). Multiple-prism grating solid-state dye laser oscillator: Optimized architecture. *Appl. Opt.* **38**, 6347–6349.

Duarte, F. J. (2003). *Tunable Laser Optics*. Elsevier-Academic, New York.

Duarte, F. J. (2008). Coherent electrically excited organic semiconductors: Coherent or laser emission?, *Appl. Phys. B* **90**, 101–108.

Duarte, F. J. (2009). Interferometric imaging. In *Tunable Laser Applications*, 2nd edn. (Duarte, F. J., ed.) CRC Press, New York, Chapter 12.

Feynman, R. P., Leighton, R. B., and Sands, M. (1965). *The Feynman Lectures on Physics*, Vol. III, Addison-Wesley, Reading, MA.

Hanbury Brown, R. and Twiss, R. Q. (1956). A test of a new type of stellar interferometer on Sirius, *Nature* **178**, 1046–1048.

Haroche, S., Brune, M., and Raimond, J-.M. (2013). Atomic clocks for controlling light Fields. *Phys. Today* **66** (1), 27–32.

Meaburn, J. (1976). *Detection and Spectrometry of Faint Light*. Reidel, Boston, MA.

Michelson, A. A. (1927). *Studies in Optics*. University of Chicago, Chicago, IL.

Ramsey, N. F. (1950). A molecular beam resonance method with separated oscillating fields. *Phys. Rev.* **78**, 695–699.

Ramsey, N. F. (1990). Experiments with separated oscillatory fields and hydrogen Masers. *Rev. Mod. Phys.* **62**, 541–552.

Steel, W. H. (1967). *Interferometry*. Cambridge University, Cambridge, U.K.

Wallenstein, R. and Hänsch, T. W. (1974). Linear pressure tuning of a multielement dye laser spectrometer. *Appl. Opt.* **13**, 1625–1628.

11

Secure Interferometric
Communications in Free Space

11.1 Introduction

The N-slit laser interferometer (NSLI) was first introduced as an alternative applicable to secure free-space optical communications in 2002 (Duarte, 2002). Albeit its initial propagation distance in the laboratory was only 15 cm, it was envisioned as an interferometric tool ideally suited for propagation in the vacuum, or outer space (Duarte, 2002). Subsequently, NSLI experiments have also been conducted over hundreds of meters, in the field, via open atmosphere (Duarte et al., 2010, 2011).

The idea is that an N-slit interferometric signal, designated as an *interferometric character* or a series of interferometric characters, can be used to transmit information securely from one point in space to another point in space. The interferometric character propagates from its origin (s) to its destiny, the interference plane (x) (see Figure 11.1), while the integrity of the character itself is secured by interferometric principles. In other words, attempts to optically intersect the interferometric character severely distort its spatial and intensity profile thus informing the receiver that the message has been compromised. Hence, the security of this free-space communication method rests simply on the principle that any optical attempt to intercept causes the collapse of the signal. This means that, in its most basic form, no security key, or code, is necessary. However, a code could be easily added as an extra layer of security (see Figure 11.1).

Here, we should add that there has been a marked recent increase in attention toward optical communications in free space, both in a terrestrial environment and for deep-space optical communications. This interest appears to be driven by saturation in the spectrum of traditional radio frequency communications (Hogan, 2013). A brief historical review of free-space optical communications is given by Duarte (2002).

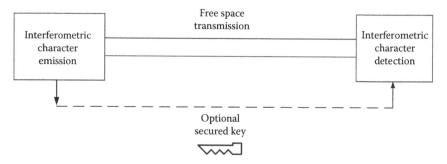

FIGURE 11.1
Cryptographic diagram applicable to N-slit interferometric communications in free space.

11.2 Theory

The probability amplitude for propagation from the source (s) to the inter-ferometric plane (x), via the slit array (j), as illustrated in Figure 11.2, can be expressed using the Dirac notation (Duarte, 1993)

$$\langle x \mid s \rangle = \sum_{j=1}^{N} \langle x \mid j \rangle \langle j \mid s \rangle \tag{11.1}$$

Assigning each probability amplitude a "wave function of ordinary wave optics" as taught by Dirac (1978), and following some algebra (see Chapter 4), leads to the generalized 1D N-slit interferometric equation (Duarte, 1991, 1993)

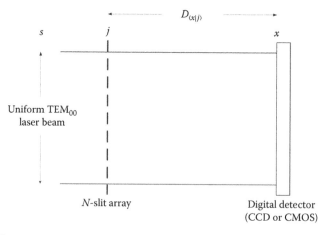

FIGURE 11.2
Top view of the N-slit laser interferometer highlighting the intra-interferometric path $D_{\langle x \mid j \rangle}$.

$$|\langle x \mid s \rangle|^2 = \sum_{j=1}^{N} \Psi(r_j)^2 + 2 \sum_{j=1}^{N} \Psi(r_j) \left(\sum_{m=j+1}^{N} \Psi(r_m) \cos(\Omega_m - \Omega_j) \right) \quad (11.2)$$

where

$\Psi(r_j)$ are wave functions (Dirac 1978; Duarte, 2004)
the term in parenthesis represents the phase that describes the exact geometry of the N-slit interferometer (Duarte, 1991, 1993)

As explained by Duarte (2004) the measured intensity is proportional to the probability $|\langle x \mid s \rangle|^2$, and it is this probability that provides the spatial distribution of the observed intensity (see Chapter 8).

At this stage, it is important to emphasize that this equation was originally derived for *single-photon propagation* (Duarte, 1993, 2004) albeit in practice it also applies to the propagation of an ensemble of indistinguishable photons, as in the case of narrow-linewidth laser emission (Duarte, 2003). The generalized N-slit interferometric equation accurately describes measurements performed in a macroscopic apparatus using quantum principles, and as such it neatly follows the van Kampen (1988) criteria.

11.3 N-Slit Interferometer for Secure Free-Space Optical Communications

The NSLI used for free-space optical communications is depicted in Figure 11.3. The essence of this method consists in the expanded beam illumination (*s*) of an N-slit array, or grating (*j*), where the interferometric characters are produced.

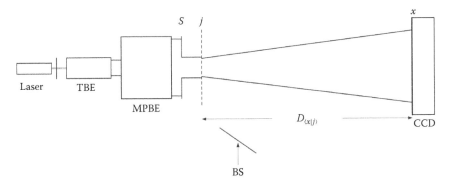

FIGURE 11.3
Top perspective of the optical architecture of the N-slit laser interferometer; see text for details. Also included in this diagram is a thin BS, inserted at the Brewster angle (relative to the optical axis), to intercept the propagating interferogram or the interferometric characters.

The interferometric character then propagates via the intra-interferometric distance $(D_{\langle x|j\rangle})$ until it reaches the interferometric plane at x. In other words, the generation, propagation, and detection of the interferometric characters take place within the NSLI, thus highlighting the conceptual and configurational simplicity of the interferometric approach.

The illumination section of the NSLI requires a single-transverse-mode narrow-linewidth laser. For the experiments described here, the coherent source is a He–Ne laser yielding 2 mW in a TEM_{00} laser beam at $\lambda = 632.8$ nm. The laser is followed by neutral density filters. Following the attenuation stage, a 2D telescopic beam expander $(M \approx 10)$ includes a 25 µm spatial filter. The expanded beam then undergoes 1D multiple-prism beam expansion thus yielding an elongated near-Gaussian beam of ~10 mm maximum height by ~50 mm maximum width. Following the prismatic beam expander, a high-precision chromium master grating (j in Figure 11.3) is positioned.

The high-precision master gratings have rather large dimensions. For instance, one of the gratings utilized has 570 µm slits separated by 570 µm isles, and the other grating has 1000 µm slits separated by 1000 µm isles. The tolerances in the slit dimensions are quoted by the manufacturer as 0.5 µm. The 570 µm grating has a total of 44 slits and the 1000 µm grating has a total of 25 slits. The overall ruled area is 50 mm × 50 mm (Duarte et al., 2010).

The digital detector, deployed at the interferometric plane (x), is a digital detector (either CCD or CMOS) with pixels ~20 µm in width. In the experiments described here, the detector was not cooled and there was no subtraction of background noise from the measurements.

11.4 Interferometric Characters

The concept of *interferometric characters* was introduced when the NSLI was disclosed as an alternative for secure free-space optical communications (Duarte, 2002). In this approach there are an infinite number of possible slit combinations that can lead to a set of interferometric characters. The simplest one consists of two-slit interferogram resulting in the interferometric character a, that is, $N = 2 \rightarrow a$, $N = 3 \rightarrow b$, $N = 4 \rightarrow c...N = 26 \rightarrow z$ (Duarte, 2002). Calculated interferograms for the interferometric characters a, b, c, and z are given in Figure 11.4.

In the N-slit interferometric approach, once the emitter, controlling the illumination of the N-slit array at j and the receiver at x, decides on an interferometric alphabet, the communications can begin immediately with the receiver reading directly the interferometric characters send by the emitter.

Since all these characters have also a theoretical counterpart, the received characters can be compared with the calculated character to verify its fidelity. As demonstrated by Duarte (2002, 2005) any attempt to optically intercept an interferometric character results in a catastrophic collapse of the interferometric signal and is immediately noticed by the receiver. The collapse sequence, and displacement, of the interferometric character a, due to the insertion of a very thin high-optical surface quality beam splitter (BS)

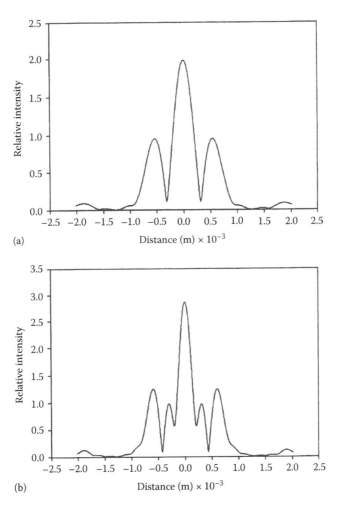

(a)

(b)

FIGURE 11.4
Interferometric characters a ($N = 2$) (a), b ($N = 3$) (b) for $\lambda = 632.8$ nm and $D_{(x|j)} = 10$ cm. The slit width here is 50 μm and uniformly separated by 50 μm. Note that the detailed and exact contour, as compared to measurements, of these interferograms (specially at the minima) depends on the choice of function to represent the radiation at the individual slits.

(continued)

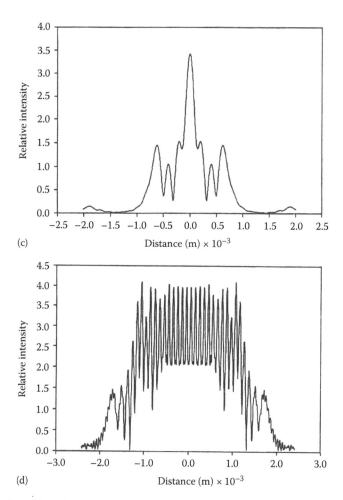

FIGURE 11.4 (continued)

Interferometric characters c ($N = 4$) (c), and z ($N = 26$) (d) for $\lambda = 632.8$ nm and $D_{\langle x|j\rangle} = 10$ cm. The slit width here is 50 μm and uniformly separated by 50 μm. Note that the detailed and exact contour, as compared to measurements, of these interferograms (specially at the minima) depends on the choice of function to represent the radiation at the individual slits. (Reproduced from Duarte, F.J., *Opt. Commun.* 205, 313, 2002, with permission from Elsevier.)

in the intra-interferometric optical path ($D_{\langle x|j\rangle}$), is illustrated in Figure 11.5. An additional sequence of measurements, using the interferometric character c ($N = 4$), is given in Figures 11.6 through 11.9.

The violent distortion of the interferometric characters, following insertion of a thin BS, as illustrated in Figures 11.5 and 11.7, can be explained in reference to Equation 11.2, which interferometrically entangles the probability amplitudes $\langle x|j\rangle$ originating at each slit (j). This entanglement is a function of photon wavelength, the number of slits illuminated at N, slit geometry, and propagation geometry (see Chapter 4).

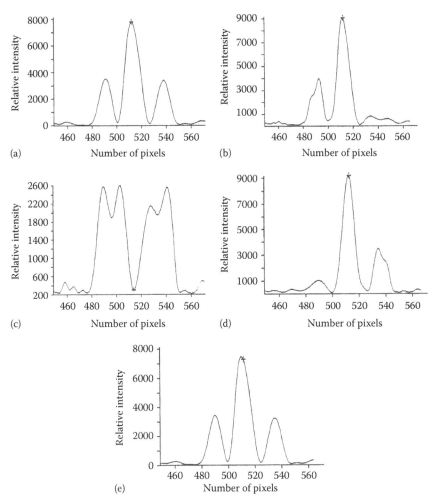

FIGURE 11.5
Collapse sequence of the interferometric characters *a*, following insertion of a thin BS into the intra-interferometric path $D_{\langle x|j\rangle}$. (a) Interferometric character *a* prior to insertion of the BS, (b–d) collapse of the interferometric character *a* during insertion, and (e) displaced interferometric character *a* once insertion is completed. Here, $N = 2$, the slit width is 50 μm, slit separation is 50 μm, and $D_{\langle x|j\rangle} = 10$ cm. (Reproduced from Duarte, F.J., *Opt. Commun.* 205, 313, 2002, with permission from Elsevier.)

This entanglement is violently disrupted by the insertion of an optical edge as provided by a classical, or macroscopic, BS regardless of the finesse of that BS.

A complementary way to think about this effect is that the mechanics of the overall probability amplitude

$$\langle x \mid s \rangle = \sum_{j=1}^{N} \langle x \mid j \rangle \langle j \mid s \rangle$$

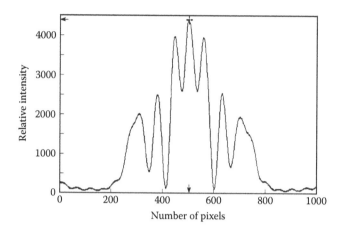

FIGURE 11.6

The interferometric character c, generated with $N = 4$ (570 μm slits separated by 570 μm), $\lambda = 632.8$ nm, at an intra-interferometric distance of $D_{\langle x|j \rangle} = 7.235$ m. (Reproduced from Duarte, F.J., *J. Opt. A: Pure Appl. Opt.* 7, 73, 2005, with permission from the Institute of Physics.)

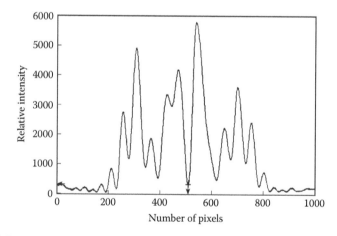

FIGURE 11.7

The interferometric character c, as in Figure 11.6, severely collapsed due to the insertion of a thin BS, at Brewster angle, 2 m from the grating (j). (Reproduced from Duarte, F.J., *J. Opt. A: Pure Appl. Opt.* 7, 73, 2005, with permission from the Institute of Physics.)

allow us to observe the input and the output but it does not allow us to obtain information while the interferometric character propagated via the intra-interferometric beam path ($D_{\langle x|j \rangle}$). In other words, the integrity of the interferometric character is protected by the very essence of interference, be it single-photon interference or interference generated by an ensemble of indistinguishable photons.

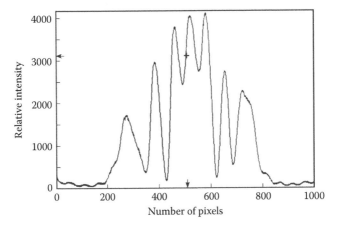

FIGURE 11.8
The interferometric character *c*, as in Figure 11.6, clearly distorted due to the static presence of a thin BS, at Brewster angle, 2 m from the grating (*j*). (Reproduced from Duarte, F.J., *J. Opt. A: Pure Appl. Opt.* 7, 73, 2005, with permission from the Institute of Physics.)

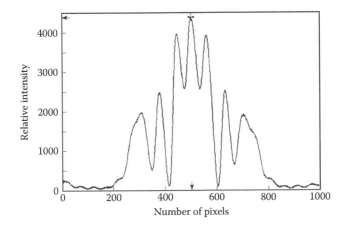

FIGURE 11.9
The interferometric character *c*, as in Figure 11.6, restored due to the removal of the thin BS that caused the distortions depicted in the previous figures. (Reproduced from Duarte, F.J., *J. Opt. A: Pure Appl. Opt.* 7, 73, 2005, with permission from the Institute of Physics.)

Indeed, as soon as the BS is introduced, Equation 11.2, that is,

$$|\langle x \mid s \rangle|^2 = \sum_{j=1}^{N} \Psi(r_j)^2 + 2 \sum_{j=1}^{N} \Psi(r_j) \left(\sum_{m=j+1}^{N} \Psi(r_m) \cos(\Omega_m - \Omega_j) \right)$$

ceases to describe the experimental situation illustrated in Figures 11.2 and 11.3, which originates in the probability amplitude expressed by

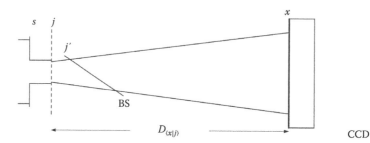

FIGURE 11.10
The N-slit laser interferometer with a thin BS inserted in the intra-interferometric path $D_{\langle x|j\rangle}$. For the propagation at the top of the figure, the probability amplitude is modified from $\langle x|j\rangle$ $\langle j|s\rangle$ to $\langle x|j'\rangle\langle j'|j\rangle\langle j|s\rangle$. And the latter probability amplitude varies continuously, as a function of distance, along the BS.

Equation 11.1. The new experimental situation, illustrated in Figure 11.10, is accounted for by a probability amplitudes of the form

$$\langle x|s\rangle = \sum_{k=1}^{N}\sum_{j=1}^{N}\langle x|j'\rangle\langle j'|j\rangle\langle j|s\rangle \qquad (11.3)$$

where the $\langle j'|j\rangle$ term represents the probability amplitude of transmission via the BS. In reality this is an undetermined spatially unsymmetric transmission that results in the destruction of the original interferometric pattern.

11.5 Propagation in Terrestrial Free Space

As mentioned at the introduction, the first experiment on the use of the NSLI as a tool for secured free-space optical communications took place in the laboratory over a distance of $D_{\langle x|j\rangle} = 0.10$ m (Duarte, 2002). Further series of measurements took place over intra-interferometric distances of $D_{\langle x|j\rangle} = 7.235$ m (Duarte, 2005), $D_{\langle x|j\rangle} = 35$ m (Duarte et al., 2010), and $D_{\langle x|j\rangle} = 527$ m (Duarte et al., 2011).

For $D_{\langle x|j\rangle} = 0.10$ m, and propagation in homogeneous laboratory air, the interferometric character a is depicted in Figure 11.5. For $D_{\langle x|j\rangle} = 7.235$ m, and propagation in homogeneous laboratory air, the interferometric character c is depicted in Figure 17.6. For $D_{\langle x|j\rangle} = 35$ m, and propagation in near-homogeneous air (at $T \approx 30°C$), the measured interferometric character c is depicted in Figure 11.11, while its calculated version is displayed in Figure 11.12. For $D_{\langle x|j\rangle} = 527$ m, and propagation in near-homogeneous air (at $T \approx 24°C$ and 66% humidity), the measured interferometric characters c ($N = 4$) and d ($N = 5$) are shown in Figures 11.13 and 11.14, respectively.

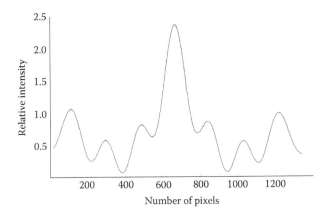

FIGURE 11.11
The interferometric character c ($N = 4$, 1000 μm slits separated by 1000 μm) for $D_{\langle x|j \rangle} = 35$ m recorded outside the laboratory. (Reproduced from Duarte, F.J. et al., *J. Opt.* 12, 015705, 2010, with permission from the Institute of Physics.)

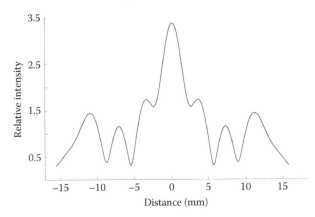

FIGURE 11.12
The calculated version of the interferometric character c ($N = 4$, 1000 μm slits separated by 1000 μm) for $D_{\langle x|j \rangle} = 35$ m. (Reproduced from Duarte, F.J. et al., *J. Opt.* 12, 015705, 2010, with permission from the Institute of Physics.)

FIGURE 11.13
The interferometric character c ($N = 4$, 1000 μm slits separated by 1000 μm) for $D_{\langle x|j \rangle} = 527$ m. (Reproduced from Duarte, F.J. et al., *J. Opt.* 13, 035710, 2011, with permission from the Institute of Physics.)

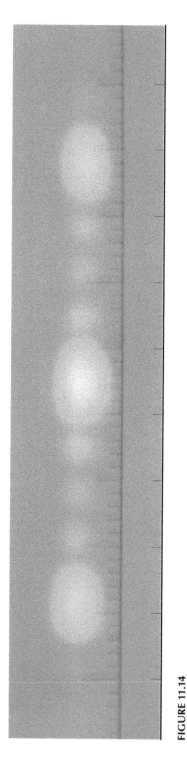

FIGURE 11.14

The interferometric character d ($N = 5$, 1000 μm slits separated by 1000 μm) for $D_{\langle x|j \rangle} = 527$ m. (Reproduced from Duarte, F.J. et al., *J. Opt.* 13, 035710, 2011, with permission from the Institute of Physics.)

11.5.1 Clear-Air Turbulence

As mentioned previously the original propagation space envisioned for communications via interferometric characters was vacuum, or outer space, where the interferometric characters can propagate free of refractive index distortions (Duarte, 2002). That is not the case for propagation in a terrestrial environment where the propagation space is subject to refractive index variations due to thermal gradients and straightforward air turbulence.

The question then becomes: can we differentiate between interferometric character distortions due to turbulence and distortions due to third-party intrusion into the intra-interferometric optical path $D_{\langle x|j\rangle}$? This question can be answered empirically. To do this, we call attention to Figures 11.5b through d, 11.7, and 11.15, all of which illustrate the catastrophic collapse of the given interferometric character due to the insertion of a BS in the $D_{\langle x|j\rangle}$ path. Now, by contrast we consider Figure 11.16 where the distortions due to mild clear-air turbulence detected at $D_{\langle x|j\rangle} = 7.235$ m, for the c ($N = 4$) character, are illustrated (Duarte, 2009). Further, the effect of severe clear-air turbulence detected at $D_{\langle x|j\rangle} = 7.235$ m, for the c ($N = 4$) character, is shown in Figure 11.17 (Duarte et al., 2010). As can be seen, the clear-air turbulence distortions can be relatively mild (Figure 11.16) and, in their severe format, tend to smooth over (Figure 11.17) the original profile of the interferometric character. In both cases it appears that the effect of clear-air turbulence should be statistically predictable. This is very different to the massive distortions and catastrophic collapse (see Figures 11.5b through d, 11.7, and 11.15) induced by BS incursions into the $D_{\langle x|j\rangle}$ optical path.

Finally, it should be mentioned that clear-air turbulence is a phenomenon of significant importance to aviation safety. And this class of turbulence is difficult to detect with traditional radar methods. In this regard, the ability of the NSLI to detect clear-air turbulence offers a practical and demonstrated

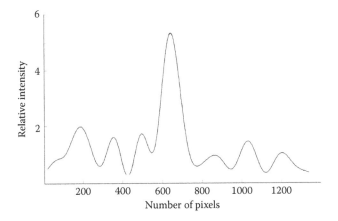

FIGURE 11.15
The interferometric character c ($N = 4$, 1000 μm slits separated by 1000 μm) for $D_{\langle x|j\rangle} = 30$ m, while a thin BS is inserted at a distance 10 m from the N-slit array. (Reproduced from Duarte, F.J. et al., *J. Opt.* 12, 015705, 2010, with permission from the Institute of Physics.)

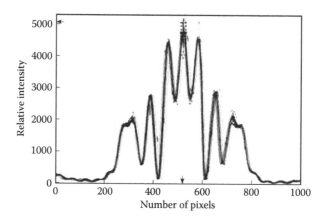

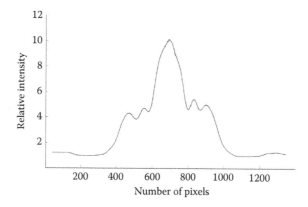

avenue of detection, specially by installing infrared-laser-based N-slit interferometers at airports near the runway thresholds.

11.6 Discussion

The N-slit interferometer, in conjunction with the N-slit interferometric equation, has been used to generate interferometric characters that have proven extremely sensitive to interception with classical, or macroscopic, optical methods.

Thus, an interferogram or interferometric character can be generated according to an established interferometric alphabet and used for free-space communications knowing that attempts to intercept the optical character will lead to its catastrophic collapse. This implies that open communications, without the use of a classified key, can proceed through the open space.

However, a cryptographic key could be easily added and implemented, if desired. In Chapter 20 we describe how microscopic optical methods can be used to intercept the interferometric character causing minimal distortions. However, from a practical perspective, it is not clear how this approach could be used to extract information. In Chapter 20 a more detailed interplay between measured and calculated interferograms is also descried.

As already mentioned the N-slit interferometric method was originally conceived for outer space communications in vacuum (Duarte, 2002). However, here we have seen that the deployment in a terrestrial environment, devoid of clear-air turbulence, should be practically feasible. Also, given the different nature of the distortions due to clear-air turbulence, as compared with macroscopic optical interception, the deployment in atmospheric environments including some degree of clear-air turbulence should also be possible.

The main advantages of the N-slit interferometric method for optical communications in free space are

1. The extraordinary simplicity of the optical architecture.
2. The use of single-transverse-mode, narrow-linewidth lasers as illumination sources that neutralize significantly signal-to-noise problems.
3. The ease with which additional security features could be implemented into its optical configuration: tunable infrared lasers and variable interferometric characters, among others.

Problems

11.1 Use the data provided in Appendix A to select an optimum laser for use in a secure interferometric character generating configuration. Explain the reasons for your selection.

11.2 Use the data provided in Appendix A to select an optimum laser for use in an NSLI for the detection of clear-air turbulence. Explain the differences with the selection of Problem 11.1.

References

Dirac, P. A. M. (1978). *The Principles of Quantum Mechanics*, 4th edn. Oxford, London, U.K.
Duarte, F. J. (1991). Dispersive dye lasers. In *High Power Dye Lasers*, (Duarte, F. J., ed.). Springer-Verlag, Berlin, Germany, Chapter 2.

Duarte, F. J. (1993). On a generalized interference equation and interferometric measurements. *Opt. Commun.* **103**, 8–14.

Duarte, F. J. (2002). Secure interferometric communications in free space. *Opt. Commun.* **205**, 313–319.

Duarte, F. J. (2003). *Tunable Laser Optics*. Elsevier-Academic, New York.

Duarte, F. J. (2004). Comment on reflection, refraction and multislit interference. *Eur. J. Phys.* **25**, L57–L58.

Duarte, F. J. (2005). Secure interferometric communications in free space: Enhanced sensitivity for propagation in the metre range. *J. Opt. A: Pure Appl. Opt.* **7**, 73–75.

Duarte, F. J. (2009). Interferometric imaging. In *Tunable Laser Applications*, 2nd edn. (Duarte, F. J., ed.) CRC Press, New York, Chapter 12.

Duarte, F. J., Taylor, T. S., Black, A. M., Davenport, W. E., and Varmette, P. G. (2011). N-slit interferometer for secure free-space optical communications: 527 m intra interferometric path length. *J. Opt.* **13**, 035710.

Duarte, F. J., Taylor, T. S., Clark, A. B., and Davenport, W. E. (2010). The *N*-slit interferometer: An extended configuration. *J. Opt.* **12**, 015705.

Hogan, F. (2013). Data demands free-space optics. *Photon. Spec.* **47**(2), 38–41.

van Kampen, N. G.(1988). Ten theorems about quantum mechanical measurements. *Physica A* **153**, 97–113.

12

Schrödinger's Equation

12.1 Introduction

As mentioned in Chapter 1, Schrödinger's equation provides one of the three main avenues to quantum mechanics. It also has important applications in atomic and molecular physics (Herzberg, 1950; Feynman et al., 1965). Given the importance of this equation, here we examine several derivational approaches. The principal aim is to gain an understanding on the physics behind this quantum wave equation. Since the application of this equation is widely and extensively treated in many textbooks (see, e.g., Saleh and Teich, 1991; Silfvast, 2008; Hooker and Webb, 2010), here we only briefly describe a few applications relevant to semiconductor lasers.

12.2 Schrödinger's Mind

As its name suggests, Schrödinger's equation is the brain child of Schrödinger and was first disclosed in 1926 in a paper entitled *An Undulatory Theory of the Mechanics of Atoms and Molecules* (Schrödinger, 1926). The word undulatory is based on the word *onde* that means wave. In that extraordinary paper, Schrödinger begins with a discussion of classical kinetic energy and work. Schrödinger's paper is not all that transparent, and it includes some imaginative arguments that are crucial to the successful development of his theory.

Next, we present Schrödinger's argument in a very abbreviated format and with some changes in notation to improve transparency.

Schrödinger begins by considering the dynamics of a particle moving through a force field. Thus he first introduces the kinetic energy as

$$T = \frac{1}{2} m \left(v_x^2 + v_y^2 + v_z^2 \right) \tag{12.1}$$

and he also introduces the Hamiltonian of action W

$$W = \int_0^t (T - V)\,dt \tag{12.2}$$

which satisfies the Hamiltonian partial differential equation

$$\frac{\partial W}{\partial t} + \frac{1}{2m}(\nabla W)^2 + V(x, y, z) = 0 \tag{12.3}$$

To solve this differential equation, he then tries a solution of the form

$$W = -Et + S(x, y, z) \tag{12.4}$$

Differentiating Equation 12.4, he arrives at $\partial W/\partial t = -E$, and substitution into Equation 12.3 yields

$$|\nabla W| = \left(2m(E - V)\right)^{1/2} \tag{12.5}$$

As it will be seen later, the appearance of the $2m(E - V)$ term is crucial to the final Schrödinger result.

Next, Schrödinger uses Equations 12.4 and 12.5 in conjunction with a geometrical argument to obtain a "phase" velocity u:

$$u = E\left(2m(E - V)\right)^{-1/2} \tag{12.6}$$

Equation 12.6 is an essential component of the classical part of the Schrödinger's argument. He then introduces the quantum component via Planck's quantum energy equation (Planck, 1901)

$$E = h\nu \tag{12.7}$$

and de Broglie's quantum momentum expression (de Broglie, 1923)

$$p = \frac{h}{\lambda} \tag{12.8}$$

which is also based on $E = h\nu$. This would enable him to later define his wave function ψ in terms of the quantum energy.

At this stage, Schrödinger introduces an ordinary wave equation of the form

$$\Delta\psi - u^{-2}\frac{\partial^2\psi}{\partial t^2} = 0 \tag{12.9}$$

where Δ is the Laplacian operator more readily recognized as ∇^2

$$\nabla^2 = \left(\frac{\partial^2}{\partial x^2}\right) + \left(\frac{\partial^2}{\partial y^2}\right) + \left(\frac{\partial^2}{\partial z^2}\right) \tag{12.10}$$

thus Equation 12.9 can be expressed in its familiar form

$$\nabla^2\psi - u^{-2}\frac{\partial^2\psi}{\partial t^2} = 0 \tag{12.11}$$

In this equation Schrödinger uses a wave function ψ that is only time dependent:

$$\psi = \psi_0 e^{-i2\pi E t/h} \tag{12.12}$$

Here, to facilitate the description we provide the first and second derivatives of ψ:

$$\frac{\partial\psi}{\partial t} = \frac{-i2\pi E}{h}\psi_0 e^{-i2\pi E t/h} \tag{12.13}$$

$$\frac{\partial^2\psi}{\partial t^2} = \frac{-4\pi^2 E^2}{h^2}\psi_0 e^{-i2\pi E t/h} \tag{12.14}$$

Insertion of Equations 12.6 and 12.14 into Equation 12.11 yields

$$\nabla^2\psi + \frac{8\pi^2 m}{h^2}(E-V)\psi = 0 \tag{12.15}$$

This wave equation is crucial to Schrödinger's development, but so far it does not exhibit much of a resemble to his final creation. In fact, in his paper, he explicitly asks: "Now what are we to do with Equation (16)?" (i.e., Equation 12.15) (Schrödinger, 1926).

He then proceeds with an *eight-page* argument where he relates Equation 12.15 with the *Hamiltonian principle* and eventually produces a generalized version of

$$\nabla^2\psi + \frac{8\pi^2 m}{h^2}(E-V)\psi = 0$$

which replaces the $\nabla^2\psi$ by a more extensive term, but the physics remains the same; thus we continue to use Equation 12.15 in our description. Also, in the original Schrödinger paper, the particle mass term m was omitted, from the generalized version of Equation 12.15, due to a misprint.

Next, about three pages later, Schrödinger goes back to Equation 12.13 to obtain an expression for E and substitutes that back into Equation 12.15 so that

$$\nabla^2\psi - \frac{8\pi^2 m}{h^2} V\psi + \frac{i4\pi m}{h}\frac{\partial\psi}{\partial t} = 0 \tag{12.16}$$

This is the original Schrödinger's equation (which in his paper appears with a missing m in the second and third terms). This original version of Schrödinger's equation can be expressed in a more familiar form using $\hbar = (h/2\pi)$, so that

$$\nabla^2\psi - \frac{2m}{\hbar^2} V\psi + i\hbar\frac{2m}{\hbar^2}\frac{\partial\psi}{\partial t} = 0 \tag{12.17}$$

Multiplying by $\hbar^2/2m$ and rearranging it, we get the Schrödinger's equation in its well-known format

$$i\hbar\frac{\partial\psi}{\partial t} = -\frac{\hbar^2}{2m}\nabla^2\psi + V\psi \tag{12.18}$$

From a physics perspective, the significant point here is that Schrödinger is using a classical argument to arrive at a phase velocity that is a function of $2m(E - V)$ and utilizes this phase velocity in the classical wave equation. The quantum aspect of this approach is introduced via the wave function

$$\psi = \psi_0 e^{-i2\pi Et/h}$$

Again, the principal quantum concept in Schrödinger's argument leading to his celebrated quantum wave equation is Planck's energy equation $E = h\nu$. In the next section an explicit heuristic derivation of Schrödinger's equation is provided.

12.3 Heuristic Explicit Approach to Schrödinger's Equation

Here, the Schrödinger's equation is arrived at in an alternative heuristic path modeled after Haken (1981). This approach relays on the use of the complex wave function $\psi(x, t)$. The advantage of this approach is that the argument is much simpler and clearer.

A free particle moves with classical kinetic energy according to

$$E = \frac{1}{2}mv^2 \tag{12.19}$$

which, using $p = mv$, can be restated as

$$E = \frac{p^2}{2m} \tag{12.20}$$

Now, using Planck's quantum energy (Planck, 1901)

$$E = h\nu$$

in conjunction with $\lambda = c/\nu$ and $E = mc^2$, the well-known de Broglie's expression for momentum can be arrived to (de Broglie, 1923)

$$p = \hbar k \tag{12.21}$$

The classical "wave functions of ordinary wave optics" (Dirac, 1978) can be written as

$$\psi(x,t) = \psi_0 e^{-i(\omega t - kx)} \tag{12.22}$$

whose derivative with respect to time becomes

$$\frac{\partial \psi(x,t)}{\partial t} = (-i\omega)\psi_0 e^{-i(\omega t - kx)} \tag{12.23}$$

Similarly the first and second derivatives with respect to displacement are

$$\frac{\partial \psi(x,t)}{\partial x} = (+ik)\psi_0 e^{-i(\omega t - kx)} \tag{12.24}$$

$$\frac{\partial^2 \psi(x,t)}{\partial x^2} = (-k^2)\psi_0 e^{-i(\omega t - kx)} \tag{12.25}$$

Multiplying the first time derivative by $(-i\hbar)$ yields

$$-i\hbar \frac{\partial \psi(x,t)}{\partial t} = (-\hbar\omega)\psi_0 e^{-i(\omega t - kx)} \tag{12.26}$$

and multiplying the second displacement derivative by $(-\hbar^2/2m)$ yields

$$-\frac{\hbar^2}{2m} \frac{\partial^2 \psi(x,t)}{\partial x^2} = \frac{\hbar^2 k^2}{2m} \psi_0 e^{-i(\omega t - kx)} \tag{12.27}$$

Recognizing that $E = \hbar^2 k^2/2m$, allow us to write

$$-\frac{\hbar^2}{2m} \frac{\partial^2 \psi(x,t)}{\partial x^2} = +i\hbar \frac{\partial \psi(x,t)}{\partial t} \tag{12.28}$$

which is the basic form of Schrödinger's equation. The Schrödinger's equation is a wave equation that incorporates classical particle concepts, and classical wave function concepts, in its derivation. In other words, a heuristic approach to Schrödinger's equation utilizes Planck's quantum energy, classical kinetic energy of a free particle, and the classical "wave functions of ordinary wave optics." In this approach it is clear that Schrödinger's equation refers to a free particle propagating in a wave motion according to ordinary wave optics. Once again, the central role of classical wave equation of the form

$$\psi(x,t) = \psi_0 e^{-i(\omega t - kx)}$$

is highlighted.

12.4 Schrödinger's Equation via the Dirac Notation

Here, once again the Feynman approach (Feynman et al., 1965) is adopted. In Chapter 8 we introduced the Hamiltonian H_{ij} and showed that the time dependence of the amplitude C_i is given by (Dirac, 1978)

$$i\hbar \frac{dC_i}{dt} = \sum_j H_{ij} C_i \qquad (12.29)$$

For $C_i = \langle i | \psi \rangle$ this equation can be rewritten as

$$i\hbar \frac{d\langle i | \psi \rangle}{dt} = \sum_j \langle i | H | j \rangle \langle j | \psi \rangle \qquad (12.30)$$

which, for $i = x$, can be written as

$$i\hbar \frac{d\langle x | \psi \rangle}{dt} = \sum_j \langle x | H | x' \rangle \langle x' | \psi \rangle \qquad (12.31)$$

Since $\langle x | \psi \rangle = \psi(x)$ this equation can be reexpressed as

$$i\hbar \frac{d\psi(x)}{dt} = \int H(x,x')\psi(x')dx' \qquad (12.32)$$

The integral on the right-hand side is given by

$$\int H(x,x')\psi(x')dx' = -\frac{\hbar^2}{2m} \frac{d^2\psi(x)}{dx^2} + V(x)\psi(x) \qquad (12.33)$$

About this stage, Feynman poses and asks the question: "Where did we get that from? Nowhere… it came from the mind of Schrödinger, invented in his struggle to find an understanding of the experimental observations of the real world" (Feynman et al., 1965).

Here, our discussion in Section 12.2 becomes quite relevant.

Next, combining Equations 12.32 and 12.33, we get Schrödinger's equation

$$+i\hbar \frac{d\psi(x)}{dt} = -\frac{\hbar^2}{2m}\frac{d^2\psi(x)}{dx^2} + V(x)\psi(x) \tag{12.34}$$

In three dimensions we use $\psi(x, y, z)$ and $V(x, y, z)$ and

$$\nabla^2 = \left(\frac{\partial^2}{\partial x^2}\right) + \left(\frac{\partial^2}{\partial y^2}\right) + \left(\frac{\partial^2}{\partial z^2}\right) \tag{12.35}$$

so that Schrödinger's equation in three dimensions can be expressed in the succinct form

$$+i\hbar \frac{\partial\psi}{\partial t} = -\frac{\hbar^2}{2m}\nabla^2\psi + V\psi \tag{12.36}$$

It is clear that even in this approach, using the quantum tools provided by Dirac and Feynman, the derivation of Schrödinger's equation still depends on Schrödinger's classical concepts. Defining the *Hamiltonian operator* $\hat{H}$ as

$$\hat{H} = \left(\frac{-\hbar^2}{2m}\right)\nabla^2 + V \tag{12.37}$$

the Schrödinger's equation can be expressed as

$$+i\hbar \frac{\partial\psi}{\partial t} = \hat{H}\psi \tag{12.38}$$

Finally, for a large number of particles, Feynman restates the Schrödinger's equation as

$$+i\hbar\left(\frac{\partial\psi(r_1, r_2, r_3...)}{\partial t}\right) = -\frac{\hbar^2}{2}\sum_i m_i^{-1}\nabla_i^2\psi + V(r_1, r_2, r_3...)\psi \tag{12.39}$$

where

$$\nabla_i^2 = \left(\frac{\partial^2}{\partial x_i^2}\right) + \left(\frac{\partial^2}{\partial y_i^2}\right) + \left(\frac{\partial^2}{\partial z_i^2}\right) \tag{12.40}$$

This is the type of Schrödinger's equation applied in the description of molecular physics (Herzberg, 1950).

12.5 Time-Independent Schrödinger's Equation

As suggested by Feynman, using a solution of the form

$$\psi(x,t) = \Psi(x)e^{-iEt/\hbar} \tag{12.41}$$

Equation 12.36, in one dimension, takes the form of

$$\frac{\partial^2 \Psi(x)}{\partial x^2} - \frac{2m}{\hbar^2}(V(x) - E)\Psi(x) = 0 \tag{12.42}$$

or

$$\frac{\partial^2 \Psi(x)}{\partial x^2} = \frac{2m}{\hbar^2}(V(x) - E)\Psi(x) \tag{12.43}$$

which is known as a 1D time-independent Schrödinger's equation. This simple form of the Schrödinger's equation is of enormous significance to semiconductor physics and semiconductor lasers. Before going any further, the reader should notice that this equation has exactly the same form as Schrödinger's inspirational Equation 12.15:

$$\nabla^2 \psi + \frac{8\pi^2 m}{h^2}(E - V)\psi = 0$$

12.5.1 Quantized Energy Levels

Let us consider a static potential well as described by the function $V(x)$ in Figure 12.1. From Equation 12.43, $\Psi(x)$ can be evaluated along x as E is varied, along the vertical, by small amounts. This evaluation indicates that $\Psi(x)$ shows an oscillatory behavior within the well. For certain definite

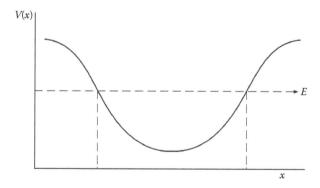

FIGURE 12.1
Static potential energy well $V(x)$.

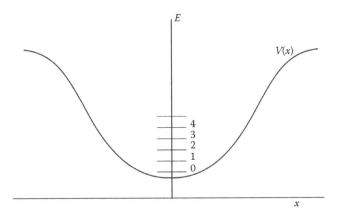

FIGURE 12.2
Potential energy well depicting a series of discrete energy levels $E = 0, 1, 3, 4$.

values of E, the shape of the curve is symmetrical as x increases pass the well boundary. However, for other slightly different values of E, $\Psi(x)$ diverges toward large positive or large negative values. In other words, as indicated in Figure 12.2, within a potential well, the particle is only bound for defi-nite discrete values of $E = 0, 1, 3, 4...$ (Feynman et al., 1965). The behavior of $\Psi(x)$, as a function of E, illustrates the phenomenon of quantized energy levels within a potential well.

12.5.2 Semiconductor Emission

Going back to the time-independent Schrödinger's equation

$$\frac{\partial^2 \psi(x)}{\partial x^2} - \frac{2m}{\hbar^2}\left(V(x) - E\right)\psi(x) = 0 \qquad (12.44)$$

and using the spatial component of the wave function as solution

$$\psi(x) = \psi_0 e^{-ikx} \qquad (12.45)$$

leads directly to

$$E = \frac{k^2 \hbar^2}{2m} + V(x) \qquad (12.46)$$

This energy expression is the sum of kinetic and potential energy so that

$$E_K + E_P = \frac{k^2 \hbar^2}{2m} + V(x) \qquad (12.47)$$

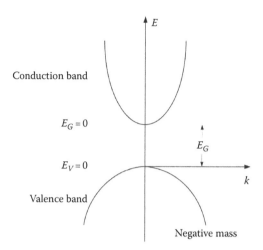

FIGURE 12.3
Conduction and valence bands according to $E_K = \pm(k^2\hbar^2/2m)$.

and obviously

$$E_K = \frac{k^2\hbar^2}{2m} \tag{12.48}$$

The kinetic energy E_K as a function of $k = 2\pi/\lambda$ is shown graphically in Figure 12.3.

The graph is a positive parabola for positive values of m and a negative parabola for negative values of m. In a semiconductor the positive parabola is known as the *conduction* band and the negative parabola is known as the *valence* band. The separation between the two bands is known as the *band gap*, E_G. If an electron is excited and transitions from the valence band to the conduction band, it is said to leave behind a vacancy or *hole*.

Electrons can transition from the bottom of the conduction band to the top of valence band by recombining with holes. In a material like gallium arsenide, the band gap is $E_G \approx 1.43$ eV and the recombination emission occurs around 870 nm (Silfvast, 2008).

Under certain conditions radiation might also occur higher from the conduction band as suggested in Figure 12.4; however, that process is undermined by fast phonon relaxation.

12.5.3 Quantum Wells

Starting with a potential well as described by Silfvast (2008), $V(x) = 0$ for $0 < x < L$, and $V(x) = \infty$ for $x = 0$ or $x = L$, as illustrated in Figure 12.5, then Equation 12.44

$$\frac{\partial^2 \psi(x)}{\partial x^2} - \frac{2m}{\hbar^2}\left(V(x) - E\right)\psi(x) = 0$$

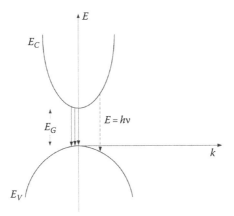

FIGURE 12.4
Emission due to recombination transitions from the bottom of the conduction band to the top of the valence band.

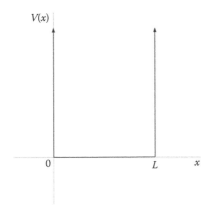

FIGURE 12.5
Potential well: $V(x) = 0$ for $0 < x < L$, and $V(x) = \infty$ for $x = 0$ or $x = L$.

becomes

$$\frac{\partial^2 \psi(x)}{\partial x^2} + \frac{2m}{\hbar^2} E \psi(x) = 0 \tag{12.49}$$

for $0 < x < L$. This is a wave equation of the form

$$\frac{d^2 \psi(x)}{dx^2} + k_x^2 \psi(x) = 0 \tag{12.50}$$

with

$$k_x^2 = \frac{2m_c}{\hbar^2} E \tag{12.51}$$

The solution to Equation 12.50 is

$$\psi(x) = \sin k_x x \tag{12.52}$$

Since $\psi(x) = 0$ at $x = 0$ or $x = L$, we have

$$k_x = \frac{n\pi}{L} \tag{12.53}$$

for $n = 1, 2, 3....$ Substituting Equation 12.53 into 12.51 leads to

$$E = \frac{n^2\pi^2\hbar^2}{2m_c L^2} \tag{12.54}$$

which should be labeled as E_n to account for the quantized nature of the energy, that is,

$$E_n = \frac{n^2\pi^2\hbar^2}{2m_c L^2} \tag{12.55}$$

This quantized energy E_n indicates a series of possible discrete energy levels above the lowest point of the conduction band so that the total energy above the valence band becomes (Silfvast, 2008)

$$E = E_c + E_n \tag{12.56}$$

12.5.4 Quantum Cascade Lasers

These lasers operate via transitions between quantized levels, *within* the conduction band, of multiple-quantum well structures. The carriers involved are electrons generated in an n-doped material. A single stage includes an injector and an active region. The electron is injected into the active region at $n = 3$, and the transition occurs down to $n = 2$ (see Figure 12.6). Following emission, the electron continues into the next injector region.

Practical devices include a series of such stages. From Equation 12.54, the energy difference between the two levels can be expressed as (Silfvast, 2008)

$$\Delta E = (3^2 - 2^2)\frac{\pi^2\hbar^2}{2m_c L^2} \tag{12.57}$$

where L is the thickness of the well. Using $\Delta E = h\nu$, it follows that the wavelength of emission is given by

$$\lambda = (3^2 - 2^2)^{-1}\frac{8m_c c L^2}{h} \tag{12.58}$$

Quantum cascade lasers (Faist et al., 1994) are tunable sources emitting in the infrared from a few micrometers to beyond 20 μm (see Appendix A).

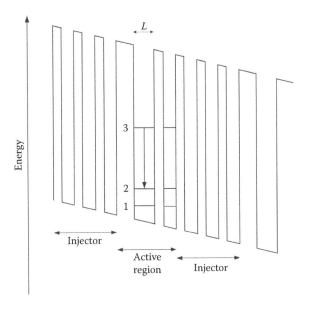

FIGURE 12.6
Simplified illustration of a multiple-quantum well structure relevant to quantum cascade lasers. An electron is injected from the "injector region" into the active region at $n = 3$. Thus a photon is emitted via the $3 \rightarrow 2$ transition. The electron continues to the next region where the process is repeated. By configuring a series of such stages, one electron can generate the emission of numerous photons.

12.5.5 Quantum Dots

Besides the multiple-quantum well configurations, other interesting semiconductor geometries include the quantum wire and the quantum dot. The quantum wire narrowly confines the electrons and holes in two directions (x, y). The quantum dot geometry severely confines the electrons in three dimensions (x, y, z). Under these circumstances the quantized energy can be expressed as

$$E = E_c + \frac{k_x^2 \hbar^2}{2m_c} + \frac{k_y^2 \hbar^2}{2m_c} + \frac{k_x^2 \hbar^2}{2m_c} \tag{12.59}$$

where k_x, k_y, k_z are defined according to Equation 12.53.

The concept of quantum dot is not limited to semiconductor materials; it also applies to nanoparticle gain media and nanoparticle core–shell gain media. For instance, for nanoparticle core–shells the physics can be described with a Schrödinger's equation of the form

$$\nabla^2 \psi(r) - \frac{2m}{\hbar^2} \left(V(r) - E \right) \psi(r) = 0 \tag{12.60}$$

with the potential $V(r)$ defined by the core–shell geometry (Dong et al., 2013).

12.6 Introduction to the Hydrogen Equation

We have already seen that Schrödinger's equation can be expressed as

$$+i\hbar \frac{\partial \psi}{\partial t} = \hat{H}\psi$$

where the Hamiltonian is given by

$$\hat{H} = \left(-\frac{\hbar^2}{2m} \right)\nabla^2 + V(r)$$

For an electron of mass m, under a potential $V(r)$ described by

$$V(r) = -\frac{e^2}{r} \tag{12.61}$$

Schrödinger's equation becomes

$$+i\hbar \frac{\partial \psi}{\partial t} = -\frac{\hbar^2}{2m}\nabla^2\psi - \frac{e^2}{r}\psi \tag{12.62}$$

Using a wave function of the form

$$\psi(r,t) = e^{-iEt/\hbar}\psi(r) \tag{12.63}$$

Equation 12.62 can be written as

$$-\frac{\hbar^2}{2m}\nabla^2\psi(r) = \left(E + \frac{e^2}{r} \right)\psi(r) \tag{12.64}$$

Using spherical polar coordinates

$$x = r\sin\theta\cos\phi$$
$$y = r\sin\theta\sin\phi \tag{12.65}$$
$$z = r\cos\theta$$

the Laplacian

$$\nabla^2 = \left(\frac{\partial^2}{\partial x^2} \right) + \left(\frac{\partial^2}{\partial y^2} \right) + \left(\frac{\partial^2}{\partial z^2} \right)$$

operating on a function $\vartheta(r, \theta, \phi)$ can be written as (see, e.g., Flanders et al., 1970)

$$\nabla^2 \vartheta(r,\theta,\phi) = r^{-1}\frac{\partial^2 (rf)}{\partial r^2} + r^{-2}\left(\sin^{-1}\theta\frac{\partial}{\partial\theta}\left(\sin\theta\frac{\partial f}{\partial\theta}\right)+\sin^{-2}\theta\frac{\partial^2 f}{\partial\phi^2}\right) \qquad (12.66)$$

Thus, for a wave function $\psi(r, \theta, \phi)$, Equation 12.64 can be expressed as

$$r^{-1}\frac{\partial^2 (r\psi)}{\partial r^2} + r^{-2}\left(\sin^{-1}\theta\frac{\partial}{\partial\theta}\left(\sin\theta\frac{\partial\psi}{\partial\theta}\right)+\sin^{-2}\theta\frac{\partial^2\psi}{\partial\phi^2}\right) = -\frac{2m}{\hbar^2}\left(E+\frac{e^2}{r}\right)\psi \qquad (12.67)$$

This is the Schrödinger's equation applicable to the description of the hydrogen atom. The proper solution to this equation is mathematically rather lengthy and involves a number of tricks that are cleverly described by Feynman. Thus, the mathematically inclined is invited to read Feynman et al. (1965) on this subject. Also, the hydrogen atom is treated in various forms and depths by numerous textbooks on quantum mechanics (see, e.g., Schiff et al., 1968). Our main purpose here has been accomplished by introducing the type of notation necessary to describe the hydrogen atom and other small atoms.

In this regard, the Schrödinger's equation has been use to describe small atoms such as hydrogen and helium in detail. However, as Feynman said: "in principle, Schrödinger's equation is capable of explaining all atomic phenomena except those involving magnetism and relativity. It explains the energy levels of an atom and all the facts of chemical bonding" (Feynman et al., 1965). When Feynman first made this statement, this was only true in principle; however, the enormous advances in modern computational techniques have converted that statement into a reality.

Problems

12.1 Obtain Equation 12.5 by differentiating Equation 12.4 and substituting into Equation 12.3.

12.2 Show that Equation 12.15 follows from substitution of Equations 12.6 and 12.14 into Equation 12.11.

12.3 Use Equations 12.13 and 12.15 to obtain the Schrödinger's equation in its well-known form, that is, Equation 12.18.

12.4 For a particle in a potential well $V(x)$, assume that the wave function describing the state of the particle is given by $\psi(x, t) = \phi(x)\vartheta(t)$. Given that

$$\vartheta(t) = Ce^{-i\omega t}$$

show that $\phi(x)$ and $\vartheta(t)$ satisfy the equation

$$\frac{\partial^2 \psi(x)}{\partial x^2} - \frac{2m}{\hbar^2}(V(x) - E)\phi(x) = 0$$

12.5 Find the stationary solutions to the Schrödinger's equation for the three regions (I, II, and III) defined in Figure 12.7. Hint: Starting from

$$+i\hbar \frac{\partial \psi}{\partial t} = -\frac{\hbar^2}{2m}\nabla^2\psi + V\psi$$

use a solution of the form

$$\psi(x,t) = \phi(x)\vartheta(t)$$

to obtain

$$\frac{\partial^2 \psi(x)}{\partial x^2} - \frac{2m}{\hbar^2}(V(x) - E)\phi(x) = 0$$

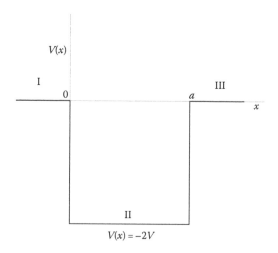

FIGURE 12.7

Potential well $V(x)$ with regions I, II, and III, applicable to Problem 12.5. $V(x) = 0$ for $x < 0$, $V(x) = -2V$ for $0 \leq x \leq a$, $V(x) = 0$ for $x > a$.

In region I, $V(x) = 0$, thus

$$\frac{\partial^2 \phi(x)}{\partial x^2} + \frac{2m}{\hbar^2} E\phi(x) = 0$$

$$\frac{\partial^2 \phi(x)}{\partial x^2} + k_1^2 \phi(x) = 0$$

$$k_1 = \left(\frac{2mE}{\hbar^2}\right)^{1/2}$$

As already seen, this equation has a solution of the form

$$\phi(x) = C_1 \sin k_1 x$$

12.6 Find the stationary solutions to the Schrödinger's equation for the three regions (I, II, and III) defined in Figure 12.8.

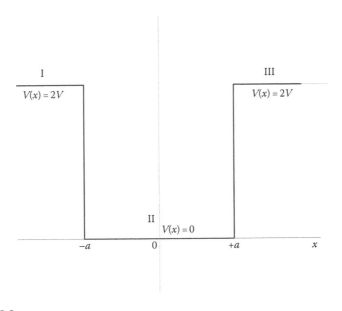

FIGURE 12.8
Potential well $V(x)$ with regions I, II, and III, applicable to Problem 12.6. $V(x) = 2V$ for $x < -a$, $V(x) = 0$ for $a \le x \le -a$, $V(x) = 2V$ for $x > a$.

References

de Broglie, L. (1923). Waves and quanta. *Nature* **112**, 540.

Dirac, P. A. M. (1978). *The Principles of Quantum Mechanics*, 4th edn. Oxford, London, U.K.

Dong, L., Sugunan, A., Hu, J., Zhou, S., Li, S., Popov, S., Toprak, M. S., Friberg, A. T., and Muhammed, M. (2013). Photoluminescence from quasi-type-II spherical CdSe-CdS core-shell quantum dots. *Appl. Opt.* **52**, 105–109.

Faist, J., Capasso, F., Sivco, D. L., Sirtori, C., Hutchinson, A. L., and Cho A. Y. (1994). Quantum cascade laser. *Science* **264**, 553–556.

Feynman, R. P., Leighton, R. B., and Sands, M. (1965). *The Feynman Lectures on Physics*, Vol. III. Addison-Wesley, Reading, MA.

Flanders, H., Korfhage, R. R., and Price, J. J. (1970). *Calculus*. Academic, New York.

Haken, H. (1981). *Light*. North-Holland, Amsterdam, the Netherlands.

Herzberg, G. (1950). *Spectra of Diatomic Molecules*. Van Nostrand Reinhold, New York.

Hooker, S. and Webb, E. (2010). *Laser Physics*. Oxford University, Oxford, U.K.

Planck, M (1901). Ueber das gesetz der energieverteilung im normalspectrum. *Ann. Phys.* **309**(3), 553–563.

Saleh, B. E. A. and Teich, M. C. (1991). *Fundamentals of Photonics*, Wiley, New York.

Schiff, L. I. (1968). *Quantum Mechanics*. McGraw-Hill, New York.

Schrödinger, E. (1926). An undulatory theory of the mechanics of atoms and molecules. *Phys. Rev.* **28**, 1049–1070.

Silfvast, W. T. (2008). *Laser Fundamentals*, 2nd edn. Cambridge University, Cambridge, U.K.

13

Introduction to Feynman Path Integrals

13.1 Introduction

The Feynman path integrals provide an alternative version of quantum mechanics (Feynman and Hibbs, 1965). This approach is outlined here since Feynman applied it to describe macroscopic beam divergence. The presentation is brief and limited since this subject was mainly selected as an example of an alternative version of quantum mechanics and is not utilized elsewhere in the book.

From a historical perspective, it should further be mentioned that *the quantum analogue of the classical action* was first introduced, as a concept, by Dirac in the early editions of his book *The Principles of Quantum Mechanics* where he introduced a section entitled "The Action Principle" (Dirac, 1978).

13.2 Classical Action

Feynman was fascinated by the *principle of least action* and he gave a lecture on this topic in the *Feynman Lectures of Physics* (Feynman et al., 1965). This principle can be used to describe the possible paths of a particle from an initial to a final point. There is a quantity S that can be computed for each path, and that S is a minimum. Feynman explains further that S is an extremum, that is, the value of S is unchanged to a first order if the path is changed slightly (Feynman and Hibbs, 1965). In his lecture Feynman states: "the average kinetic energy less the average potential energy is as little as possible for the path of an object." Then he introduces the integral

$$S = \int_{t_1}^{t_2} \left(\frac{1}{2} m \left(\frac{dx}{dt} \right)^2 - mgx \right) dt \tag{13.1}$$

which he also writes as

$$S = \int_{t_1}^{t_2} (KE - PE) dt \qquad (13.2)$$

or

$$S = \int_{t_a}^{t_b} L(\dot{x}, x, t) dt \qquad (13.3)$$

where

$$L = \left(\frac{m}{2}\right) \dot{x}^2 - V(x, t) \qquad (13.4)$$

is known as the *Lagrangian*.

In summary, for the true path, between an initial point x_a and a final point x_b, S is a minimum. This concept is illustrated in Figure 13.1 where S_1 and S_2 are two possible paths, S_1 is the true path, and $S_1 < S_2$.

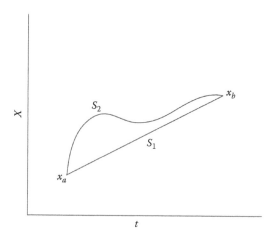

FIGURE 13.1
Least action: S_1 and S_2 are two possible paths between an initial x_a and a final point x_b. S_1 represents the direct *true path*. S_2 represents a longer convoluted path: $S_1 < S_2$.

13.3 Quantum Link

Feynman and Hibbs (1965) introduce a notation in which a square bracket represents a path between two points in a 2D space. Thus, the action S between a and b is expressed as $S[b, a]$. For the amplitudes due to successive events, Feynman writes

$$S[b,a] = S[b,c] + S[c,a] \tag{13.5}$$

and then he defines the kernel as a *path integral*

$$K(b,a) = \int_a^b e^{(i/\hbar)S[b,a]} D\, x(t) \tag{13.6}$$

This is the crucial step where he introduces the concept of quantum energy into his otherwise classical approach. This is done via de Broglie's relation

$$p = k\hbar = \frac{\partial S}{\partial x} \tag{13.7}$$

which dimensionally leads to an exponent of the form

$$e^{(i/\hbar)S[b,a]} \rightarrow e^{(i/2\hbar)px} \rightarrow e^{-(mx^2/2i\hbar)} \tag{13.8}$$

Here, it should be mentioned that equations of the form

$$\langle q'_t \mid q'' \rangle = e^{iS/\hbar}$$

were first introduced by Dirac (1978) in his discussion of the action principle. Evaluation of the kernel, via integration, leads to (Feynman and Hibbs, 1965)

$$K(x,t;0,0) = \left(\frac{2\pi i\hbar t}{m} \right)^{-1/2} e^{(imx^2/2\hbar t)} \tag{13.9}$$

In terms that are familiar with concepts already introduced, where else in this book, the probability $P(b, a)$ to go from x_a, at time t_a, to x_b, at time t_b, is given by

$$P(b,a) = \mid K(b,a) \mid^2 \tag{13.10}$$

and an alternative way to express the probability amplitude is

$$K(b,a) = \sum_{a \to b} \phi[x(t)] = \sum_{a \to b} ce^{(i/\hbar)S[x(t)]} \tag{13.11}$$

where
c is a constant
$a \to b$ means all possible paths from a to b

Thus, Equations 13.9 and 13.11 refer to probability amplitudes and thus relate to the usual wave function.

13.4 Propagation Through a Slit and the Uncertainty Principle

Now we consider Feynman's description of propagation through a slit and we do this in reference to Figure 13.2. The event to be described here can be summarized in the following set of steps:

1. The particle is at $x = 0$ at $t = 0$.
2. The particle passes between $(x_0 - w)$ and $(x_0 + w)$ at $t = T$.

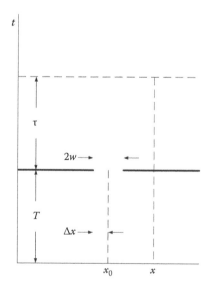

FIGURE 13.2

Propagation through a slit of width $2w$. The particle starts at $x = 0$ at $t = 0$ and passes through $2w$ at $t = T$. This passage position can also be designated as lying between $x_0 - w$ and $x_0 + w$. First, one can calculate the probability amplitude $\psi(x)$ of finding the particle at a position x at some time later denoted by $t = T + \tau$. Once $\psi(x)$ is established, then the probability $|\psi(x)|^2$ can be calculated.

3. The problem is to calculate the probability of finding the particle at a position x at a later time $t = T + \tau$.
4. The width of the slit, from $-w$ to $+w$, is $2w$

The correct quantum mechanical answer must consider all possible paths so the wave function depends on the sum of all possible paths in the range $-w$ to $+w$ or the integral in that range. Thus, Feynman and Hibbs (1965) write

$$\psi(x) = \int_{-w}^{+w} K(x + x_0, T + \tau; \ x_0 + y, T) K(x_0 + y, T; \ 0, 0) dy \qquad (13.12)$$

which, using Equation 13.9, can be expressed as

$$\psi(x) = \int_{-b}^{+b} \left(\frac{2\pi i \hbar t}{m} \right)^{-1/2} e^{(im(x-y)^2 / 2\hbar\tau)} \left(\frac{2\pi i \hbar T}{m} \right)^{-1/2} e^{(im(x_0 + y)^2 / 2\hbar T)} dy \qquad (13.13)$$

Feynman then argues that a Gaussian function $G(y)$ can be introduced in the integrand while extending the range of integration to $\pm\infty$. The function $G(y)$ has a value of unity in the $-w \le y \le +w$ range and zero elsewhere. The introduction of $G(y)$ with

$$G(y) = e^{-y^2 / 2w^2}$$

modifies Equation 13.13 to

$$\psi(x) = \int_{-\infty}^{+\infty} \left(\frac{m}{2\pi i \hbar} \right) (\tau T)^{-1/2} e^{(im(x-y)^2 / 2\hbar\tau)} e^{(im(x_0 + y)^2 / 2\hbar T)} e^{-y^2 / 2w^2} dy \qquad (3.14)$$

Expanding the exponentials, this equation can be written as

$$\psi(x) = \int_{-\infty}^{+\infty} \left(\frac{m}{2\pi i \hbar} \right) (\tau T)^{-1/2} e^{\alpha + \beta y + \gamma y^2} dy \qquad (13.15)$$

where

$$\alpha = \frac{im}{2\hbar}\left(\frac{x^2}{\tau} + \frac{x_0^2}{T}\right) \tag{13.16}$$

$$\beta = \frac{im}{2\hbar}\left(-\frac{2x}{\tau} + \frac{2x_0}{T}\right) \tag{13.17}$$

$$\gamma = \frac{im}{2\hbar}\left(\frac{1}{\tau} + \frac{1}{T} + \frac{i\hbar}{mw^2}\right) \tag{13.18}$$

This integral, via the exponent described by Equation 13.18, links the quantum quantity $i\hbar$ to the width of the slit $2w$, the interesting physics is in the term $(i\hbar/mw^2)$. Feynman goes on to perform the integral and ends up with the quantity

$$\frac{im}{2\hbar}\left(\frac{1}{\tau} + \frac{1}{T} + \frac{i\hbar}{mw^2}\right)$$

as a component of the amplitude of $\psi(x)$. However, rather than performing the integral and following the steps of Feynman and Hibbs (1965), we return to our previous observation that the physics of interest is contained in the term $(i\hbar/mw^2)$ and that this term is the inverse of a temporal quantity, so that

$$\frac{1}{t} \approx \frac{i\hbar}{mw^2} \tag{13.19}$$

The absolute value of this quantity is obtained by multiplying it with its complex conjugate and taking its square root, so that

$$\left|\frac{1}{t}\right| \approx \left|\frac{\hbar}{mw^2}\right| \tag{13.20}$$

If we abstract the absolute notation and notice that the quantity w is a segment of x, that is, $w \rightarrow \Delta x$ (see Figure 13.1), then

$$\frac{\Delta x}{t} \approx \frac{\hbar}{m\Delta x} \tag{13.21}$$

$$\Delta p \approx \frac{\hbar}{\Delta x} \tag{13.22}$$

and we arrive to an expression of the form

$$\Delta p \, \Delta x \approx \hbar \tag{13.23}$$

which is a reduced expression of the uncertainty principle as compared to

$$\Delta p \, \Delta x \approx h$$

In Chapter 3, we describe how to use the uncertainty principle to arrive to the diffraction-limited beam divergence

$$\Delta \theta \approx \frac{\lambda}{\pi w}$$

where $2w$ is the width of the slit.

13.4.1 Discussion

Beyond discussing the propagation through a slit and its link to the Heisenberg uncertainty principle, Feynman goes on to derive Schrödinger's equation and to discuss many other applications to quantum mechanics using his path integral approach. In regard to our optics perspective, in this brief introduction, we have learned that

1. The path integral approach can be used to describe propagation of particles through macroscopic slits.
2. This approach leads to the uncertainty principle that can be used to describe diffraction-limited beam divergence (see Chapter 3).
3. The path integral approach is mathematically more involved and not as succinct and elegant as the straightforward Dirac notation to describe basic optics phenomena such as beam divergence.

13.5 Feynman Diagrams in Optics

Since we are on the subject of Feynman, perhaps this is the best place to mention the Feynman diagrams. An introduction to this subject is provided by Feynman himself in his book *QED: The Strange Theory of Light and Matter* (Feynman, 1985). Feynman is said to have invented these diagrams "as a bookkeeping device for wading through complicated calculations" (Kaiser, 2005) in his particle physics renormalization work.

Here, we provide a brief pictorial visit to the subject simply to illustrate one of the applications of these diagrams in optics. In doing so, we refer to the works of Yariv (1977) and Butcher and Cotter (1991).

A Feynman diagram in its simplest form depicts a sequence of events, from left to right, as a function of time. Other diagrams are 2D with space on the horizontal axis and time on the vertical axis. The diagram that we are going to consider depicts the absorption and emission of light as a function of time. Of the many rules that apply to the Feynman diagrams, the principal rule observed here is that the absorption of light is depicted by a wavy line moving from left to right, on one side of the diagram, while the emission of light is also depicted by a wavy line, moving from left to right, at the other side of the diagram. The quantum states involved are also included in the diagram.

The first example we consider is the case of Raman scattering (Yariv, 1977) including the ground state (1), an upper state (2), and the lower state (3), as well as a virtual state (see Figure 13.3). It should be pointed out that the equations describing this process are rather crowded and the Feynman diagram is a practical avenue to illustrate the mechanism.

An additional example involves the pictorial description, using the Feynman diagrams, of the third harmonic generation given by Butcher and Cotter (1991). These authors derive an extensive equation describing the third harmonic generation. Here we describe one of the terms of the equation using a Feynman diagram as illustrated in Figure 13.4. The third harmonic generation involves four states, a, b, c, and d. In the diagram shown, the molecule in state a emits a 3ω photon while making a virtual transition to the state b. Via the absorption of a photon, with frequency ω, the molecule transitions from b to c. Via the absorption of two more photons, with frequency ω,

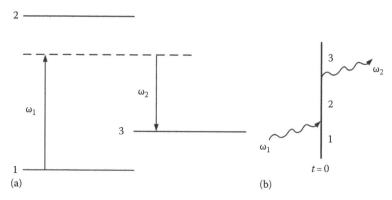

FIGURE 13.3
(a) Energy diagram of Raman scattering. (b) Corresponding Feynman diagram.

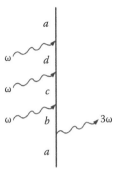

FIGURE 13.4
Feynman diagram of the third harmonic generation.

the molecule transitions to state d and finally to its initial state a. The other three complementary processes involve the emission of 3ω photons from b, c, and d while absorbing ω photons at their respective non-emitting states.

Problems

13.1 Show that the exponential forms in Equation 13.8 are equivalent.
13.2 Show that Equation 13.14 can be expressed as Equation 13.15 in conjunction with Equations 13.16 through 13.18.
13.3 Draw a Feynman diagram for a two-photon absorption process involving a ground state, an intermediate virtual state, and an upper state.

References

Butcher, P. N. and Cotter D. (1991). *The Elements of Nonlinear Optics*. Cambridge University, Cambridge, U.K.

Dirac, P. A. M. (1978). *The Principles of Quantum Mechanics*, 4th edn. Oxford, London, U.K.

Feynman, R. P. (1985). *QED: The Strange Theory of Light and Matter*. Princeton University, Princeton, NJ.

Feynman, R. P. and Hibbs, A. R. (1965). *Quantum Mechanics and Path Integrals*. McGraw-Hill, New York.

Feynman, R. P., Leighton, R. B., and Sands, M. (1965). *The Feynman Lectures on Physics*, Vol. II. Addison-Wesley, Reading, MA.

Kaiser, D. (2005). *Drawing Theories Apart: The Dispersion of Feynman Diagrams in Postwar Physics*. Chicago University, Chicago, IL.

Yariv, A. (1977). The application of time evolution operators and Feynman diagrams to nonlinear optics. *IEEE J. Quant. Electron.* **QE-13**, 943–950.

14

Matrix Aspects of Quantum Mechanics

14.1 Introduction

As mentioned in Chapter 1, the Heisenberg matrix mechanics provides one of the three main avenues to quantum mechanics. This approach to quantum mechanics was disclosed in three papers authored by Heisenberg (1925), Born and Jordan (1925), and Born et al. (1926). An iconic result from the Heisenberg–Born–Jordan contribution was the *commutation rule*

$$pq - qp = \frac{h}{2\pi i} \tag{14.1}$$

or

$$pq - qp = -i\hbar \tag{14.2}$$

Here, we examine the origin of the commutation rule, using the Feynman approach, and provide a brief pragmatic introduction to some salient aspects of matrix quantum mechanics with a focus on Pauli matrices. We begin with a review preamble on vector and matrix algebra.

14.2 Introduction to Vector and Matrix Algebra

Here, a few of the salient, and useful, features of vector and matrix algebra are introduced in a very pragmatic approach without derivation or proof.

14.2.1 Vector Algebra

A vector in three dimensions (x, y, z) is depicted in Figure 14.1. The sum of two vectors $r + s$, illustrated in Figure 14.2, is defined as

$$\begin{pmatrix} r_1 \\ r_2 \\ r_3 \end{pmatrix} + \begin{pmatrix} s_1 \\ s_2 \\ s_3 \end{pmatrix} = \begin{pmatrix} r_1 + s_1 \\ r_2 + s_2 \\ r_3 + s_3 \end{pmatrix} \tag{14.3}$$

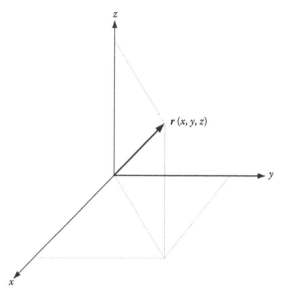

FIGURE 14.1
Vector in three dimensions.

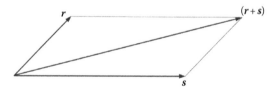

FIGURE 14.2
Vector addition $r + s$.

Subtraction of two vectors $r - s$, illustrated in Figure 14.3, is defined as

$$\begin{pmatrix} r_1 \\ r_2 \\ r_3 \end{pmatrix} - \begin{pmatrix} s_1 \\ s_2 \\ s_3 \end{pmatrix} = \begin{pmatrix} r_1 - s_1 \\ r_2 - s_2 \\ r_3 - s_3 \end{pmatrix} \tag{14.4}$$

Multiplication of a vector r with a scalar number a, thus creating a new vector ar, is defined as

$$a \begin{pmatrix} r_1 \\ r_2 \\ r_3 \end{pmatrix} = \begin{pmatrix} ar_1 \\ ar_2 \\ ar_3 \end{pmatrix} \tag{14.5}$$

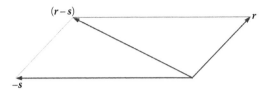

FIGURE 14.3
Vector subtraction $r - s$.

The length of a vector s is defined as $|s|$

$$|s|^2 = \begin{vmatrix} s_1 \\ s_2 \\ s_3 \end{vmatrix}^2 = \left(s_1^2 + s_2^2 + s_3^2 \right) \tag{14.6}$$

and the dot product of two vectors $r \cdot s$, see Figure 14.4, is a scalar defined as

$$\begin{pmatrix} r_1 \\ r_2 \\ r_3 \end{pmatrix} \cdot \begin{pmatrix} s_1 \\ s_2 \\ s_3 \end{pmatrix} = \left(r_1 s_1 + r_2 s_2 + r_3 s_3 \right) \tag{14.7}$$

In reference to Figure 14.4, the angle between the two vectors is defined as θ and applying the law of cosines

$$r \cdot s = |r||s| \cos \theta$$

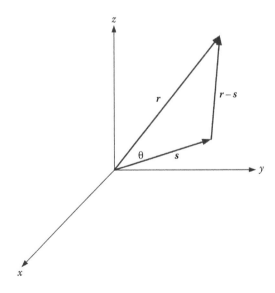

FIGURE 14.4
Dot product $r \cdot s$. The angle between the two vectors is θ.

The cross product of two vectors $r \times s$ is a new vector defined as

$$\begin{pmatrix} r_1 \\ r_2 \\ r_3 \end{pmatrix} \times \begin{pmatrix} s_1 \\ s_2 \\ s_3 \end{pmatrix} = \begin{pmatrix} r_2 s_3 - r_3 s_2 \\ r_3 s_1 - r_1 s_3 \\ r_1 s_2 - r_2 s_1 \end{pmatrix} \tag{14.8}$$

The magnitude of $r \times s$ is the area of the parallelogram formed by r and s and its direction is determined by the right-hand rule as illustrated in Figure 14.5. Some useful vector identities involve the *derivative operators* ∇ and ∇^2

$$\nabla(AB) = (\nabla A)B + A(\nabla B) = B\nabla A + A\nabla B \tag{14.9}$$

$$\nabla \times (AB) = B(\nabla \times A) - A \times \nabla B \tag{14.10}$$

$$\nabla \times (\nabla \times C) = \nabla(\nabla \cdot C) - (\nabla \cdot \nabla)C \tag{14.11}$$

where

$$\nabla = \begin{pmatrix} \partial / \partial x \\ \partial / \partial y \\ \partial / \partial z \end{pmatrix} \tag{14.12}$$

and

$$\nabla^2 = \left(\frac{\partial^2}{\partial x^2} + \frac{\partial^2}{\partial y^2} + \frac{\partial}{\partial z^2} \right) \tag{14.13}$$

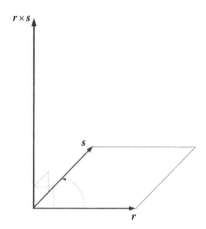

FIGURE 14.5
Cross product $r \times s$ (vector diagram not to scale).

These identities are useful in dealing with wave equations (see Chapter 12) and a range of other applications including electromagnetism (see Chapter 15).

14.2.2 Matrix Algebra

For simplicity we consider mainly 2×2 matrices albeit the algebra is also applicable to 3×3 and higher order matrices. We begin by defining the 2×2 matrix A

$$A = \begin{pmatrix} a_{11} & a_{12} \\ a_{21} & a_{22} \end{pmatrix} \tag{14.14}$$

and the 2×2 matrix B

$$B = \begin{pmatrix} b_{11} & b_{12} \\ b_{21} & b_{22} \end{pmatrix} \tag{14.15}$$

thus the matrix addition $A + B$ yields

$$A + B = \begin{pmatrix} a_{11} + b_{11} & a_{12} + b_{12} \\ a_{21} + b_{21} & a_{22} + b_{22} \end{pmatrix} \tag{14.16}$$

and the subtraction $A - B$ yields

$$A - B = \begin{pmatrix} a_{11} - b_{11} & a_{12} - b_{12} \\ a_{21} - b_{21} & a_{22} - b_{22} \end{pmatrix} \tag{14.17}$$

Multiplication of the matrix A by (-1) is equivalent to multiplying each individual component of the matrix by (-1)

$$(-1)A = \begin{pmatrix} -a_{11} & -a_{12} \\ -a_{21} & -a_{22} \end{pmatrix} \tag{14.18}$$

Similarly, multiplication of the matrix A by the quantity i is equivalent to multiplying each individual component of the matrix by i

$$iA = \begin{pmatrix} ia_{11} & ia_{12} \\ ia_{21} & ia_{22} \end{pmatrix} \tag{14.19}$$

The simple product of the two matrices AB is

$$AB = \begin{pmatrix} a_{11} & a_{12} \\ a_{21} & a_{22} \end{pmatrix} \begin{pmatrix} b_{11} & b_{12} \\ b_{21} & b_{22} \end{pmatrix} = \begin{pmatrix} a_{11}b_{11} + a_{12}b_{21} & a_{11}b_{12} + a_{12}b_{22} \\ a_{21}b_{11} + a_{22}b_{21} & a_{21}b_{12} + a_{22}b_{22} \end{pmatrix} \tag{14.20}$$

and the simple product of the two matrices BA is

$$BA = \begin{pmatrix} b_{11} & b_{12} \\ b_{21} & b_{22} \end{pmatrix} \begin{pmatrix} a_{11} & a_{12} \\ a_{21} & a_{22} \end{pmatrix} = \begin{pmatrix} b_{11}a_{11} + b_{12}a_{21} & b_{11}a_{12} + b_{12}a_{22} \\ b_{21}a_{11} + b_{22}a_{21} & b_{21}a_{12} + b_{22}a_{22} \end{pmatrix} \quad (14.21)$$

Notice that

$$AB - BA \neq 0 \quad (14.22)$$

or

$$AB \neq BA$$

which is a very important, and even iconic, result in matrix algebra. However, with certain special matrices, the condition described by Equation 14.22 might not hold as we shall see next.

Unitary matrices are defined as follows for a 2 × 2 matrix

$$1 = \begin{pmatrix} 1 & 0 \\ 0 & 1 \end{pmatrix} \quad (14.23)$$

a 3 × 3 matrix

$$1 = \begin{pmatrix} 1 & 0 & 0 \\ 0 & 1 & 0 \\ 0 & 0 & 1 \end{pmatrix} \quad (14.24)$$

and a 4 × 4 matrix

$$1 = \begin{pmatrix} 1 & 0 & 0 & 0 \\ 0 & 1 & 0 & 0 \\ 0 & 0 & 1 & 0 \\ 0 & 0 & 0 & 1 \end{pmatrix} \quad (14.25)$$

$$1A = A1 = A \quad (14.26)$$

where 1 refers either to the number 1 or the unitary matrix U, as defined in Equations 14.23 through 14.25

$$UA = AU = A \quad (14.27)$$

$$U = 1 \quad (14.28)$$

Obviously, $UA - AU = 0$ which is an exception to the condition described in Equation 14.22.

The inverse matrix I^{-1} is a unique matrix that when multiplied with the original matrix I yields unity, so that

$$II^{-1} = I^{-1}I = 1 \tag{14.29}$$

$$(I^{-1})^{-1} = I \tag{14.30}$$

If matrices I and H have inverses, I^{-1} and H^{-1}, then

$$(IH)^{-1} = I^{-1}H^{-1} \tag{14.31}$$

and

$$I^{-1}H^{-1}IH = IHI^{-1}H^{-1} = 1 \tag{14.32}$$

The *determinant* of a 2×2 matrix is defined as

$$|A| = \begin{vmatrix} a_{11} & a_{12} \\ a_{21} & a_{22} \end{vmatrix} = a_{11}a_{22} - a_{21}a_{12} \tag{14.33}$$

For a 3×3 matrix, the determinant is defined as

$$|A| = \begin{vmatrix} a_{11} & a_{12} & a_{13} \\ a_{21} & a_{22} & a_{23} \\ a_{31} & a_{32} & a_{33} \end{vmatrix} = a_{11}\begin{vmatrix} a_{22} & a_{23} \\ a_{32} & a_{33} \end{vmatrix} - a_{12}\begin{vmatrix} a_{21} & a_{23} \\ a_{31} & a_{33} \end{vmatrix} + a_{13}\begin{vmatrix} a_{21} & a_{22} \\ a_{31} & a_{32} \end{vmatrix} \tag{14.34}$$

and so on for $N \times N$ matrices.

Finally, in this section we define the *trace* of a matrix: this quantity Tr is defined as the sum of the diagonal elements of a matrix

$$Tr(A) = \sum_j a_{jj} \tag{14.35}$$

so that for our 3×3 matrix

$$Tr(A) = \sum_{j=1}^{N=3} a_{jj} = a_{11} + a_{22} + a_{33} \tag{14.36}$$

For example, if

$$I = \begin{pmatrix} 1 & 1-i \\ 1+i & -1 \end{pmatrix} \tag{14.37}$$

$$I^{-1} = 3^{-1} \begin{pmatrix} 1 & 1-i \\ 1+i & -1 \end{pmatrix}$$ (14.38)

$$II^{-1} = \begin{pmatrix} 1 & 1-i \\ 1+i & -1 \end{pmatrix} 3^{-1} \begin{pmatrix} 1 & 1-i \\ 1+i & -1 \end{pmatrix} = 3^{-1} \begin{pmatrix} 3 & 0 \\ 0 & 3 \end{pmatrix}$$ (14.39)

moreover, the determinant of this matrix $|I|$ (see Equation 14.33) is

$$|I| = \begin{vmatrix} 1 & 1-i \\ 1+i & -1 \end{vmatrix} = -1 + (1+i)(1-i) = -1 + 2 = 1$$ (14.40)

and its trace is

$$Tr(I) = 1 + (-1) = 0$$ (14.41)

14.3 Quantum Operators

There are three types operators in quantum mechanics: the position operator, the momentum operator, and the energy operator. We are already familiar with the Hamiltonian that is the energy operator (see Chapters 8 and 12). Here, the position, momentum, and energy operators are introduced using Feynman's style and notation (Feynman et al., 1965). Note: given the similarity of the equations involved, the Feynman approach appears to be inspired in Dirac's discussion on momentum (Dirac, 1978).

14.3.1 Position Operator

The position operator is related to the average position of a particle. More specifically, it is related to the average value of x in a state $|\psi\rangle$. Following a probabilistic argument, Feynman defines the average of x using an integral first disclosed by Dirac (1978) (see note at the end of this section):

$$\bar{x} = \int \langle \psi | x \rangle x \langle x | \psi \rangle dx$$ (14.42)

Then, Feynman identifies the average of x with the probability amplitude

$$\bar{x} = \langle \psi | \alpha \rangle$$ (14.43)

where

$$|\alpha\rangle = \hat{x} | \psi \rangle$$ (14.44)

Hence, combining Equations 14.42 and 14.43, we get

$$\bar{x} = \langle \psi \,|\, \alpha \rangle = \int \langle \psi \,|\, x \rangle x \langle x \,|\, \psi \rangle dx \tag{14.45}$$

and since, by definition,

$$\langle \psi \,|\, \alpha \rangle = \int \langle \psi \,|\, x \rangle \langle x \,|\, \alpha \rangle dx \tag{14.46}$$

then comparison of Equations 14.45 and 14.46 yields

$$\langle x \,|\, \alpha \rangle = x \langle x \,|\, \psi \rangle \tag{14.47}$$

$$|\, \alpha \rangle = x \,|\, \psi \rangle \tag{14.48}$$

and from the definition in Equation 14.44

$$|\, \alpha \rangle = \hat{x} \,|\, \psi \rangle = x \,|\, \psi \rangle \tag{14.49}$$

Hence, the position operators $\hat{x}$, $\hat{y}$, $\hat{z}$ are related to the coordinates x, y, z according to

$$\hat{x} \,|\, \psi \rangle = x \,|\, \psi \rangle$$
$$\hat{y} \,|\, \psi \rangle = y \,|\, \psi \rangle \tag{14.50}$$
$$\hat{z} \,|\, \psi \rangle = z \,|\, \psi \rangle$$

Note: equations of the form

$$\int \langle \psi \,|\, x \rangle x \langle x \,|\, \psi \rangle dx$$

were first introduced by Dirac in his discussion of the momentum representation (Dirac, 1978).

14.3.2 Momentum Operator

Again, following Feynman's style, as with the position operator, the process begins by relating the average momentum $\bar{p}$ to a state $|\beta\rangle$

$$\bar{p} = \langle \psi \,|\, \beta \rangle \tag{14.51}$$

$$|\, \beta \rangle = p \,|\, \psi \rangle \tag{14.52}$$

$$|\, \beta \rangle = \hat{p} \,|\, \psi \rangle \tag{14.53}$$

Next, in analogy with the position operator approach, Feynman defines

$$\bar{p} = \langle \psi | \beta \rangle = \int \langle \psi | x \rangle \langle x | \beta \rangle dx \tag{14.54}$$

and

$$\langle p | \beta \rangle = \int \langle p | x \rangle \langle x | \beta \rangle dx \tag{14.55}$$

where, following Dirac (1978), we define

$$\langle p | x \rangle = e^{-ipx/\hbar} \tag{14.56}$$

Then, using the definitions given in Equations 14.52 and 14.56, with $\langle x | \psi \rangle = \psi(x)$

$$\langle p | \beta \rangle = p \langle p | \psi \rangle = p \int \langle p | x \rangle \langle x | \psi \rangle dx = \int e^{-ipx/\hbar} p\psi(x) dx \tag{14.57}$$

which can be integrated by parts (Feynman et al., 1965) to yield

$$\langle p | \beta \rangle = -i\hbar \int e^{-ipx/\hbar} \frac{d\psi(x)}{dx} dx \tag{14.58}$$

Comparison of Equations 14.55 and 14.58 renders

$$\langle x | \beta \rangle = -i\hbar \frac{d\psi(x)}{dx} \tag{14.59}$$

which is equivalent to

$$\langle x | \beta \rangle = -i\hbar \frac{d}{dx} \langle x | \psi \rangle \tag{14.60}$$

$$| \beta \rangle = -i\hbar \frac{d}{dx} | \psi \rangle \tag{14.61}$$

Now, using the definition given in Equation 14.53

$$| \beta \rangle = \hat{p} | \psi \rangle = -i\hbar \frac{d}{dx} | \psi \rangle \tag{14.62}$$

In this manner, the three momentum operators become (Feynman et al., 1965; Dirac, 1978)

$$\hat{p}_x \rightarrow \hat{\wp}_x = -i\hbar \frac{\partial}{\partial x}$$

$$\hat{p}_y \rightarrow \hat{\wp}_y = -i\hbar \frac{\partial}{\partial y} \quad (14.63)$$

$$\hat{p}_z \rightarrow \hat{\wp}_z = -i\hbar \frac{\partial}{\partial z}$$

In three dimensions and using vector notation, these results are nicely summarized as (Feynman et al., 1965)

$$\hat{p} \rightarrow \hat{\wp} = -i\hbar \nabla \quad (14.64)$$

14.3.3 Example

Letting the operator $\hat{p}_x$ operate on some function $\psi(x)$, analogously to $xp_x - p_x x$, yields

$$x\hat{p}_x\psi(x) - \hat{p}_x x\psi(x) = x(-i\hbar)\frac{\partial\psi(x)}{\partial x} - (-i\hbar)\frac{\partial(x\psi(x))}{\partial x}$$

$$x(-i\hbar)\frac{\partial\psi(x)}{\partial x} + i\hbar\psi(x) + xi\hbar\frac{\partial\psi(x)}{\partial x} = +i\hbar\psi(x)$$

$$x\hat{p}_x - \hat{p}_x x = i\hbar \quad (14.65)$$

as given by Dirac (1978). If both sides of this equation are multiplied by –1, we get

$$\hat{p}_x x - x\hat{p}_x = -i\hbar \quad (14.66)$$

which is the all important result discovered by Heisenberg and colleagues in 1926 (see Equation 14.1). Notice that in the classical analogy $xp_x - p_x x = 0$.
In general, any two operators, $\hat{A}$ and $\hat{B}$, exhibiting the condition

$$\hat{A}\hat{B} - \hat{B}\hat{A} \neq 0 \quad (14.67)$$

are said not to commute. In other words, "the operators do not commute" (Feynman et al., 1965). Also, Equation 14.65 is referred to as the *commutation rule*. Note that this result is analogous to that encountered in classical matrix algebra (see Equation 14.22).

14.3.4 Energy Operator

The energy operator is the *Hamiltonian* and has already been introduced in Chapters 8 and 12 as (Dirac, 1978)

$$\hat{H} \rightarrow \hat{\mathcal{H}} = -\frac{\hbar^2}{2m}\nabla^2 + V(r) \tag{14.68}$$

which in vector notation becomes

$$\hat{H} \rightarrow \hat{\mathcal{H}} = \frac{1}{2m}\hat{\wp}\cdot\hat{\wp} + V(r) \tag{14.69}$$

14.3.5 Heisenberg Equation of Motion

Feynman defines an operator $\hat{A}$ that he eventually links to the operator $\hat{H}$ via an equation of the form (Dirac, 1987)

$$\dot{\hat{A}} = \frac{i}{\hbar}[\hat{H}, \hat{A}] \tag{14.70}$$

where $[\hat{H}, \hat{A}] = (\hat{H}\hat{A} - \hat{A}\hat{H})$ is the *Poisson bracket* introduced by Dirac (1978).

Here we'll do things a little bit differently since we are interested in the Heisenberg equation of motion. To do this, we consider Dirac's definition of $A(t)$ (Dirac, 1978):

$$A(t) = e^{iHt/\hbar}Ae^{-iHt/\hbar} \tag{14.71}$$

Differentiation of this function leads to

$$\frac{dA(t)}{dt} = \frac{iH}{\hbar}A + \frac{\partial A}{\partial t} - A\frac{iH}{\hbar}$$

$$\frac{dA(t)}{dt} = \frac{i}{\hbar}(HA - AH) + \frac{\partial A}{\partial t}$$

$$\frac{dA(t)}{dt} = \frac{i}{\hbar}[H, A] + \frac{\partial A}{\partial t} \tag{14.72}$$

which is known as the *Heisenberg equation of motion*.

Finally, we should mention that operators provide the entrance stage to the subfield of "second quantization" where a creation operator a_ξ^t creates a quantum state ξ and an annihilation operator a_η annihilates a quantum state η. Bosons satisfy several commutation rules including $[a_\xi^t, a_\eta^t] = 0$. For an introduction to this subject, see Judd (2006).

14.4 Pauli Matrices

Pauli matrices are widely used in quantum polarization (see Chapter 16); thus, here we provide an introduction to the subject via the Hamiltonian following Feynman's style.

The time-dependent equation for two-state system is (Dirac, 1978)

$$i\hbar \frac{dC_i}{dt} = \sum_j H_{ij}C_j \tag{14.73}$$

or

$$i\hbar \frac{dC_1}{dt} = H_{11}C_1 + H_{12}C_2 \tag{14.74}$$

$$i\hbar \frac{dC_2}{dt} = H_{21}C_1 + H_{22}C_2 \tag{14.75}$$

and for a spin one-half particle, such as an electron, under the influence of a magnetic field, the Hamiltonian becomes (Feynman et al., 1965)

$$\begin{pmatrix} H_{11} & H_{12} \\ H_{21} & H_{22} \end{pmatrix} = \begin{pmatrix} -\mu B_z & -\mu(B_x - iB_y) \\ -\mu(B_x + iB_y) & +\mu B_z \end{pmatrix} \tag{14.76}$$

In general, the Hamiltonian for a spin one-half particle can be defined as (Feynman et al., 1965)

$$H_{ij} = -\mu(\sigma_{ij}^x B_x + \sigma_{ij}^y B_y + \sigma_{ij}^z B_z) \tag{14.77}$$

While observing the definitions given in Equations 14.76 and 14.77, we see that for the z component, that is B_z, we have

$$H_{11} = -\mu\sigma_{11}^z B_z = -\mu B_z$$

$$H_{12} = -\mu\sigma_{12}^z B_z = 0$$

$$H_{21} = -\mu\sigma_{21}^z B_z = 0$$

$$H_{22} = -\mu\sigma_{22}^z B_z = +\mu B_z$$

which implies that

$$\sigma_{11}^z = 1$$

$$\sigma_{12}^z = 0$$

$$\sigma_{21}^z = 0$$

$$\sigma_{22}^z = -1$$

so that in matrix form (Dirac, 1978)

$$\sigma_{ij}^z = \sigma_z = \begin{pmatrix} 1 & 0 \\ 0 & -1 \end{pmatrix} \qquad (14.78)$$

Similarly, we find for the x and y components (Dirac, 1978)

$$\sigma_{ij}^x = \sigma_x = \begin{pmatrix} 0 & 1 \\ 1 & 0 \end{pmatrix} \qquad (14.79)$$

$$\sigma_{ij}^y = \sigma_y = \begin{pmatrix} 0 & -i \\ i & 0 \end{pmatrix} \qquad (14.80)$$

The 2×2 matrices σ_x, σ_y, and σ_z, are known as Pauli matrices (Dirac, 1978) and are very important to spin and magnetic moment computations. Some properties of these matrices are (Dirac, 1978; Jordan, 1986)

$$\sigma_x^2 = \sigma_y^2 = \sigma_z^2 = 1$$

$$\sigma_x \sigma_y = i\sigma_z \quad \sigma_y \sigma_x = -i\sigma_z$$

$$\sigma_y \sigma_z = i\sigma_x \quad \sigma_z \sigma_y = -i\sigma_x \qquad (14.81)$$

$$\sigma_z \sigma_x = i\sigma_y \quad \sigma_x \sigma_z = -i\sigma_y$$

$$\sigma_x \sigma_y \sigma_z = i$$

The σ_y matrix is also known to be *Hermitian*, that is, this matrix is identical to its own *conjugate transpose*. In other words, the conjugate of σ_y is

$$\sigma_y^* = \begin{pmatrix} 0 & -i \\ +i & 0 \end{pmatrix}^* = \begin{pmatrix} 0 & +i \\ -i & 0 \end{pmatrix} \qquad (14.82)$$

and the transpose (rows → column) of σ_y^* is $(\sigma_y^*)^T$

$$(\sigma_y^*)^T = \begin{pmatrix} 0 & +i \\ -i & 0 \end{pmatrix}^T = \begin{pmatrix} 0 & -i \\ +i & 0 \end{pmatrix} \tag{14.83}$$

Comparison of Equation 14.82 and 14.83 indicates that

$$\sigma_y = (\sigma_y)^\dagger \tag{14.84}$$

where the symbol† represents the combined conjugate transpose operation, that is, the Hermitian property.

14.4.1 Pauli Matrices for Spin One-Half Particles

Electrons and protons are defined as fermions and have an intrinsic angular momentum of $\hbar/2$ (Feynman et al., 1965). Thus, the electron is said to be a spin one-half, or spin $-1/2$, particle. Also, the electron can have its spin-up $|+\rangle$ or it can have its spin down $|-\rangle$. Pauli introduced spin operators to describe the condition of the electron as

$$\hat{S} = \tfrac{1}{2}\hbar\hat{\sigma} \tag{14.85}$$

where

$$\hat{S}_x = \tfrac{1}{2}\hbar \begin{pmatrix} 0 & 1 \\ 1 & 0 \end{pmatrix} \tag{14.86}$$

$$\hat{S}_y = \tfrac{1}{2}\hbar \begin{pmatrix} 0 & -i \\ i & 0 \end{pmatrix} \tag{14.87}$$

$$\hat{S}_z = \tfrac{1}{2}\hbar \begin{pmatrix} 1 & 0 \\ 0 & -1 \end{pmatrix} \tag{14.88}$$

These matrices satisfy (Robson, 1974)

$$\hat{S}_x^2 + \hat{S}_y^2 + \hat{S}_z^2 = \tfrac{3}{4}\hbar^2 \begin{pmatrix} 1 & 0 \\ 0 & 1 \end{pmatrix} \tag{14.89}$$

and the angular momentum commutation rules become

$$\hat{S}_x\hat{S}_y - \hat{S}_y\hat{S}_x = \frac{\hbar^2}{4}\left\{ \begin{pmatrix} 0 & 1 \\ 1 & 0 \end{pmatrix} \begin{pmatrix} 0 & -i \\ i & 0 \end{pmatrix} - \begin{pmatrix} 0 & -i \\ i & 0 \end{pmatrix} \begin{pmatrix} 0 & 1 \\ 1 & 0 \end{pmatrix} \right\} = \frac{i\hbar^2}{2} \begin{pmatrix} 1 & 0 \\ 0 & -1 \end{pmatrix}$$

$$\tag{14.90}$$

so that

$$\hat{S}_x\hat{S}_y - \hat{S}_y\hat{S}_x = i\hbar\hat{S}_z \tag{14.91}$$

and similarly

$$\hat{S}_z\hat{S}_x - \hat{S}_x\hat{S}_z = i\hbar\hat{S}_y \tag{14.92}$$

$$\hat{S}_y\hat{S}_z - \hat{S}_z\hat{S}_y = i\hbar\hat{S}_x \tag{14.93}$$

as can be verified by expansion. Equations 14.86 through 14.89 are useful in the description of polarized beam of electrons (Robson, 1974).

The eigenvalues of Equation 14.89 are

$$\tfrac{3}{4}\hbar^2 = s(s+1)\hbar^2 \tag{14.94}$$

where $s = 1/2$. By inspection, eigenvalues of $\hat{S}_z$ are

$$\pm\tfrac{1}{2}\hbar = \pm s\hbar \tag{14.95}$$

Now, without derivation, we introduce the quantum angular momentum

$$|J| = [j(j+1)]^{1/2}\hbar \tag{14.96}$$

where

$$j = l \pm s = l \pm \tfrac{1}{2} \tag{14.97}$$

Here, we already know that s is related to the spin angular momentum and can take the values $\pm\tfrac{1}{2}$. On the other hand, l is related to the orbital angular momentum and can take the values $l = 0, 1, 2, 3\dots$. Equation 14.97 in conjunction with the Pauli exclusion principle can be used to gain a glimpse of the energy level structure of hydrogen. This is an alternative way to describe the hydrogen atom to the formal Schrödinger equation path outlined in Chapter 12.

Pauli exclusion principle: "two electrons are never in the same state" (Dirac, 1978). This means that no two electrons can have the same quantum numbers.

14.5 Introduction to the Density Matrix

Here we provide a brief introduction to the concept of density matrices. The description given here is aimed at illustrating the density matrix formalism applicable to the description of multiple-level transitions (see Chapter 8). First, we express the Schrödinger equation as (Dirac, 1978)

$$i\hbar \frac{d\,|\,m\rangle}{dt} = H\,|\,m\rangle \tag{14.98}$$

$$i\hbar \frac{d\langle m\,|}{dt} = -\langle m\,|\,H \tag{14.99}$$

and define the "quantum density" as (Dirac, 1978)

$$\rho = \sum_m |\,m\rangle P_m \langle m\,| \tag{14.100}$$

where P_m is an mth state. Differentiating the quantum density with respect to time yields

$$i\hbar \frac{d\rho}{dt} = \sum_m i\hbar \left(\frac{d\,|\,m\rangle}{dt} P_m \langle m\,| + |\,m\rangle P_m \frac{d\langle m\,|}{dt} \right) \tag{14.101}$$

and using Equations 14.98 and 14.99

$$i\hbar \frac{d\rho}{dt} = \sum_m \left(H\,|\,m\rangle P_m \langle m\,| - |\,m\rangle P_m \langle m\,|\,H \right)$$

which can be written as (Dirac, 1978)

$$i\hbar \frac{d\rho}{dt} = (H\rho - \rho H) \tag{14.102}$$

$$\frac{d\rho}{dt} = -\frac{i}{\hbar}[H,\rho] \tag{14.103}$$

Dirac argues that the sum of the P_m states can be equal to unity thus allowing Equation 14.100 to be expressed simply as

$$\rho = |\,m\rangle \langle m\,| \tag{14.104}$$

which can be expanded in explicit form as

$$\rho = |\,m\rangle \langle m\,| = \begin{pmatrix} m_1 \\ m_2 \end{pmatrix} \begin{pmatrix} m_1^* & m_2^* \end{pmatrix} = \begin{pmatrix} m_1 m_1^* & m_1 m_2^* \\ m_2 m_1^* & m_2 m_2^* \end{pmatrix} = \begin{pmatrix} \rho_{11} & \rho_{12} \\ \rho_{21} & \rho_{22} \end{pmatrix} \tag{14.105}$$

Characteristics of the density matrix include $\rho = \rho^\dagger$, the Hermitian property, $Tr(\rho) = 1$, and $\rho^2 = \rho$.

14.5.1 Examples

If a state is described by the vector

$$|m\rangle = \frac{1}{\sqrt{2}}\begin{pmatrix} e^{-i\varphi} \\ 1 \end{pmatrix} \tag{14.106}$$

let us find out the corresponding density matrix. Using Equation 14.104,

$$|m\rangle\langle m| = \begin{pmatrix} 2^{-1/2}e^{-i\varphi} \\ 2^{-1/2} \end{pmatrix}\begin{pmatrix} 2^{-1/2}e^{i\varphi} & 2^{-1/2} \end{pmatrix} = \begin{pmatrix} 1/2 & e^{-i\varphi}/2 \\ e^{i\varphi}/2 & 1/2 \end{pmatrix} \tag{14.107}$$

Next, we can easily verify that all the conditions for a normalized density matrix, that is, $\rho = \rho^\dagger$, $Tr(\rho) = 1$, and $\rho^2 = \rho$, are met (see Problem 8).

If $\phi = \pi/2$, then this equation becomes the matrix for right-handed circularly polarized light

$$|m\rangle\langle m| = \begin{pmatrix} 1/2 & -i/2 \\ i/2 & 1/2 \end{pmatrix} \tag{14.108}$$

and if $\phi = -\pi/2$, then this equation becomes the matrix for left-handed circularly polarized light (see Chapter 16)

$$|m\rangle\langle m| = \begin{pmatrix} 1/2 & i/2 \\ -i/2 & 1/2 \end{pmatrix} \tag{14.109}$$

Dirac's definition given in Equation 14.100 suggests that density matrices can be added thus creating mixed states. If we take linearly polarized light in the x direction

$$|1\rangle\langle 1| = \begin{pmatrix} 1 \\ 0 \end{pmatrix}\begin{pmatrix} 1 & 0 \end{pmatrix} = \begin{pmatrix} 1 & 0 \\ 0 & 0 \end{pmatrix} \tag{14.110}$$

and mix it with light linearly polarized in the y direction

$$|0\rangle\langle 0| = \begin{pmatrix} 0 \\ 1 \end{pmatrix}\begin{pmatrix} 0 & 1 \end{pmatrix} = \begin{pmatrix} 0 & 0 \\ 0 & 1 \end{pmatrix} \tag{14.111}$$

we get a mixed state described by the density matrix

$$\rho = \frac{1}{2}(|1\rangle\langle 1| + |0\rangle\langle 0|) = \frac{1}{2}\begin{pmatrix} 1 & 0 \\ 0 & 1 \end{pmatrix} \tag{14.112}$$

which is the density matrix for unpolarized light (Robson, 1974). The density matrix description of polarization in photons is considered in more detail in Chapter 16, including Stokes parameters and Pauli matrices.

14.5.2 Transitions via the Density Matrix

Equation 14.104 is the starting point of the description of excitation of multi-level systems. For a two-level system, this equation can be written in vector form with wave function components

$$\rho = \begin{pmatrix} \Psi_a \\ \Psi_b \end{pmatrix} \begin{pmatrix} \Psi_a^* & \Psi_b^* \end{pmatrix} = \begin{pmatrix} \Psi_a \Psi_a^* & \Psi_a \Psi_b^* \\ \Psi_b \Psi_a^* & \Psi_b \Psi_b^* \end{pmatrix} \tag{14.113}$$

Using the waveform definitions given by Demtröder (2008), we write

$$\Psi_a(r,t) = \alpha(t) u_a e^{-iE_a t/\hbar} \tag{14.114}$$

$$\Psi_b(r,t) = \beta(t) u_b e^{-iE_b t/\hbar + i\varphi} \tag{14.115}$$

and the density matrix becomes

$$\begin{pmatrix} \rho_{aa} & \rho_{ab} \\ \rho_{ba} & \rho_{bb} \end{pmatrix} = \begin{pmatrix} |\alpha(t)|^2 & \alpha(t)\beta(t)e^{-i[(E_a - E_b)t/\hbar + \phi]} \\ \alpha(t)\beta(t)e^{+i[(E_a - E_b)t/\hbar + \phi]} & |\beta(t)|^2 \end{pmatrix} \tag{14.116}$$

since the u factors normalize to unity (Demtröder, 2008). The non-diagonal terms in this matrix (ρ_{ab} and ρ_{ba}) describe the coherence of the system. If ρ_{ab} and ρ_{ba} average to zero, then the matrix

$$\rho_{in} = \begin{pmatrix} |\alpha(t)|^2 & 0 \\ 0 & |\beta(t)|^2 \end{pmatrix} \tag{14.117}$$

describes the incoherently excited system. Equation 14.103 can also be used to describe the relaxation of coherently excited systems (Demtröder, 2008). For this, the total Hamiltonian is separated into three components:

$$H = H_0 + H_I(t) + H_R \tag{14.118}$$

where

$$H_0 = \begin{pmatrix} E_a & 0 \\ 0 & E_b \end{pmatrix} \tag{14.119}$$

$$H_I = \begin{pmatrix} 0 & -\mu_{ab} \cdot E_0(t) \\ -\mu_{ba} \cdot E_0(t) & 0 \end{pmatrix} \cos(\omega t + \phi) \tag{14.120}$$

$$H_R = \hbar \begin{pmatrix} \gamma_a & \gamma_a^\varphi \\ \gamma_b^\varphi & \gamma_b \end{pmatrix} \tag{14.121}$$

are the internal (H_0), interaction (H_I), and relaxation (H_R) Hamiltonians, respectively (Demtröder, 2008). For example, the population relaxation of

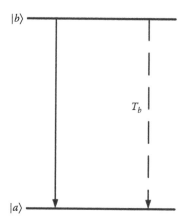

FIGURE 14.6
Spontaneous emission from level $|b\rangle$ to level $|a\rangle$.

state $|b\rangle$ to state $|a\rangle$, with a decay rate $\gamma_b = (1/T_b)$ (see Figure 14.6), can be described, using Equation 14.103, as

$$i\hbar\,\frac{\rho_{bb}}{T_b} = -[H_R, \rho]_{bb} \qquad (14.122)$$

$$i\hbar\,\frac{\rho_{bb}}{T_b} = +[H_R, \rho]_{aa} \qquad (14.123)$$

Furthermore, using Equations 14.103 and 14.113, a complete set of motion equations known as "master equations" can be derived. The density matrix formalism is particularly apt to describe quantum mechanically the transition mechanics of n-level systems. For example, this has been done in detail for lithium, a five-level system, by Olivares et al. (1998, 2002). Note that n-level transitions are considered, using rate equations, in Chapter 8.

Problems

14.1 Perform the integration by parts of the integral in Equation 14.57 and show that the result corresponds to Equation 14.58. Hint: $\psi(x) \rightarrow 0$ at $\pm\infty$.

14.2 Use Equations 14.76 and 14.77 to find the Pauli matrices σ_{ij}^x and σ_{ij}^y.

14.3 Using Equations 14.78 through 14.80, prove the following Pauli matrix identities:

$$\sigma_x^2 = \sigma_y^2 = \sigma_z^2 = 1.$$

14.4 Using Equations 14.78 through 14.80, prove the following Pauli matrix identities:

$$\sigma_z\sigma_x = i\sigma_y \quad \text{and} \quad \sigma_x\sigma_z = -i\sigma_y.$$

14.5 Evaluate the determinant and the trace of σ_x, σ_y, and σ_y.

14.6 Use the definitions for the spin one-half particles given in Equations 14.86 through 14.88 to verify Equation 14.89.

14.7 Use the definitions for the spin one-half particles given in Equations 14.86 through 14.88 to verify the commutation rules given in Equations 14.92 and 14.93.

14.8 Verify the conditions $\rho = \rho^\dagger$, $Tr(\rho) = 1$, and $\rho^2 = \rho$, for the normalized density matrix given in Equation 14.107.

References

Born, M., Heisenberg, W., and Jordan, P. (1926). Zur quantenmechanik II. *Z. Phys.* **35**, 557–617.

Born, M. and Jordan, P. (1925). Zur quantenmechanik. *Z. Phys.* **34**, 858–888.

Demtröder, W. (2008). *Laser Spectroscopy*, 4th edn. Springer, Berlin, Germany.

Dirac, P. A. M. (1978). *The Principles of Quantum Mechanics*, 4th edn. Oxford, London, U.K.

Feynman, R. P., Leighton, R. B., and Sands, M. (1965). *The Feynman Lectures on Physics*, Vol. III, Addison-Wesley, Reading, MA.

Heisenberg, W. (1925). Uber quantenthoretische umdeutung kinematischer und mechanischer beziehungen. *Z. Phys.* **33**, 879–893.

Jordan, T. F. (1986). *Quantum Mechanics in Simple Matrix Form*. Wiley, New York.

Judd, B. R. (2006). Second quantization. In *Springer Handbook of Atomic, Molecular, and Optical Physics* (Drake, G. W. F., ed.). Springer, Berlin, Germany, Part A Chapter 6.

Olivares, I. E., Duarte, A. E., Lokajczyk, T., Dinklage, A., and Duarte, F. J. (1998). Doppler-free spectroscopy and collisional studies with tunable diode lasers of isotopes in a heat-pipe oven. *J. Opt. Soc. Am. B* **15**, 1932–1939.

Olivares, I. E., Duarte, A. E., Saravia, E. A., and Duarte, F. J. (2002). Lithium isotope separation with tunable diode lasers. *App. Opt.* **41**, 2973–2977.

Robson, B. A. (1974). *The Theory of Polarization Phenomena*. Clarendon, Oxford, U.K.

15

Classical Polarization

15.1 Introduction

Here, we provide a revision of concepts and techniques in classical polarization as a background to Chapter 16 that deals with quantum polarization. Also, various elements and techniques considered here are used as tools in optical systems in quantum optics experiments. This exposition is based on a revised version of a review given by Duarte (2003) and includes new and updated material.

15.2 Maxwell Equations

The Maxwell equations are of fundamental importance since they describe the whole of classical electromagnetic phenomena. From a classical perspective, light can be described as waves of electromagnetic radiation. As such, the Maxwell equations are very useful to illustrate a number of the characteristics of light including polarization. It is customary to just state these equations without derivation. Since our goal is simply to apply them, the usual approach will be followed. However, for those interested, it is mentioned that a derivation by Dyson (1990) attributed to Feynman is available in the literature. The *Maxwell equations* in the rationalized metric system are given by

$$\nabla \cdot \boldsymbol{B} = 0 \tag{15.1}$$

$$\nabla \cdot \boldsymbol{E} = \frac{\rho}{\varepsilon_0} \tag{15.2}$$

$$c^2 \nabla \times \boldsymbol{B} = \frac{\partial \boldsymbol{E}}{\partial t} + \frac{\boldsymbol{j}}{\varepsilon_0} \tag{15.3}$$

$$\nabla \times \boldsymbol{E} = -\frac{\partial \boldsymbol{B}}{\partial t} \tag{15.4}$$

Feynman et al. (1965). These equations illustrate, with succinct beauty, the unique coexistence of the electric field and the magnetic field in nature. The first two equations give the value of the given flux through a closed surface, while the second two equations give the value of a line integral around a loop. In this notation,

$$\nabla = \left(\frac{\partial}{\partial x}, \frac{\partial}{\partial y}, \frac{\partial}{\partial z} \right)$$

where
  E is the electric vector
  B is the magnetic induction
  ρ is the electric charge density
  j is the electric current density
  ε_0 is the *permittivity of free space*
  c is the speed of light (see Appendix K)

In addition to the Maxwell equations, the following identities are useful:

$$j = \sigma E \tag{15.5}$$

$$D = \varepsilon E \tag{15.6}$$

$$B = \mu H \tag{15.7}$$

where
  D is the electric displacement
  H is the magnetic vector
  σ is the specific conductivity
  ε is the dielectric constant (or permittivity)
  μ is the magnetic permeability

In the Gaussian systems of units, the Maxwell equations are given in the form of

$$\nabla \cdot B = 0 \tag{15.8}$$

$$\nabla \cdot E = 4\pi\rho \tag{15.9}$$

$$\nabla \times H = \frac{1}{c}\left(\frac{\partial D}{\partial t} + 4\pi\, j \right) \tag{15.10}$$

$$\nabla \times E = -\frac{1}{c}\frac{\partial B}{\partial t} \tag{15.11}$$

(see, e.g., Born and Wolf, 1999). It should be noted that many authors in the field of optics prefer to use the Maxwell equations in the Gaussian system of units. As explained by Born and Wolf (1999) in this system, E, D, j, and ρ are measured in electrostatic units while H and B are measured in electromagnetic units.

For the case of no charges or currents, that is, $j = 0$ and $\rho = 0$, and a homogeneous medium, the Maxwell equations and the given identities can be applied in conjunction with the vector identity

$$\nabla \times \nabla \times E = \nabla \nabla \cdot E - \nabla^2 E \tag{15.12}$$

to obtain wave equations of the form (Born and Wolf, 1999)

$$\nabla^2 E - \frac{\varepsilon \mu}{c^2} \frac{\partial^2 E}{\partial t^2} = 0 \tag{15.13}$$

This leads to an expression for the velocity of propagation

$$v = c(\varepsilon \mu)^{-1/2} \tag{15.14}$$

Comparison of this expression with the law of positive refraction, derived in Chapter 5, leads to what is known as the Maxwell formula (Born and Wolf, 1999)

$$n = (\varepsilon \mu)^{1/2} \tag{15.15}$$

where n is the refractive index. It is useful to note that in vacuum

$$c^2 = (\varepsilon_0 \mu_0)^{-1} \tag{15.16}$$

in the rationalized metric system, where μ_0 is the *permeability of free space* (Lorrain and Corson, 1970). The values of fundamental constants are listed in Appendix K.

15.3 Polarization and Reflection

Following the convention described by Born and Wolf (1999), we consider a reflection boundary, depicted in Figure 15.1, and a plane of incidence established by the incidence ray the and the normal to the reflection surface. Here, the reflected component $\mathcal{R}_{\parallel}$ is parallel to the plane of incidence, and the reflected component $\mathcal{R}_{\perp}$ is perpendicular to the plane of incidence.

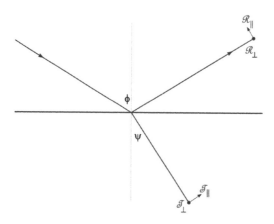

FIGURE 15.1
Reflection boundary defining the plane of incidence.

For the case of $\mu_1 = \mu_2 = 1$, Born and Wolf (1999) consider the electric, and magnetic, vectors as complex plane waves. In this approach, the incident electric vector is represented by equations of the form

$$E_x^{(i)} = -A_\| \cos\phi \left(e^{-i\tau_i} \right)$$ (15.17)

$$E_y^{(i)} = -A_\perp \left(e^{-i\tau_i} \right)$$ (15.18)

$$E_z^{(i)} = -A_\| \sin\phi \left(e^{-i\tau_i} \right)$$ (15.19)

where
$A_\|$ and $A_\perp$ are complex amplitudes
τ_i is the usual plane wave phase factor

Using corresponding equations for E and H for transmission and reflection in conjunction with the Maxwell relation, with $\mu = 1$, and the law or positive refraction, the Fresnel formulae can be derived (Born and Wolf, 1999). Using the Fresnel formulae, the transmissivity and reflectivity, for both polarizations, can be expressed as

$$\mathcal{T}_\| = \left(\frac{(\sin 2\phi \sin 2\psi)}{\sin^2(\phi+\psi)\cos^2(\phi-\psi)} \right)$$ (15.20)

$$\mathcal{T}_\perp = \left(\frac{(\sin 2\phi \sin 2\psi)}{\sin^2(\phi+\psi)} \right)$$ (15.21)

$$\mathcal{R}_{\parallel} = \left(\frac{\tan^2(\phi - \psi)}{\tan^2(\phi + \psi)} \right) \tag{15.22}$$

$$\mathcal{R}_{\perp} = \left(\frac{\sin^2(\phi - \psi)}{\sin^2(\phi + \psi)} \right) \tag{15.23}$$

and

$$\mathcal{R}_{\parallel} + \mathcal{T}_{\parallel} = 1 \tag{15.24}$$

$$\mathcal{R}_{\perp} + \mathcal{T}_{\perp} = 1 \tag{15.25}$$

Using these expressions for transmissivity and reflectivity, the degree of polarization, $\mathcal{P}$, is defined as (Born and Wolf, 1999)

$$\mathcal{P} = \frac{(\mathcal{R}_{\parallel} - \mathcal{R}_{\perp})}{(\mathcal{R}_{\parallel} + \mathcal{R}_{\perp})} \tag{15.26}$$

The usefulness of these equations is self-evident once $\mathcal{R}_{\parallel}$ is calculated, as a function of angle of incidence (Figure 15.2), for fused silica at $\lambda \approx 590$ nm ($n = 1.4583$). Here we see that $\mathcal{R}_{\parallel} = 0$ at 55.5604°. At this angle $(\phi + \psi)$ becomes 90° so that $\tan(\phi + \psi)$ approaches infinity thus causing $\mathcal{R}_{\parallel} = 0$. This particular ϕ is known as the *Brewster angle* (ϕ_B) and has a very important role in laser optics. At $\phi = \phi_B$ the angle of refraction becomes $\psi = (90 - \phi)$ degrees and the law of refraction takes the form of

$$\tan \phi_B = n \tag{15.27}$$

For orthogonal, or normal, incidence, the difference between the two polarizations vanishes. Using the law of positive refraction and the appropriate trigonometric identities, in Equations 15.20 through 15.23, it can be shown that (Born and Wolf, 1999)

$$\mathcal{R} = \left(\frac{(n-1)}{(n+1)} \right)^2 \tag{15.28}$$

and

$$\mathcal{T} = \left(\frac{4n}{(n+1)} \right)^2 \tag{15.29}$$

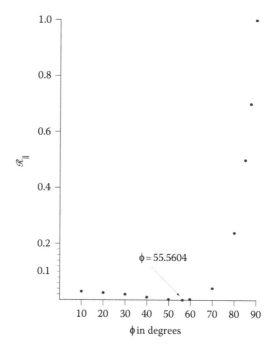

FIGURE 15.2
Reflection intensity as a function of angle of incidence. The angle at which the reflection vanishes is known as the Brewster angle.

15.3.1 Plane of Incidence

The discussion in the preceding section uses parameters such as $\mathcal{R}_{||}$ and $\mathcal{R}_{\perp}$. In this convention, $||$ means parallel to the plane of incidence and $\perp$ means perpendicular, orthogonal, or normal to the plane of incidence. The plane of incidence is defined, following Born and Wolf (1999), in Figure 15.1. However, on more explicit terms, let us consider a laser beam propagating on a plane parallel to the surface of an optical table. If that beam is made to illuminate the hypotenuse of a right-angle prism, whose triangular base is parallel to the surface of the table, then the plane of incidence is established by the incident laser beam and the perpendicular to the hypotenuse of the prism. In other words, in this case, the plane of incidence is parallel to the surface of the optical table.

Moreover, if that prism is allowed to expand the transmitted beam, as discussed later in this chapter, then the beam expansion is parallel to the plane of incidence.

The linear polarization of a laser can often be orthogonal to an external plane of incidence. When that is the case, and maximum transmission of the laser through external optics is desired, the laser is rotated by $\pi/2$ about its axis of propagation as will be discussed later in this chapter.

15.4 Jones Calculus

The Jones calculus is a matrix approach to describe, in a unified form, both linear and circular polarization. It was introduced by Jones (1947) and a good review of the subject is given by Robson (1974). Here, salient features of the Jones calculus are described without derivation.

A more general approach to express the electric field in complex terms in x and y coordinates is in vector form

$$\begin{pmatrix} E_{0x} \\ E_{0y} \end{pmatrix} = \begin{pmatrix} E_0 e^{i\phi_x} \\ E_0 e^{i\phi_y} \end{pmatrix} \tag{15.30}$$

In this notation, linear polarization in the x direction is represented by

$$\begin{pmatrix} E_{0x} \\ E_{0y} \end{pmatrix} = \begin{pmatrix} 1 \\ 0 \end{pmatrix} \tag{15.31}$$

while linear polarization in the y direction is described by

$$\begin{pmatrix} E_{0x} \\ E_{0y} \end{pmatrix} = \begin{pmatrix} 0 \\ 1 \end{pmatrix} \tag{15.32}$$

Subsequently,

$$\begin{pmatrix} E_{0x} \\ E_{0y} \end{pmatrix} = \frac{1}{\sqrt{2}} \begin{pmatrix} 1 \\ \pm 1 \end{pmatrix} \tag{15.33}$$

describes diagonal (or oblique) polarization at a $\pi/4$ angle, relative to the x axis (+), or relative to the y axis (−).

Circular polarization is described by the vector

$$\begin{pmatrix} E_{0x} \\ E_{0y} \end{pmatrix} = \frac{1}{\sqrt{2}} \begin{pmatrix} 1 \\ \pm i \end{pmatrix} \tag{15.34}$$

where
$+i$ applies to right circularly polarized light
$-i$ to left circularly polarized light (see Appendix E for a description of i)

Figure 15.3 illustrates the various polarization alternatives.

The Jones calculus introduces 2×2 matrices to describe optical elements transforming the polarization of the incidence radiation in the following format:

$$\begin{pmatrix} a_{11} & a_{12} \\ a_{21} & a_{22} \end{pmatrix} \begin{pmatrix} A_x \\ A_y \end{pmatrix} = \begin{pmatrix} a_{11}A_x + a_{12}A_y \\ a_{21}A_x + a_{22}A_y \end{pmatrix} \tag{15.35}$$

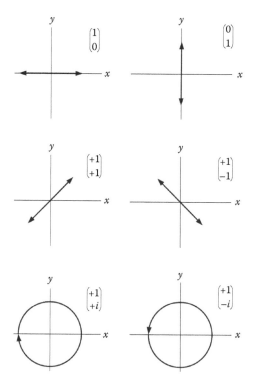

FIGURE 15.3
The various forms of polarization.

where the 2×2 matrix represents the optical element, the polarization vector multiplying this matrix corresponds to the incident radiation, and the resulting vector describes the polarization of the resulting radiation.

Useful Jones matrices include the matrix for transmission of linearly polarized light in the x direction

$$\begin{pmatrix} a_{11} & a_{12} \\ a_{21} & a_{22} \end{pmatrix} = \begin{pmatrix} 1 & 0 \\ 0 & 0 \end{pmatrix} \tag{15.36}$$

and the y direction

$$\begin{pmatrix} a_{11} & a_{12} \\ a_{21} & a_{22} \end{pmatrix} = \begin{pmatrix} 0 & 0 \\ 0 & 1 \end{pmatrix} \tag{15.37}$$

For light linearly polarized at a $\pi/4$ angle, the matrix becomes

$$\begin{pmatrix} a_{11} & a_{12} \\ a_{21} & a_{22} \end{pmatrix} = \frac{1}{2}\begin{pmatrix} 1 & 1 \\ 1 & 1 \end{pmatrix} \tag{15.38}$$

The generalized polarization Jones matrix for linearly polarized light, at an angle θ to the x axis, is given by

$$\begin{pmatrix} a_{11} & a_{12} \\ a_{21} & a_{22} \end{pmatrix} = \begin{pmatrix} \cos^2\theta & \sin\theta\cos\theta \\ \sin\theta\cos\theta & \sin^2\theta \end{pmatrix} \tag{15.39}$$

The right circular polarizer is described as

$$\begin{pmatrix} a_{11} & a_{12} \\ a_{21} & a_{22} \end{pmatrix} = \frac{1}{2}\begin{pmatrix} 1 & -i \\ i & 1 \end{pmatrix} \tag{15.40}$$

while the left circular polarizer is described as

$$\begin{pmatrix} a_{11} & a_{12} \\ a_{21} & a_{22} \end{pmatrix} = \frac{1}{2}\begin{pmatrix} 1 & i \\ -i & 1 \end{pmatrix} \tag{15.41}$$

In Chapter 16 we see how these matrices can be obtained from the density matrix formalism.

The generalized rotation matrix for birefringent rotators is given by (Robson, 1974)

$$\begin{pmatrix} a_{11} & a_{12} \\ a_{21} & a_{22} \end{pmatrix} = \begin{pmatrix} \cos\theta & \sin\theta \\ -e^{i\delta}\sin\theta & e^{i\delta}\cos\theta \end{pmatrix} \tag{15.42}$$

where
  δ is the phase angle
  α is the rotation angle about the z axis

For a quarter-wave plate, $\delta = \pi/2$, the rotation matrix becomes

$$\begin{pmatrix} a_{11} & a_{12} \\ a_{21} & a_{22} \end{pmatrix} = \begin{pmatrix} \cos\theta & \sin\theta \\ -i\sin\theta & i\cos\theta \end{pmatrix} \tag{15.43}$$

15.4.1 Example

A laser beam polarized in the x direction is sent through an optical element that allows the transmission of y polarization only; thus, using Equations 15.31 and 15.37

$$\begin{pmatrix} 0 & 0 \\ 0 & 1 \end{pmatrix}\begin{pmatrix} 1 \\ 0 \end{pmatrix} = \begin{pmatrix} 0 \\ 0 \end{pmatrix} \tag{15.44}$$

we find that no light is transmitted as can be demonstrated by a simple experiment.

15.5 Polarizing Prisms

There are two avenues to induce polarization using prisms. The first involves simple reflection as characterized by the equations of reflectivity and straightforward refraction.

  This approach is valid for windows, prisms, or multiple-prism arrays, made from homogeneous optical materials such as optical glass or fused silica. The second approach involves double refraction in crystalline transmission media exhibiting birefringence.

15.5.1 Transmission Efficiency in Multiple-Prism Arrays

For a generalized multiple-prism array, as shown in Figure 15.4, the cumulative reflection losses at the incidence surface of the mth prism are given by (Duarte et al., 1990)

$$L_{1,m} = L_{2,(m-1)} + (1 - L_{2,(m-1)})\mathcal{R}_{1,m} \tag{15.45}$$

while the losses at the mth exit surface are given by

$$L_{2,m} = L_{1,m} + (1 - L_{1,m})\mathcal{R}_{2,m} \tag{15.46}$$

where $\mathcal{R}_{1,m}$ and $\mathcal{R}_{2,m}$ are given by either $\mathcal{R}_{||}$ or $\mathcal{R}_{\perp}$. In practice, the optics is deployed so that the polarization of the propagation beam is parallel to the plane of incidence meaning that the reflection coefficient is given by $\mathcal{R}_{||}$. It should be noted that these equations apply not just to prisms but also to optical wedges and any homogeneous optical element, with an input and exit surface, used in the transmission domain.

15.5.2 Induced Polarization in a Double-Prism Beam Expander

Polarization induction in multiple-prism beam expanders should be apparent once the reflectivity equations are combined with the transmission Equations 15.45 and 15.46.

  In this section, this effect is made clear by considering the transmission efficiency, for both components of polarization, of a simple double-prism beam expander as illustrated in Figure 15.5. This beam expander is a modified version of one described by Duarte (2003) and consists of two identical prisms made of fused silica, with $n = 1.4583$ at $\lambda \approx 590$ nm, and an apex angles of $42.7098°$. Both prism are deployed to yield identical magnifications and for orthogonal beam exit. This implies that

$\phi_{1,1} = \phi_{1,2} = 81.55°, \quad \psi_{1,1} = \psi_{1,2} = 42.7098°, \quad \phi_{2,1} = \phi_{2,2} = 0°, \quad \psi_{2,1} = \psi_{2,2} = 0°.$

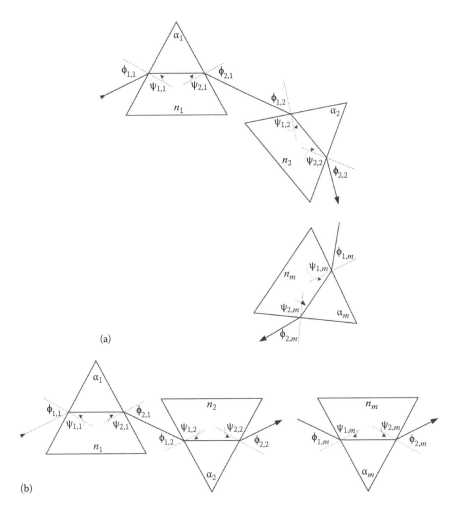

(a)

(b)

FIGURE 15.4
Generalized multiple-prism arrays. Depiction of these generalized prismatic arrays in (a) additive and (b) compensating configurations was introduced by Duarte and Piper (1983).

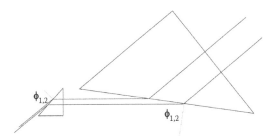

FIGURE 15.5
Double-prism expander as described in the text.

Thus for radiation polarized parallel to the plane of incidence

$$L_{1,1} = \mathscr{R}_{1,1} = 0.3008$$

$$L_{2,1} = L_{1,1}$$

$$L_{1,2} = L_{2,1} + (1 - L_{2,1})\mathscr{R}_{1,2} = 0.5111$$

$$L_{2,2} = L_{1,2}$$

while for radiation polarized perpendicular to the plane of incidence

$$L_{1,1} = \mathscr{R}_{1,1} = 0.5758$$

$$L_{2,1} = L_{1,1}$$

$$L_{1,2} = L_{2,1} + (1 - L_{2,1})\mathscr{R}_{1,2} = 0.8200$$

$$L_{2,2} = L_{1,2}$$

also

$$k_{1,1} = k_{1,2} = 5.0005$$

$$M = k_{1,1}k_{1,2} = 25.0045$$

Thus, for this particular beam expander, the cumulative reflection losses are 51.11% for light polarized parallel to the plane of incidence while they increase to 82.00% for radiation polarized perpendicular to the plane of incidence. This example helps to illustrate the fact that multiple-prism beam expanders exhibit a clear polarization preference. It is easy to see that the addition of further stages of beam magnification lead to increased discrimination. When incorporated in frequency-selective dispersive laser cavities, these beam expanders contribute significantly toward the emission of laser emission polarized parallel to the plane of propagation.

The reader should refer to Chapter 6 for a generalized description of multiple-prism dispersion. Chapter 9 describes the use of multiple-prism arrays in laser oscillators.

15.5.3 Double-Refraction Polarizers

These are crystalline prism pairs that exploit the birefringence effect in crystals. In birefringent materials, the dielectric constant, ε, is different in each of the x, y, and z directions so that the propagation velocity is different in each direction:

$$v_a = c(\varepsilon_x)^{-1/2} \tag{15.47}$$

$$v_b = c(\varepsilon_y)^{-1/2} \tag{15.48}$$

$$v_c = c(\varepsilon_z)^{-1/2} \tag{15.49}$$

Since polarization of a transmission medium is determined by the D vector, it is possible to describe the polarization characteristics in each direction. Further, it can be shown that there are two different velocities, for the refracted radiation, in any given direction (Born and Wolf, 1999). As a consequence of the law of refraction, these two velocities lead to two different propagation paths in the crystal and give origin to the *ordinary* and *extraordinary* ray. In other words, the two velocities lead to *double refraction*.

Of particular interest in this class of polarizers are those known as the Nicol prism, the Rochon prism, the Glan–Foucault prism, the Glan–Thompson prism, and the Wollaston prism. According to Bennett and Bennett (1978), a Glan–Foucault prism pair is an air-spaced Glan–Thompson prism pair. In Glan-type polarizers, the extraordinary ray is transmitted from the first to the second prism in the propagation direction of the incident beam. On the other hand, the diagonal surface of the two prisms is predetermined to induce total internal reflection for the ordinary ray (see Figure 15.6).

Glan-type polarizers are very useful since they can be oriented to discriminate in favor of either polarization component with negligible beam deviation. Normally these polarizers are made of either quartz or calcite. Commercially available calcite Glan–Thompson polarizer with a useful aperture of 10 mm provides extinction ratios of $\sim 5 \times 10^{-5}$. It should be noted that Glan-type polarizers are used in straightforward propagation applications

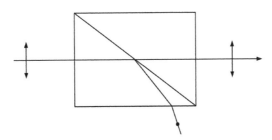

FIGURE 15.6
Generic Glan–Thompson polarizer. The beam polarized parallel to the plane of incidence is transmitted, while the complementary component is deviated (drawing not to scale). (For further details, refer to Jenkins F.A. and White H.E., *Fundamentals of Optics*, McGraw-Hill, New York, 1957.)

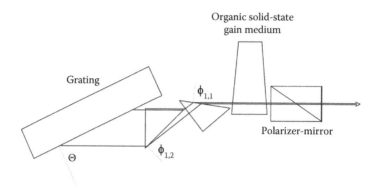

FIGURE 15.7
Solid-state multiple-prism Littrow (MPL) grating dye laser oscillator, yielding single-longitudinal-mode emission, incorporating a Glan–Thompson polarizer output coupler. The reflective coating is applied to the outer surface of the polarizer.

as well as intracavity elements. For instance, the tunable single-longitudinal-mode laser oscillator depicted in Figure 15.7 incorporates a Glan–Thompson polarizer as output coupler. In this particular polarizer, the inner window is antireflection coated while the outer window is coated for partial reflectivity to act as an output coupler mirror. The laser emission from multiple-prism grating oscillators is highly polarized parallel to the plane of incidence by the interaction of the intracavity flux with the multiple-prism expander and the grating. The function of the polarizer output coupler here is to provide further discrimination against unpolarized single-pass amplified spontaneous emission. These dispersive tunable laser oscillators yield extremely low levels of broadband amplified spontaneous emission measured to be in the 10^{-7}–10^{-6} range (Duarte, 1995, 1999).

The Wollaston prism, illustrated in Figure 15.8, is usually fabricated of either crystalline quartz or calcite. These prisms are assembled from two matched and complimentary right-angle prisms whose crystalline optical axes are oriented orthogonal to each other. The Wollaston prisms are widely used as beam splitters of beams with orthogonal polarizations. The beam separation provided by calcite is significantly greater than the beam

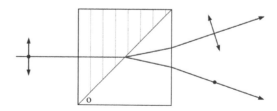

FIGURE 15.8
Generic Wollaston prism. The lines and circle represent the direction of the crystalline optical axis of the prism components (drawing not to scale).

separation achievable with crystalline quartz. Also, for both materials, the beam separation is wavelength dependent.

The use of these prisms in quantum cryptography optical configurations is outlined in Chapter 19. In those optical configurations, a Wollaston prism is used after an electro-optical polarization rotator (such as a Pockels cell) to spatially separate photons corresponding to orthogonal polarizations. For a description of electro-optical polarization rotators, see Saleh and Teich (1991).

15.5.4 Attenuation of the Intensity of Laser Beams Using Polarization

A very simple and yet powerful technique to attenuate the intensity of linearly polarized laser beams involves the transmission of the laser beam through a prism pair such as a Glan–Thompson polarizer followed by rotation of the polarizer. This technique is illustrated in Figure 15.9. In this technique for a ~100% laser beam polarized parallel to the plane of incidence, there is almost total transmission when the Glan–Thompson prism pair is oriented as in Figure 15.9a. As the prism pair is rotated about the axis of propagation, the intensity of the transmission decreases until it becomes zero once the angular displacement has reached $\pi/2$. With precision rotation of the prism pair, a scale of well-determined intensities can be easily obtained (Duarte, 2001). This has a number of applications including the generation of precise laser intensity scales for exposing instrumentation, and laser printers, used in imaging (Duarte, 2001). Also, this technique has been successfully applied to *laser cooling* experiments to independently vary the intensity of the cooling and repumping lasers as illustrated in Figure 15.10 (Olivares et al., 2009).

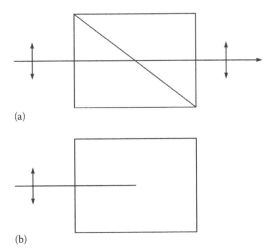

(a)

(b)

FIGURE 15.9

Attenuation of polarized laser beams using a Glan–Thompson polarizer. (a) Polarizer set for ~100% transmission. (b) Rotation of the polarizer, about the axis of propagation by $\pi/2$, yields ~0% transmission. The amount of transmitted light can be varied continuously by rotating the polarizer in the $0 \leq \theta \leq \pi/2$ range. (From Duarte, F.J., *US Patent* 6, 236, 461 B1, 2001.)

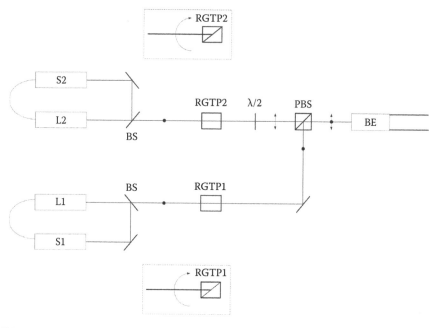

FIGURE 15.10

Top view of schematics of a laser cooling experiment including Glan–Thompson polarizers to independently control the laser intensity of the cooling lasers (L1) and the repumping laser (L2). S1 and S2 are stabilizer systems, RGTP1 and RGTP2 are the rotating Glan–Thompson polarizers, λ/2 is a half-wave plate, PBS is a polarizer beam splitter, and BE a beam expander. Polarizations perpendicular to the plane of propagation are indicated by the solid dot. Rotation by π/2 of an RGTP extinguishes the transmission of the laser beam that is polarized perpendicular to the plane of propagation. (Adapted from Olivares, I.E. et al., *J. Mod. Opt.* 56, 1780, 2009. With permission.)

15.6 Polarization Rotators

Maximum transmission efficiency is always a goal in optical systems. If the polarization of a laser is mismatched to the polarization preference of the optics, then transmission efficiency will be poor. Furthermore, efficiency can be significantly improved if the polarization of a pump laser is matched to the polarization preference of the laser being excited (Duarte, 1990). Although sometimes the efficiency can be improved, or even optimized, by the simple rotation of a laser, it is highly desirable and practical to have optical elements to perform this function. In this section we shall consider three alternatives to perform such rotation: birefringent polarization rotators and prismatic rotators. An additional alternative to rotate polarization are rhomboids (Figure 15.11) and are described in Duarte (2003).

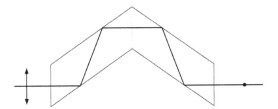

FIGURE 15.11
Side view of double Fresnel rhomb. Linearly polarized light is rotated by $\pi/2$ and exits polarized orthogonally to the original polarization.

15.6.1 Birefringent Polarization Rotators

In birefringent uniaxial crystalline materials, the ordinary and extraordinary rays propagate at different velocities. The generalized matrix for birefringent rotators is given by Equation 15.42

$$\begin{pmatrix} a_{11} & a_{12} \\ a_{21} & a_{22} \end{pmatrix} = \begin{pmatrix} \cos\theta & \sin\theta \\ -e^{i\delta}\sin\theta & e^{i\delta}\cos\theta \end{pmatrix}$$

For a quarter-wave plate $\delta = \pi/2$, the phase term is $e^{i\pi/2} = +i$, and the rotation matrix becomes

$$\begin{pmatrix} a_{11} & a_{12} \\ a_{21} & a_{22} \end{pmatrix} = \begin{pmatrix} \cos\theta & \sin\theta \\ -i\sin\theta & i\cos\theta \end{pmatrix} \tag{15.50}$$

For a half-wave plate $\delta = \pi$ and the phase term is $e^{i\pi} = -1$. Thus, the rotation matrix becomes

$$\begin{pmatrix} a_{11} & a_{12} \\ a_{21} & a_{22} \end{pmatrix} = \begin{pmatrix} \cos\theta & \sin\theta \\ \sin\theta & -\cos\theta \end{pmatrix} \tag{15.51}$$

From experiment we know that a half-wave plate causes a rotation of a linearly polarized beam by $\theta = \pi/2$ so that Equation 15.51 reduces to

$$\begin{pmatrix} a_{11} & a_{12} \\ a_{21} & a_{22} \end{pmatrix} = \begin{pmatrix} 0 & 1 \\ 1 & 0 \end{pmatrix} \tag{15.52}$$

15.6.1.1 Example

Thus, if we send a beam polarized in the x direction through a half-wave plate, the emerging beam polarization will be

$$\begin{pmatrix} 0 & 1 \\ 1 & 0 \end{pmatrix}\begin{pmatrix} 1 \\ 0 \end{pmatrix} = \begin{pmatrix} 0 \\ 1 \end{pmatrix} \tag{15.53}$$

which corresponds to a beam polarized in the y direction as observed experimentally and as illustrated in Figure 15.10 (Duarte, 2003; Olivares et al., 2009).

15.6.2 Broadband Prismatic Polarization Rotators

An alternative to frequency-selective polarization rotators are prismatic rotators (Duarte, 1989). These devices work at normal incidence and apply the principle of total internal reflection. The basic operation of polarization rotation, by $\pi/2$, due to total internal reflection is shown in Figure 15.12. This operation, however, reflects the beam into a direction that is orthogonal to the original propagation. Furthermore, the beam is not in the same plane. In order to achieve collinear polarization rotation, by $\pi/2$, the beam must be displaced upward and then be brought into alignment with the incident beam while conserving the polarization rotation achieved by the initial double-reflection operation. A collinear prismatic polarization rotator that performs this task using seven total internal reflections is depicted in Figure 15.13. For high-power laser applications, this rotator is best assembled using a high-precision mechanical mount that allows air interfaces between the individual prisms. The useful aperture in this rotator is about 10 mm and its physical length is 30 mm.

It should be noted that despite the apparent complexity of this collinear polarization rotator, the transmission efficiency is relatively high using antireflection coatings. In fact, using broadband (425–675 nm) antireflection coatings with a nominal loss of 0.5% per surface, the measured transmission efficiency becomes 94.7% at $\lambda = 632.8$ nm.

The predicted transmission losses using

$$L_r = 1-(1-L)^r \tag{15.54}$$

are 4.9%, with $L = 0.5\%$, as compared to a measured value of 5.3%. Equation 15.54 is derived combining Equations 15.45 and 15.46 for the special case of identical reflection losses. Here, r is the total number of reflection surfaces. For this particular collinear rotator, $r = 10$. A further parameter of interest is the transmission fidelity of the rotator since it is also important to keep spatial distortions of the rotated beam to a minimum.

The integrity of the beam due to transmission and rotation is quantified in Figure 15.14 where a very slight beam expansion, of ~3.2% at FWHM, is evident (Duarte, 1992).

15.6.2.1 Example

The $\pi/2$ prismatic polarization rotator just described rotates linearly x polarized radiation into linearly y polarized radiation and vice versa. Here we will find the Jones matrix that describes its rotational capability. Considering first the case of $x \to y$ and using the Jones matrix formalism, we can write

$$\begin{pmatrix} a_{11} & a_{12} \\ a_{21} & a_{22} \end{pmatrix}\begin{pmatrix} 1 \\ 0 \end{pmatrix} = \begin{pmatrix} 1a_{11}+0a_{12} \\ 1a_{21}+0a_{22} \end{pmatrix} = \begin{pmatrix} 0 \\ 1 \end{pmatrix} \tag{15.55}$$

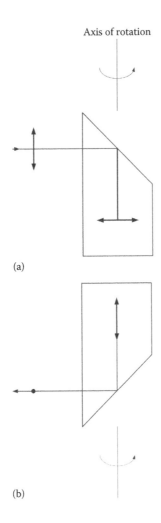

Axis of rotation

(a)

(b)

FIGURE 15.12
Basic prism operator for polarization rotation using two reflections. This can be composed of two 45° prisms adjoined $\pi/2$ to each other (note that it is also manufactured as one piece). (a) Side view of the rotator illustrating the basic rotation operation due to one reflection. The beam with the rotated polarization exits the prism into the plane of the figure. (b) The prism rotator is itself rotated anticlockwise by $\pi/2$ about the rotation axis (as indicated) thus providing an alternative perspective of the operation: the beam is now incident into the plane of the figure and it is reflected downward with is polarization rotated by $\pi/2$ relative to the original orientation. (From Duarte, F.J., Optical device for rotating the polarization of a light beam, *US Patent* 4822150, 1989.)

which means that $a_{11} = 0$ and $a_{21} = 1$. To find the other two components, we use the complementary rotation $y \rightarrow x$ that can be described as

$$\begin{pmatrix} a_{11} & a_{12} \\ a_{21} & a_{22} \end{pmatrix} \begin{pmatrix} 0 \\ 1 \end{pmatrix} = \begin{pmatrix} 0a_{11} + 1a_{12} \\ 0a_{21} + 1a_{22} \end{pmatrix} = \begin{pmatrix} 1 \\ 0 \end{pmatrix} \qquad (15.56)$$

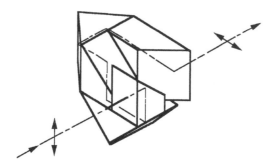

FIGURE 15.13
Broadband collinear prism polarization rotator. (From Duarte, F.J., Optical device for rotating the polarization of a light beam, *US Patent* 4822150, 1989.)

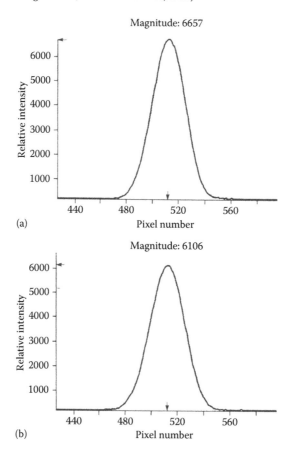

FIGURE 15.14
Transmission fidelity of the broadband collinear polarization rotator: (a) intensity profile of incident beam, prior to rotation, and (b) intensity profile of transmitted beam with rotated polarization. (Reproduced from Duarte, F.J., *Appl. Opt.* 31, 3377, 1992, with permission from the Optical Society of America.)

which implies that $a_{12} = 1$ and $a_{22} = 0$. Thus, the Jones matrix for $\pi/2$ rotation, that applies directly to the prismatic rotator described in Figure 15.13, becomes

$$R = \begin{pmatrix} a_{11} & a_{12} \\ a_{21} & a_{22} \end{pmatrix} = \begin{pmatrix} 0 & 1 \\ 1 & 0 \end{pmatrix} \tag{15.57}$$

Thus, we have again arrived to the $\pi/2$ rotation matrix Equation 15.52, by observation, using simple linear algebra.

Problems

15.1 Design a single right-angle prism, made of fused silica, to expand a laser beam by a factor of two with orthogonal beam exit. Calculate $R_{\parallel}$ and $R_{\perp}$ (use $n = 1.4583$ at $\lambda \approx 590$ nm).

15.2 For a four-prism beam expander, with orthogonal beam exit, using fused silica prisms with an apex angle of 41°, calculate the overall beam magnification factor M.

   Also, calculate and the overall transmission efficiency for a laser beam polarized parallel to the plane of incidence (use $n = 1.4583$ at $\lambda \approx 590$ nm).

15.3 Use the Maxwell equations in the Gaussian system, for the $j = 0$ and $\rho = 0$ case, to derive the wave equations

$$\nabla^2 E - (\varepsilon\mu)c^{-2} \frac{\partial^2 E}{\partial t^2} = 0$$

$$\nabla^2 H - (\varepsilon\mu)c^{-2} \frac{\partial^2 H}{\partial t^2} = 0$$

15.4 If a linearly polarized beam in the x direction is sent through a rotator plate represented by the matrix

$$\begin{pmatrix} a_{11} & a_{12} \\ a_{21} & a_{22} \end{pmatrix} = \begin{pmatrix} 0 & 1 \\ -i & 0 \end{pmatrix}$$

What will be the polarization of the transmitted beam? What kind of plate would that be?

References

Bennett, J. M. and Bennett, H. E. (1978). Polarization. In *Handbook of Optics* (Driscoll, W. G. and Vaughan, W., eds.). McGraw-Hill, New York.

Born, M. and Wolf, E. (1999). *Principles of Optics*, 7th edn. Cambridge University, Cambridge, U.K.

Duarte, F. J. (1989). Optical device for rotating the polarization of a light beam. *US Patent* 4,822,150.

Duarte, F. J. (1990). Technology of pulsed dye lasers. In *Dye Laser Principles* (Duarte, F. J. and Hillman, L. W., eds.). Academic, New York, Chapter 6.

Duarte, F. J. (1992). Beam transmission characteristics of a collinear polarization rotator. *Appl. Opt.* **31**, 3377–3378.

Duarte, F. J. (1995). Solid-state dispersive dye laser oscillator: Very compact cavity. *Opt. Commun.* **117**, 480–484.

Duarte, F. J. (1999). Multiple-prism grating solid-state dye laser oscillator: Optimized architecture. *Appl. Opt.* **38**, 6347–6349.

Duarte, F. J. (2001). Laser sensitometer using a multiple-prism beam expander and a Polarizer. *US Patent* 6, 236, 461 B1.

Duarte, F. J. (2003). *Tunable Laser Optics.* Elsevier-Academic, New York.

Duarte, F. J., Ehrlich, J. J., Davenport, W. E., and Taylor, T. S. (1990). Flashlamp pumped narrow-linewidth dispersive dye laser oscillators: Very low amplified spontaneous emission levels and reduction of linewidth instabilities. *Appl. Opt.* **29**, 3176–3179.

Duarte, F. J. and Piper, J. A. (1983). Generalized prism dispersion theory. *Am. J. Phys.* **51**, 1132–1134.

Dyson, F. J. (1990). Feynman's proof of Maxwell equations. *Am. J. Phys.* **58**, 209–211.

Feynman, R. P., Leighton, R. B., and Sands, M. (1965). *The Feynman Lectures on Physics*, Vol. II. Addison-Wesley, Reading, MA.

Jenkins, F. A. and White, H. E. (1957). *Fundamentals of Optics*, McGraw-Hill, New York.

Jones, R. C. (1947). A new calculus for the treatment of optical systems, *J. Opt. Soc. Am.* **37**, 107–110.

Lorrain, P. and Corson, D. (1970). *Electromagnetic Fields and Waves.* Freeman, San Francisco, CA.

Olivares, I. E., Cuadra, J. A., Aguilar, F. A., Aguirre Gomez, J. G., and Duarte, F. J. (2009). Optical method using rotating Glan-Thompson polarizers to independently vary the power of the excitation and repumping lasers in laser cooling experiments. *J. Mod. Opt.* **56**, 1780–1784.

Robson, B. A. (1974). *The Theory of Polarization Phenomena*, Clarendon, Oxford, U.K.

Saleh, B. E. A. and Teich, M. C. (1991). *Fundamentals of Photonics.* Wiley, New York.

16

Quantum Polarization

16.1 Introduction

In this chapter we examine the quantum aspects of polarization primarily via the Dirac notation (Dirac, 1978) and also using density matrices. The approach follows the style of Feynman (Feynman et al., 1965). Classical polarization is examined in Chapter 15.

16.2 Linear Polarization

Linear polarization in the x direction represented in the Jones calculus by (Jones, 1947)

$$\begin{pmatrix} E_{0x} \\ E_{0y} \end{pmatrix} = \begin{pmatrix} 1 \\ 0 \end{pmatrix} \tag{16.1}$$

is expressed simply as $|x\rangle$ in the *bra–ket* representation. Linear polarization in the y direction described in the Jones calculus by

$$\begin{pmatrix} E_{0x} \\ E_{0y} \end{pmatrix} = \begin{pmatrix} 0 \\ 1 \end{pmatrix} \tag{16.2}$$

is expressed simply as $|y\rangle$ in the *bra–ket* representation.

Rotation of axes, $x \to x'$ and $y \to y'$, as illustrated in Figure 16.1, leads directly to the following rotation relations:

$$\frac{x}{x'} = \cos\theta \tag{16.3}$$

$$\frac{x}{y'} = \sin\theta \tag{16.4}$$

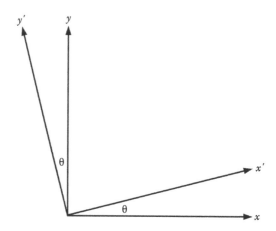

FIGURE 16.1
x and y axes and rotating x' and y' axes.

$$\frac{y}{x'} = -\sin\theta \tag{16.5}$$

$$\frac{y}{y'} = \cos\theta \tag{16.6}$$

which can be expressed in matrix form as

$$R = \begin{pmatrix} \cos\theta & \sin\theta \\ -\sin\theta & \cos\theta \end{pmatrix} \tag{16.7}$$

Equation 16.7 is consistent with the generalized rotation matrix introduced by Robson (1974):

$$\begin{pmatrix} a_{11} & a_{12} \\ a_{21} & a_{22} \end{pmatrix} = \begin{pmatrix} \cos\theta & \sin\theta \\ -e^{i\delta}\sin\theta & e^{i\delta}\cos\theta \end{pmatrix} \tag{16.8}$$

which includes additional phase terms.

From the axes transformations described in Equations 16.3 through 16.6, we can write

$$\langle x \,|\, x'\rangle = \cos\theta \tag{16.9}$$

$$\langle y \,|\, x'\rangle = \sin\theta \tag{16.10}$$

$$\langle x \,|\, y'\rangle = -\sin\theta \tag{16.11}$$

$$\langle y \,|\, y'\rangle = \cos\theta \tag{16.12}$$

Thus, the probability amplitude $\langle x|x'\rangle$ can be expanded into

$$\langle x|x'\rangle = \langle x|x\rangle\langle x|x'\rangle + \langle x|y\rangle\langle y|x'\rangle \qquad (16.13)$$

$$\langle x|x'\rangle = \cos\theta\langle x|x\rangle + \sin\theta\langle x|y\rangle \qquad (16.14)$$

and in abstract form it becomes as given by Feynman (Feynman et al., 1965)

$$|x'\rangle = \cos\theta|x\rangle + \sin\theta|y\rangle \qquad (16.15)$$

Since $\langle x|x\rangle = 1$ and $\langle x|y\rangle = 0$, then the probability is

$$|\langle x|x'\rangle|^2 = \cos^2\theta \qquad (16.16)$$

In abstract form, the complete set of transformation identities can be written as

$$|x'\rangle = \cos\theta|x\rangle + \sin\theta|y\rangle \qquad (16.17)$$

$$|y'\rangle = \cos\theta|y\rangle - \sin\theta|x\rangle \qquad (16.18)$$

$$|x\rangle = \cos\theta|x'\rangle + \sin\theta|y'\rangle \qquad (16.19)$$

$$|y\rangle = \cos\theta|y'\rangle - \sin\theta|x'\rangle \qquad (16.20)$$

16.2.1 Example

Consider the polarization configuration described in Figure 16.2: a source s is followed by a polarizer deployed to allow y polarization only and is followed by a polarizer deployed to allow x polarization only. The light is then detected at detector d. The probability amplitude of this transmission configuration can be described as

$$\langle d|s\rangle = \langle d|x\rangle\langle x|y\rangle\langle y|s\rangle \qquad (16.21)$$

and assuming $\langle y|s\rangle = \langle d|x\rangle = 1$, the probability amplitude reduces to $\langle x|y\rangle$

$$\langle d|s\rangle = \langle x|y\rangle \qquad (16.22)$$

which, by definition, is $\langle x|y\rangle = 0$, so that

$$\langle d|s\rangle = 0 \qquad (16.23)$$

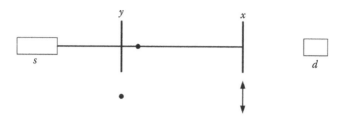

FIGURE 16.2
Top view of polarization configuration including a polarizer set for *y* transmission followed by a polarized set for transmission in the *x* direction. No light reaches the detector *d*. The *y* direction is perpendicular to the plane of incidence and is indicated by a dot.

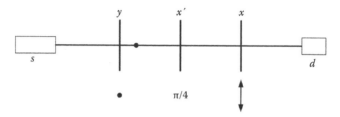

FIGURE 16.3
Top view of polarization configuration including a polarizer set for *y* transmission followed by a polarizer set for transmission in the *x* direction. A polarizer set for $\pi/4$ is inserted between the two polarizers. As explained in the text, some light is now allowed to reach the detector *d*.

If instead we now add a $\pi/4$ polarizer in between *x* and *y* polarizers, as illustrated in Figure 16.3, the probability amplitude becomes

$$\langle d \mid s \rangle = \langle d \mid x \rangle \langle x \mid x' \rangle \langle x' \mid y \rangle \langle y \mid s \rangle \tag{16.24}$$

Assuming, as before, that $\langle y|s \rangle = \langle d|x \rangle = 1$

$$\langle d \mid s \rangle = \langle x \mid x' \rangle \langle x' \mid y \rangle \tag{16.25}$$

and using Equation 16.9,

$$\langle d \mid s \rangle = \cos\theta \langle x' \mid y \rangle \tag{16.26}$$

Now, substituting Equation 16.20 for $\langle x'|y \rangle$ yields

$$\langle d \mid s \rangle = -\sin\theta \cos\theta \tag{16.27}$$

so that the probability of transmission at $\theta = \pi/4$ becomes

$$|\langle d \mid s \rangle|^2 = (-\sin\theta\cos\theta)^2 = \frac{1}{4} \tag{16.28}$$

16.3 Polarization as a Two-State System

From the solution to the two-state system described by (Feynman et al., 1965)

$$ i\hbar \frac{dC_i}{dt} = \sum H_{ij} C_j \tag{16.29} $$

where H_{ij} is the Hamiltonian, we obtain (see Chapter 8)

$$ C_{II} = \frac{1}{\sqrt{2}} (C_1 + C_2) \tag{16.30} $$

and

$$ C_I = \frac{1}{\sqrt{2}} (C_1 - C_2) \tag{16.31} $$

which eventually lead to probability amplitudes of the form

$$ |s\rangle = \frac{1}{\sqrt{2}} (|B\rangle + |A\rangle) \tag{16.32} $$

and

$$ |s\rangle = \frac{1}{\sqrt{2}} (|B\rangle - |A\rangle) \tag{16.33} $$

16.3.1 Diagonal Polarization

In Chapter 15 the classical vector representation for diagonally polarized light was introduces as

$$ \begin{pmatrix} E_{0x} \\ E_{0y} \end{pmatrix} = \frac{1}{\sqrt{2}} \begin{pmatrix} 1 \\ \pm 1 \end{pmatrix} \tag{16.34} $$

Linearly polarized light sustaining a $\pi/4$ angle relative to the x axis is referred to as diagonally polarized light and is described by the probability amplitude

$$ |D\rangle = \frac{1}{\sqrt{2}} (|x\rangle + |y\rangle) \tag{16.35} $$

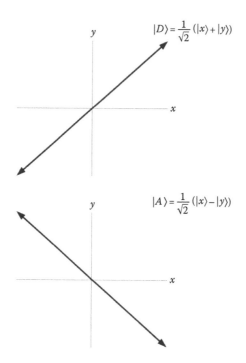

FIGURE 16.4
Diagonally polarized light depicting quantum notation.

If the angle is $-(\pi/4)$, this obliquely polarized light is represented by the probability amplitude

$$|A\rangle = \frac{1}{\sqrt{2}}\left(|x\rangle - |y\rangle\right) \tag{16.36}$$

Diagonally polarized light is described in Figure 16.4.

16.3.2 Circular Polarization

In Chapter 15 the classical vector representation for circularly polarized light was introduces as

$$\begin{pmatrix} E_{0x} \\ E_{0y} \end{pmatrix} = \frac{1}{\sqrt{2}}\begin{pmatrix} 1 \\ \pm i \end{pmatrix} \tag{16.37}$$

where the $+i$ factor applies to right-handed polarization and the $-i$ factor applies to left-handed polarization. Circularly polarized light is described in Figure 16.5.

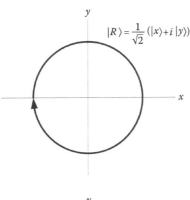

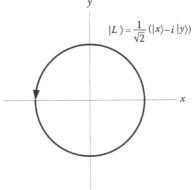

FIGURE 16.5
Circularly polarized light depicting quantum notation.

Using Equations 16.32 and 16.33, the probability amplitude representation for right-handed polarization ($|R\rangle$) and left-handed polarization ($|L\rangle$) becomes

$$|R\rangle = \frac{1}{\sqrt{2}}\left(|x\rangle + i|y\rangle\right) \tag{16.38}$$

$$|L\rangle = \frac{1}{\sqrt{2}}\left(|x\rangle - i|y\rangle\right) \tag{16.39}$$

Adding and subtracting Equations 16.38 and 16.39 yields

$$|x\rangle = \frac{1}{\sqrt{2}}\left(|R\rangle + |L\rangle\right) \tag{16.40}$$

$$|y\rangle = -\frac{i}{\sqrt{2}}\left(|R\rangle - |L\rangle\right) \tag{16.41}$$

Further, we can write

$$|R'\rangle = \frac{1}{\sqrt{2}}\left(|x'\rangle + i\,|y'\rangle\right) \tag{16.42}$$

from Equations 16.17 and 16.18, we get (Feynman et al., 1965)

$$|R'\rangle = \frac{1}{\sqrt{2}}\left(\cos\theta\,|x\rangle + \sin\theta\,|y\rangle + i\cos\theta\,|y\rangle - i\sin\theta\,|x\rangle\right) \tag{16.43}$$

$$|R'\rangle = \frac{1}{\sqrt{2}}\left((\cos\theta - i\sin\theta)(|x\rangle + i\,|y\rangle)\right) \tag{16.44}$$

$$|R'\rangle = \left(e^{-i\theta}\,|R\rangle\right) \tag{16.45}$$

so that

$$|L'\rangle = \left(e^{+i\theta}\,|R\rangle\right) \tag{16.46}$$

16.4 Density Matrix Notation

Here we return to the Jones calculus initially to consider the classical density matrix for polarization followed by the quantum description. The notation used is consistent with that of Robson (1974) in the matrix case and that of Feynman in the *bra–ket* approach.

The Jones calculus does not offer a direct representation for unpolarized or partially polarized light. However, Robson (1974) points out that a Jones vector of the form

$$\begin{pmatrix} a \\ be^{i\delta} \end{pmatrix} \tag{16.47}$$

generally describes polarized beams. If we define the earlier vector as a state $|J\rangle$

$$|J\rangle = \begin{pmatrix} a_1 \\ a_2 e^{i\delta} \end{pmatrix} \tag{16.48}$$

then, using the definition for the density matrix given in Chapter 14, we can write

$$|J\rangle\langle J| = \begin{pmatrix} a_1 \\ a_2 e^{i\delta} \end{pmatrix}\begin{pmatrix} a_1 & a_2 e^{-i\delta} \end{pmatrix} = \begin{pmatrix} a_1 a_1 & a_1 a_2 e^{-i\delta} \\ a_2 a_1 e^{i\delta} & a_2 a_2 \end{pmatrix} \tag{16.49}$$

and the resulting matrix is a 2 × 2 density matrix

$$\rho = \begin{pmatrix} \rho_{11} & \rho_{12} \\ \rho_{21} & \rho_{22} \end{pmatrix} = \begin{pmatrix} a_1 a_1 & a_1 a_2 e^{-i\delta} \\ a_2 a_1 e^{i\delta} & a_2 a_2 \end{pmatrix} \tag{16.50}$$

The trace of this matrix

$$Tr(\rho) = a_1 a_1 + a_2 a_2 \tag{16.51}$$

corresponds to the intensity of the beam and the off-diagonal terms provide information about the phase of the two components (Robson, 1974).

The *Stokes parameters* (see Appendix H) are defined in terms of combinations of the density matrix elements (Robson, 1974):

$$I = \rho_{11} + \rho_{22}$$

$$P_1 = \rho_{11} - \rho_{22} \tag{16.52}$$

$$P_2 = \rho_{12} + \rho_{21}$$

$$P_3 = i(\rho_{12} - \rho_{21})$$

Thus, for polarized light, the density matrix can be re-expressed as

$$\rho = \begin{pmatrix} \rho_{11} & \rho_{12} \\ \rho_{21} & \rho_{22} \end{pmatrix} = \frac{1}{2}\begin{pmatrix} I+P_1 & P_2-iP_3 \\ P_2+iP_3 & I-P_1 \end{pmatrix} \tag{16.53}$$

For unpolarized light $P_1 = P_2 = P_3 = 0$ so that the matrix reduces to

$$\rho_u = \frac{1}{2}\begin{pmatrix} I & 0 \\ 0 & I \end{pmatrix} \tag{16.54}$$

As seen in Chapter 14, we can write the $|J\rangle$ state in a general form, so that

$$|J\rangle\langle J| = \begin{pmatrix} j_1 \\ j_2 \end{pmatrix}\begin{pmatrix} j_1^* & j_2^* \end{pmatrix} = \begin{pmatrix} j_1 j_1^* & j_1 j_2^* \\ j_2 j_1^* & j_2 j_2^* \end{pmatrix} = \begin{pmatrix} \rho_{11} & \rho_{12} \\ \rho_{21} & \rho_{22} \end{pmatrix} = \rho_q \tag{16.55}$$

The trace of this matrix

$$Tr(\rho_q) = |j_1|^2 + |j_2|^2 \tag{16.56}$$

gives information on the relative intensities of the two linearly polarized and orthogonal components. If we define $j_1 = |x\rangle$ and $j_2 = |y\rangle$, then the trace of the matrix provides information on the relative probabilities of finding a photon in the in the $|x\rangle$ and $|y\rangle$ polarization states (Robson, 1974).

16.4.1 Stokes Parameters and Pauli Matrices

Previously in Chapter 14, we found that the identity matrix and the x, y, and z Pauli matrices are given by

$$I = \begin{pmatrix} 1 & 0 \\ 0 & 1 \end{pmatrix}$$

(16.57)

$$\sigma_x = \begin{pmatrix} 0 & 1 \\ 1 & 0 \end{pmatrix}$$

(16.58)

$$\sigma_y = \begin{pmatrix} 0 & -i \\ i & 0 \end{pmatrix}$$

(16.59)

$$\sigma_z = \begin{pmatrix} 1 & 0 \\ 0 & -1 \end{pmatrix}$$

(16.60)

Multiplication of the density matrix with these matrices yields

$$\rho I = \begin{pmatrix} \rho_{11} & \rho_{12} \\ \rho_{21} & \rho_{22} \end{pmatrix} \begin{pmatrix} 1 & 0 \\ 0 & 1 \end{pmatrix} = \begin{pmatrix} \rho_{11} & \rho_{12} \\ \rho_{21} & \rho_{22} \end{pmatrix}$$

(16.61)

$$\rho \sigma_x = \begin{pmatrix} \rho_{11} & \rho_{12} \\ \rho_{21} & \rho_{22} \end{pmatrix} \begin{pmatrix} 0 & 1 \\ 1 & 0 \end{pmatrix} = \begin{pmatrix} \rho_{12} & \rho_{11} \\ \rho_{22} & \rho_{21} \end{pmatrix}$$

(16.62)

$$\rho \sigma_y = \begin{pmatrix} \rho_{11} & \rho_{12} \\ \rho_{21} & \rho_{22} \end{pmatrix} \begin{pmatrix} 0 & -i \\ i & 0 \end{pmatrix} = \begin{pmatrix} i\rho_{12} & -i\rho_{11} \\ i\rho_{22} & -i\rho_{21} \end{pmatrix}$$

(16.63)

$$\rho \sigma_z = \begin{pmatrix} \rho_{11} & \rho_{12} \\ \rho_{21} & \rho_{22} \end{pmatrix} \begin{pmatrix} 1 & 0 \\ 0 & -1 \end{pmatrix} = \begin{pmatrix} \rho_{11} & -\rho_{12} \\ \rho_{21} & -\rho_{22} \end{pmatrix}$$

(16.64)

Computing the trace of these matrices leads directly to

$$Tr(\rho I) = \rho_{11} + \rho_{22} = I$$

(16.65)

$$Tr(\rho \sigma_z) = \rho_{11} - \rho_{22} = P_1$$

(16.66)

$$Tr(\rho\sigma_x) = \rho_{12} + \rho_{21} = P_2 \tag{16.67}$$

$$Tr(\rho\sigma_y) = i(\rho_{12} - \rho_{21}) = P_3 \tag{16.68}$$

Due to the Stoke nomenclature, Robson (1974) uses a different notation (see Appendix H):

$$\sigma_x \rightarrow \hat{\sigma}_2 = \begin{pmatrix} 0 & 1 \\ 1 & 0 \end{pmatrix} \tag{16.69}$$

$$\sigma_y \rightarrow \hat{\sigma}_3 = \begin{pmatrix} 0 & -i \\ i & 0 \end{pmatrix} \tag{16.70}$$

$$\sigma_z \rightarrow \hat{\sigma}_1 = \begin{pmatrix} 1 & 0 \\ 0 & -1 \end{pmatrix} \tag{16.71}$$

Here, Robson (1974) points out that while the σ_x, σ_y, and σ_z matrices refer to ordinary space, the matrices $\hat{\sigma}_1$, $\hat{\sigma}_2$, and $\hat{\sigma}_3$ correspond to the Poincaré space that defines P_1, P_2, and P_3 (see Appendix H). Also, notice the use of the operator hat to designate these matrices. Introducing the operator $\hat{\gamma}$ to designate the identity matrix, the set of relations just described becomes (Robson, 1974)

$$Tr(\rho\hat{\gamma}) = \rho_{11} + \rho_{22} = I \tag{16.72}$$

$$Tr(\rho\hat{\sigma}_1) = \rho_{11} - \rho_{22} = P_1 \tag{16.73}$$

$$Tr(\rho\hat{\sigma}_2) = \rho_{12} + \rho_{21} = P_2 \tag{16.74}$$

$$Tr(\rho\hat{\sigma}_3) = i(\rho_{12} - \rho_{21}) = P_3 \tag{16.75}$$

Albeit most of the applications of Pauli matrices are found in the context of spin one-half particles (Fermions), such as electrons, some applications are found for spin one particles (Bosons) such as the photon. For instance, in Chapter 14 we discovered that the matrix describing a $\pi/2$ rotator of linear polarization, for either a half-wave plate or a prismatic rotator, is given by

$$R = \begin{pmatrix} 0 & 1 \\ 1 & 0 \end{pmatrix}$$

This matrix is identical to either σ_x or $\hat{\sigma}_2$. This means that $\hat{\sigma}_2$ is a rotation operator of linear polarization by $\theta = \pi/2$. In general, $\hat{\sigma}_1$ and $\hat{\sigma}_2$ and are used in the description of linearly polarized photons, while $\hat{\sigma}_3$ is used in the description of circularly polarized photons.

16.4.2 Density Matrix and Circular Polarization

Using the probability amplitude representation for right-handed polarization ($|R\rangle$) and left-handed polarization ($|L\rangle$) given previously, we can write

$$|R\rangle = \frac{1}{\sqrt{2}}\left(|x\rangle + i|y\rangle\right)$$

$$|L\rangle = \frac{1}{\sqrt{2}}\left(|x\rangle - i|y\rangle\right)$$

which can be expressed in vector form as

$$|R\rangle = \frac{1}{\sqrt{2}}\begin{pmatrix} 1 \\ +i \end{pmatrix} \tag{16.76}$$

$$|L\rangle = \frac{1}{\sqrt{2}}\begin{pmatrix} 1 \\ -i \end{pmatrix} \tag{16.77}$$

Using these vectors the density matrix for right-handed polarization becomes (see Chapter 14)

$$|R\rangle\langle R| = \frac{1}{2}\begin{pmatrix} 1 \\ +i \end{pmatrix}\begin{pmatrix} 1 & -i \end{pmatrix} = \frac{1}{2}\begin{pmatrix} 1 & -i \\ +i & 1 \end{pmatrix} \tag{16.78}$$

and the density matrix for left-handed polarization becomes

$$|L\rangle\langle L| = \frac{1}{2}\begin{pmatrix} 1 \\ -i \end{pmatrix}\begin{pmatrix} 1 & +i \end{pmatrix} = \frac{1}{2}\begin{pmatrix} 1 & +i \\ -i & 1 \end{pmatrix} \tag{16.79}$$

16.4.3 Example

Setting $a_1 = a_2 = 2^{-1/2}$ and $\delta = \pi/2$, in Equation 16.50, gives us again

$$\begin{pmatrix} \rho_{11} & \rho_{12} \\ \rho_{21} & \rho_{22} \end{pmatrix} = \frac{1}{2}\begin{pmatrix} 1 & -i \\ +i & 1 \end{pmatrix} \tag{16.80}$$

which is the density matrix for right-handed polarized light.

Problems

16.1 Complete the set of axes transformations, that is, $\langle x'|x \rangle$, $\langle y'|x \rangle$, $\langle x'|y \rangle$, and $\langle y'|y \rangle$, and proceed to derive Equations 16.17 through 16.20.

16.2 Circular polarization: using the same approach outlined in Equations 16.42 through 16.45, derive the expression for $|L'\rangle$ given in Equation 16.46.

16.3 Use the definition of the Stokes parameters given in Equations 16.52 to express the polarization density matrix given in Equation 16.53.

16.4 For the density matrix applicable to unpolarized light, Equation 16.54, find the values of ρ_{11} and ρ_{22} so that $Tr(\rho) = 1$.

16.5 Set $a_1 = a_2 = 2^{-1/2}$ and $\delta = -\pi/2$, in Equation 16.50, to obtain the density matrix for left-handed polarized light.

References

Dirac, P. A. M. (1978). *The Principles of Quantum Mechanics*, 4th edn. Oxford, London, U.K.

Feynman, R. P., Leighton, R. B, and Sands, M. (1965). *The Feynman Lectures on Physics*, Vol. III. Addison-Wesley, Reading, MA.

Jones, R. C. (1947). A new calculus for the treatment of optical systems, *J. Opt. Soc. Am.* **37**, 107–110.

Robson, B. A. (1974). *The Theory of Polarization Phenomena*, Clarendon, Oxford, U.K.

17

Entangled Polarizations: Probability Amplitudes and Experimental Configurations

17.1 Introduction

In this chapter we derive the ubiquitous equation for the probability amplitude of polarization entanglement of two photons moving away in different directions, from a common source. The first derivation is performed from a Hamiltonian approach using the two-state infrastructure taught to us by Feynman. The second derivation utilizes an interferometric approach, while the third approach is based on the original analysis used by Ward (1949). All these approaches rely on the Dirac notation, exclusively.

17.2 Hamiltonian Approach

First, we review some of the relevant notation: following Feynman (Feynman et al., 1965) in the treatment of a two-state system (see Chapter 8), we can define

$$C_{II} = C_1 + C_2 \tag{17.1}$$

where

$$C_1 = \langle 1 \mid \phi \rangle \tag{17.2}$$

$$C_2 = \langle 2 \mid \phi \rangle \tag{17.3}$$

and the amplitude of finding $|\phi\rangle$ in a new state $|II\rangle$ is given by

$$C_{II} = \langle II \,|\, \phi\rangle = \langle 2 \,|\, \phi\rangle + \langle 1 \,|\, \phi\rangle \tag{17.4}$$

$$\langle II \,| = \langle 2 \,| + \langle 1 \,| \tag{17.5}$$

which is equivalent to (Feynman et al., 1965)

$$|II\rangle = |2\rangle + |1\rangle \tag{17.6}$$

The amplitude for state $|II\rangle$ to be in $|1\rangle$ is

$$\langle 1 \,|\, II\rangle = \langle 1 \,|\, 1\rangle + \langle 1 \,|\, 2\rangle = 1 \tag{17.7}$$

since $|1\rangle$ and $|2\rangle$ are base states. Also, from the basic principle

$$\langle i \,|\, j\rangle = \delta_{ij} \tag{17.8}$$

$$\langle I \,|\, I\rangle = \langle II \,|\, II\rangle = 1 \tag{17.9}$$

$$\langle II \,|\, I\rangle = \langle I \,|\, II\rangle = 0 \tag{17.10}$$

The dynamics of a two-state system can be described using (Feynman et al., 1965)

$$i\hbar \frac{dC_i}{dt} = \sum H_{ij} C_j \tag{17.11}$$

where H_{ij} is the Hamiltonian. Since the resulting probability has a maximum value of one, we have

$$|C_1|^2 + |C_2|^2 = 1 \tag{17.12}$$

Now, going back to the principle

$$\langle \chi \,|\, \phi\rangle = \sum_i \langle \chi \,|\, i\rangle\langle i \,|\, \phi\rangle \tag{17.13}$$

and setting $\chi = \phi = II$, for a two-state system, we get

$$\langle II \,|\, II\rangle = \langle II \,|\, 1\rangle\langle 1 \,|\, II\rangle + \langle II \,|\, 2\rangle\langle 2 \,|\, II\rangle \tag{17.14}$$

$$\langle II \,|\, II\rangle = \langle II \,|\, 1\rangle\langle II \,|\, 1\rangle^* + \langle II \,|\, 2\rangle\langle II \,|\, 2\rangle^* \tag{17.15}$$

$$\langle II \,|\, II\rangle = C_I C_I^* + C_{II} C_{II}^* \tag{17.16}$$

$$\langle II \,|\, II\rangle = |C_I|^2 + |C_{II}|^2 \tag{17.17}$$

Redefining

$$C_{II} = \frac{1}{\sqrt{2}}(C_1 + C_2) \qquad (17.18)$$

$$C_I = \frac{1}{\sqrt{2}}(C_1 - C_2) \qquad (17.19)$$

and substituting Equations 17.18 and 17.19 into (17.17), it can be verified that

$$1 = C_1^2 + C_2^2 \qquad (17.20)$$

Thus, for a two-state system, using Equations 17.2 and 17.3, we can have

$$\langle II | \phi \rangle = \frac{1}{\sqrt{2}}(\langle 1 | \phi \rangle + \langle 2 | \phi \rangle) \qquad (17.21)$$

$$\langle I | \phi \rangle = \frac{1}{\sqrt{2}}(\langle 1 | \phi \rangle - \langle 2 | \phi \rangle) \qquad (17.22)$$

Following abstraction, Equation 17.22 can be restated as

$$\langle I | = \frac{1}{\sqrt{2}}(\langle 1 | - \langle 2 |) \qquad (17.23)$$

which can also be written as

$$| I \rangle = \frac{1}{\sqrt{2}}(| 1 \rangle - | 2 \rangle) \qquad (17.24)$$

The probability amplitude given in Equation 17.24 can be reexpressed as

$$| s \rangle = \frac{1}{\sqrt{2}}(| B \rangle - | A \rangle) \qquad (17.25)$$

Now, for an assembly of particles, Dirac (1978) expresses the *ket* for the assembly as

$$| X \rangle = | a_1 \rangle | b_2 \rangle | c_3 \rangle ... | g_n \rangle \qquad (17.26)$$

In this nomenclature, the numerical subscripts (1, 2, 3 ...) refer to different individual particles. Thus, for two photons with different polarizations, we can have

$$| B \rangle = | x_1 \rangle | y_2 \rangle \qquad (17.27)$$

where x refers to one polarization and y to the alternative polarization, so that

$$|s\rangle = \frac{1}{\sqrt{2}}\left(|x\rangle_1 |y\rangle_2 - |x\rangle_2 |y\rangle_1\right) \tag{17.28}$$

and using the identity $|\phi\rangle|\psi\rangle = |\psi\rangle|\phi\rangle$

$$|s_-\rangle = \frac{1}{\sqrt{2}}\left(|x\rangle_1 |y\rangle_2 - |y\rangle_1 |x\rangle_2\right) \tag{17.29}$$

where $|s\rangle$ is replaced by $|s_-\rangle$ to highlight the sign inside the parentheses. This is the probability amplitude equation widely used, in the literature, to describe the polarization entanglement of two photons moving in opposite directions.

The same approach, starting on Equation 17.21, leads to

$$|s_+\rangle = \frac{1}{\sqrt{2}}\left(|x\rangle_1 |y\rangle_2 + |y\rangle_1 |x\rangle_2\right) \tag{17.30}$$

In Equations 17.29 and 17.30, pairs of photons representing different polarizations are represented. These probability amplitudes, describing entangled pairs of photons with different polarizations, are the result of applying the Dirac identity for the representation of a series of different particles $|X\rangle = |a_1\rangle |b_2\rangle |c_3\rangle \ldots |g_n\rangle$ that for two photons with different polarizations reduces to $|B\rangle = |x_1\rangle |y_2\rangle$. If the pair of entangled photons have the same polarization, we can have $|B\rangle = |x_1\rangle |x_2\rangle$ or $|B\rangle = |y_1\rangle |y_2\rangle$. Thus, an additional set of linear combinations for the probability amplitude are

$$|s_-\rangle = \frac{1}{\sqrt{2}}\left(|x\rangle_1 |x\rangle_2 - |y\rangle_1 |y\rangle_2\right) \tag{17.31}$$

$$|s_+\rangle = \frac{1}{\sqrt{2}}\left(|x\rangle_1 |x\rangle_2 + |y\rangle_1 |y\rangle_2\right) \tag{17.32}$$

17.2.1 Example

One further note: if instead of $|X\rangle = |a_1\rangle |b_2\rangle |c_3\rangle \ldots |g_n\rangle$ we just have

$$|X\rangle = |a\rangle |b\rangle |c\rangle \ldots |g\rangle \tag{17.33}$$

then Equation 17.29 becomes

$$|s_-\rangle = \frac{1}{\sqrt{2}}\left(|x\rangle |y\rangle - |y\rangle |x\rangle\right) \tag{17.34}$$

which, prior to normalization, can just be written as

$$|s_-\rangle = \left(|x\rangle|y\rangle - |y\rangle|x\rangle\right)$$ (17.35)

that is the original result obtained by Ward (1949).

17.3 Interferometric Approach

Consider the two-slit interference experiment as illustrated in Figure 17.1. The probability amplitude for a photon to propagate from the source s to the detector d, via apertures 1 and 2, is given by $\langle d|s\rangle$:

$$\langle d|s\rangle = \langle d|2\rangle\langle 2|s\rangle + \langle d|1\rangle\langle 1|s\rangle$$ (17.36)

With that background in mind, let us consider the geometry of an experiment with a central photon source emitting toward *identical detectors* ($d_1 = d_2 = d$), in opposite directions, via polarization analyzers p_1 and p_2, as illustrated in Figure 17.2. The corresponding probability amplitude can be described as

$$\langle d|s\rangle = \langle d|p_2\rangle\langle p_2|s\rangle + \langle d|p_1\rangle\langle p_1|s\rangle$$ (17.37)

which can be expressed in abstract form as

$$|s\rangle = |p_2\rangle\langle p_2|s\rangle + |p_1\rangle\langle p_1|s\rangle$$ (17.38)

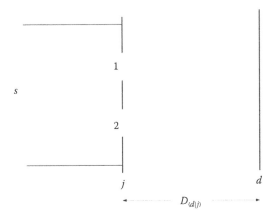

FIGURE 17.1
Schematics of double-slit interference experiment also known as two-slit interference experiment and Young's interference experiment.

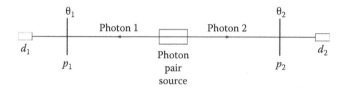

FIGURE 17.2
Basic two-photon polarization entanglement geometry assuming identical detectors (d_1 = $d_2 = d$). The θ_1 and θ_2 angles indicate the angular mobility of the respective polarizers designated as p_1 and p_2.

Now, using the identity $|\phi\rangle = |j\rangle\langle j|\phi\rangle$, we can write $|s\rangle = |p_1\rangle\langle p_1|s\rangle$. However, to differentiate between the two probability amplitudes, on the right-hand side, we write $|A\rangle = |p_1\rangle\langle p_1|s\rangle$ and $|B\rangle = |p_2\rangle\langle p_2|s\rangle$ so that

$$|s\rangle = |B\rangle + |A\rangle \tag{17.39}$$

which, using $|X\rangle = |a_1\rangle\,|b_2\rangle\,|c_3\rangle\,...|g_n\rangle$, allows us to write

$$|s\rangle = |x\rangle_1\,|y\rangle_2 + |y\rangle_1\,|x\rangle_2 \tag{17.40}$$

Once normalized, this probability amplitude becomes

$$|s_+\rangle = \frac{1}{\sqrt{2}}(|x\rangle_1\,|y\rangle_2 + |y\rangle_1\,|x\rangle_2) \tag{17.41}$$

and its linear combination is

$$|s_-\rangle = \frac{1}{\sqrt{2}}(|x\rangle_1\,|y\rangle_2 - |y\rangle_1\,|x\rangle_2) \tag{17.42}$$

which are the ubiquitous probability amplitude equations associated with counterpropagating photons with entangled polarizations (Duncan and Kleinpoppen, 1988; Mandel and Wolf, 1995).

17.4 Pryce–Ward–Snyder Probability Amplitude of Entanglement

The initial link between quantum mechanical concepts and the polarization correlation of photons propagating in opposite directions was given by Wheeler (1946): "According to the pair theory, if one of these photons is polarized in one plane, then the photon that goes off in the opposite direction with equal momentum is linearly polarized in the perpendicular plane." This fundamental idea, expressed by Wheeler, is crucial to the concept of quantum entanglement. The *pair theory* is the theory of electron–positron pairs due to Dirac (1930).

Next, a description of the Ward derivation of the entangled polarizations probability amplitude is given based on a critical review of this subject by Duarte (2012): indeed, Ward (1949) uses Wheeler's initial work on the positron–electron annihilation

$$e^+e^- \rightarrow \gamma\gamma \tag{17.43}$$

to produce counterpropagating correlated quanta as the inspiration of his work. In his thesis, the young Ward explains that Wheeler did attempt to calculate this effect *but* "through the neglect of interference terms he derived an incorrect, and in fact, far too small value for the angular correlations of the scattered quanta" (Ward, 1949).

Next we briefly describe Ward's quantum argument included in his doctoral thesis (Ward, 1949), which was used to derive the correlation equation published by Pryce and Ward (1947). First, Ward considers the polarization alternatives for the x and y polarization axes related to two counterpropagating photons. These are

$$|x,x\rangle, |x,y\rangle, |y,x\rangle, |y,y\rangle \tag{17.44}$$

Since the first coordinate refers to photon 1 and the second coordinate refers to photon 2, we can write

$$|x_1,x_2\rangle, |x_1,y_2\rangle, |y_1,x_2\rangle, |y_1,y_2\rangle$$

Next, Ward focuses on the momenta alternatives depicted in Figure 17.3 and writes (Ward, 1949)

$$|+k,-k\rangle, |-k,+k\rangle \tag{17.45}$$

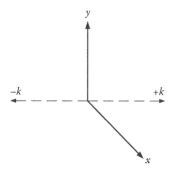

FIGURE 17.3
Momenta coordinates applicable to the PW entanglement experimental configuration.

Following a remarkable discussion that includes antisymmetrical single states, in both polarizations, and the importance of the condition of zero angular momentum, he arrives at (Ward, 1949)

$$\left(|x_1,y_2\rangle-|y_1,x_2\rangle\right)\left(|+k,-k\rangle-|-k,+k\rangle\right) \tag{17.46}$$

Ward then provides an extensive argument that selects the equation above as the likely alternative for the correct physics (Duarte, 2012). Focusing on the polarization component exclusively, we have

$$\left(|x_1,y_2\rangle-|y_1,x_2\rangle\right) \tag{17.47}$$

Using the identities given by Dirac (1978) ($|x\rangle|y\rangle = |x, y\rangle$), this can be written as

$$\left(|x\rangle_1|y\rangle_2-|y\rangle_1|x\rangle_2\right) \tag{17.48}$$

and following normalization, we get

$$\frac{1}{\sqrt{2}}\left(|x\rangle_1|y\rangle_2-|y\rangle_1|x\rangle_2\right) \tag{17.49}$$

and its linear combination is

$$\frac{1}{\sqrt{2}}\left(|x\rangle_1|y\rangle_2+|y\rangle_1|x\rangle_2\right) \tag{17.50}$$

for quanta propagating in the opposite directions as depicted in Figure 17.4. These are the widely used equations to describe the probability amplitude of

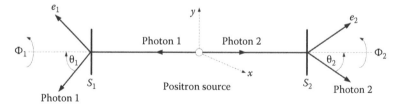

FIGURE 17.4
PW experiment for two-photon entangled polarizations. Photon 1 and photon 2 are emitted in opposite directions along the z axis, from a single source. The photons undergo Compton scattering at S_1 and S_2 thus being scattered at angles θ_1 and θ_2. The essence of this experiment consists in the following: (1) the emission of entangled photons in opposite directions, (2) the propagation of these photons in opposite directions, (3) angular selectivity at each of the propagation paths, and (4) irreversible detection at the end of each propagation path. In this regard, this experimental configuration is equivalent to that described in Figure 17.2.

entangled polarizations of quanta propagating in opposite directions (Bohm and Aharonov, 1957; Duncan and Kleinpoppen, 1988). Since probability amplitude equations of this form were also derived independently by Snyder et al. (1948), it has been proposed that they should be named Pryce–Ward–Snyder (PWS) probability amplitudes (Duarte, 2012).

In his thesis, Ward (1949) uses these probability amplitudes as an initial step in the calculation that eventually yielded the quantum ratio for perpendicular polarization over parallel polarization counting rate (Pryce and Ward, 1947).

17.5 Pryce–Ward–Snyder Probability

In order to evaluate numerically the corresponding measurable, which is the corresponding probability, we use the PWS probability amplitude

$$|s\rangle = \frac{1}{\sqrt{2}}(|x\rangle_1 |y\rangle_2 - |y\rangle_1 |x\rangle_2) \tag{17.51}$$

in conjunction with the geometrical identities involved in rotating some generic polarization axis from x to x' and so on (see Chapter 16):

$$\langle x' | x\rangle = \cos\theta \tag{17.52}$$

$$\langle y' | x\rangle = \sin\theta \tag{17.53}$$

$$\langle x' | y\rangle = -\sin\theta \tag{17.54}$$

$$\langle y' | y\rangle = \cos\theta \tag{17.55}$$

Thus, using Equation 17.51, the probability amplitude $\langle x'|s\rangle$ becomes

$$\langle x' | s\rangle = \frac{1}{\sqrt{2}}\left(\langle x' | x\rangle_1 \langle x' | y\rangle_2 - \langle x' | y\rangle_1 \langle x' | x\rangle_2\right) \tag{17.56}$$

and the corresponding probability is

$$|\langle x' | s\rangle|^2 = \frac{1}{2}\left(\langle x' | x\rangle_1 \langle x' | y\rangle_2 - \langle x' | y\rangle_1 \langle x' | x\rangle_2\right)^2 \tag{17.57}$$

Substituting the corresponding trigonometric terms into Equation 17.57, we get

$$|\langle x' | s \rangle|^2 = \frac{1}{2}(\sin \theta_1 \cos \theta_2 - \cos \theta_1 \sin \theta_2)^2$$

$$|\langle x' | s \rangle|^2 = \frac{1}{2}\sin^2(\theta_1 - \theta_2) \qquad (17.58)$$

which allow us to evaluate numerically the probability depending on the settings θ_1 and θ_2. The PWS probability is applied to evaluate Bell's inequality in Chapter 21.

17.6 Pryce–Ward Experimental Arrangement

The first experimental configuration including a central source emitting two correlated quanta in opposite directions is due to Pryce and Ward (1947) and is depicted in Figure 17.4. The two counterpropagating quanta undergo Compton scattering at S_1 and S_2 that causes the respective photons to be scattered at angles θ_1 and θ_2. The essence of this experiment consists in (1) the emission of entangled photons in opposite directions, (2) the propagation of these photons in opposite directions, (3) angular selectivity at each propagation path, and (4) irreversible detection at the end of each propagation path.

Duarte (2012) goes into a detailed discussion on the origin of the ideas that culminated in the PW configuration. In this regard, the first written, nondiagrammatic, description of this type of experiment was provided by Wheeler (1946), while Pryce and Ward (1947) also mention a proposal by R. C. Hanna. However, all participants appear to refer to Dirac himself as the source of the seminal idea (Duarte, 2012). Certainly, Snyder et al. (1948) also disclosed an experimental diagram that they based on Wheeler's suggestion.

17.7 Wu–Shaknov Experiment

Following the publications of Pryce and Ward (1947) and Snyder et al. (1948), Wu and Shaknov (1950) reported experimental results on the maximum polarization ratio of the two counterpropagating quanta (2.04 ± 0.08) that was only ~2% higher than the theoretical value, as per the PW theory. Other experimental efforts were those of Hanna (1948) and Bleuler and

Bradt (1948). The literature as presented here led Dalitz and Duarte (2000) to state explicitly that the correct quantum theory for entangled quantum traveling in opposite directions was already known in 1947 and already confirmed by experiment by 1950.

17.7.1 Relevance of the Pryce–Ward Theory and the Wu–Shaknov Experiment to EPR

Here, we quote directly from relevant publications on the importance of the PWS theory and the Wu and Shaknov (WS) experiment to the Einstein, Podolsky, and Rosen (1935) (EPR) paradox (see Chapter 21).

First, in their original paper, Wu and Shaknov (1950) refer to the theory behind their experiment: "As early as 1946 J. A. Wheeler proposed an experiment to verify a prediction of pair theory, that the two quanta emitted in the annihilation of a positron–electron pair, with zero angular momentum, are polarized at right angles to each other... The detailed theoretical investigations were reported by Pryce and Ward and Snyder et al." This is a clear and explicit statement providing a firm nexus between the PWS theory and the WS experiment.

Next is the link between the WS experiment and the EPR paradox. In this regard, Bohn and Aharanov (1957) write in reference to the WS experiment: "Thus, the paradox of EPR can equally well be tested by polarization properties of pair of photons." Via simple deduction, if the experiment of Wu and Shaknov is considered as an EPR-type experiment, equally relevant should be the theory and experimental scheme disclosed by Pryce and Ward (Duarte, 2012).

The Bohm and Aharanov (1957) opinion on the WS experiment was questioned by some authors (Peres and Singer, 1960). However, this criticism was explicitly rejected by these authors: "In a previous paper (Bohm and Aharanov, 1957) we have discussed the paradox of Einstein, Podolsky, and Rosen (1935), and we have shown that the Wu- Shaknov experiment (Wu and Shaknov, 1950)... provides an experimental confirmation of the features of quantum mechanisms which are the basis of the above paradox" (Bohm and Aharanov, 1960).

Following the publication of the famous Bell (1964) paper, other authors argued that the WS experiment did not produce "evidence against local hidden-variable theories" given the "use of Compton polarimeters" (Clauser et al., 1969). Wu and colleagues responded that "even though a Compton experiment cannot rule out hidden-variable theories, it can provide strong evidence against them" (Kasday et al., 1975). Dalitz and Duarte (2000) also argue that the PWS theory and the WS experiment provide evidence against local realism.

From a historical perspective, it should be mentioned that the 1957 Bohm and Aharanov paper became the inspiration of researchers interested in optical experiments on entangled polarizations (see, e.g., Aspect et al. 1982).

17.8 Conclusion

The theory of entangled quantum polarizations, central to EPR-type optical experiments, was established in the 1947–1949 period by Pryce and Ward (Pryce and Ward 1947; Ward 1949) and independently by Snyder et al. (1948). Thus, it would be proper to name the expression

$$| s \rangle = \frac{1}{\sqrt{2}} \left(| x \rangle_1 | y \rangle_2 - | y \rangle_1 | x \rangle_2 \right)$$

as the PW probability amplitude or the PWS probability amplitude. Here, we have established that in addition to Ward's original derivation, it is possible to arrive at this equation using the Hamiltonian approach and a direct interferometric approach.

On a broader perspective, the evidence presented here tends to indicate that the physics of quantum entanglement, as initiated by Dirac, discussed by Wheeler, and resolved by Pryce and Ward, would still be here even in the apparent absence of interpretational questions (see Chapter 21). In other words, the physics of polarization entanglement was established as a purely quantum physics result independent of interpretational efforts.

This is an observation of fundamental importance that is not widely appreciated.

Problems

17.1 Verify that substituting Equations 17.18 and 17.19 into 17.17 yields Equation 17.20.

17.2 Show that Equation 17.35 in its normalized version becomes

$$| s \rangle = \frac{1}{\sqrt{2}} \left(| x \rangle_1 | y \rangle_2 - | y \rangle_1 | x \rangle_2 \right)$$

17.3 Show that substitution of Equations 17.52 and 17.54 into Equation 17.56 leads to the PWS probability given in Equation 17.58

17.4 Using Equation 17.51, in conjunction with Equations 17.53 and 17.55, find the probability amplitude $\langle y' | s \rangle$.

17.5 For $\theta_1 = \pi/3$ and $\theta_2 = \pi/6$, evaluate the PWS probability using Equation 17.58.

References

Aspect, A., Dalibard, J., and Roger, G. (1982). Experimental test of Bell's inequalities using time-varying analyzers. *Phys. Rev. Lett.* **49**, 1804–1807.

Bell, J. S. (1964). On the Einstein-Podolsky-Rosen paradox. *Physics* **1**, 195–200.

Bleuler, E. and Bradt, H. L. (1948). Correlation between the states of polarization of the two quanta of annihilation radiation. *Phys. Rev.* **73**, 1398.

Bohm, D. and Aharonov, Y. (1957). Discussion of experimental proof for the paradox of Einstein, Rosen, and Podolsky. *Phys. Rev.* **108**, 1070–1076.

Bohm, D. and Aharonov, Y. (1960). Further discussion of possible experimental tests for the paradox of Einstein, Podolsky and Rosen. *Nouvo Cimento* **17**, 964–976.

Clauser, J. F., Horne, M. A., Shimony, A., and Holt, R. A. (1969). Proposed experiment to test hidden variable theories. *Phys. Rev. Lett.* **23**, 880–884.

Dalitz, R. H. and Duarte, F. J. (2000). John Clive Ward. *Phys. Today* **53**(10), 99–100.

Dirac, P. A. M. (1930). On the annihilation of electron and protons. *Camb. Phil. Soc.* **26**, 361–375.

Dirac, P. A. M. (1978). *The Principles of Quantum Mechanics*, 4th edn. Oxford, London, U.K.

Duarte, F. J. (2012). The origin of quantum entanglement experiments based on polarization measurements. *Eur. Phys. J. H* **37**, 311–318.

Duncan, A. J. and Kleinpoppen, H. (1988). The experimental investigation of the Einstein-Podolsky-Rosen question and Bell's inequality. In *Quantum Mechanics Versus Local Realism* (Selleri, F., ed.), Plenum, New York. Chapter 7.

Einstein, A., Podolsky, B., and Rosen, N. (1935). Can quantum mechanical description of physical reality be considered complete? *Phys. Rev.* **47**, 777–780.

Feynman, R. P., Leighton, R. B., and Sands, M. (1965). *The Feynman Lectures on Physics*, Vol. III, Addison-Wesley, Reading, MA.

Hanna, R. C. (1948). Polarization of annihilation radiation. *Nature* **162**, 332.

Kasday, L. R., Hullman, J. D., and Wu, C. S. (1975). Angular correlation of Compton-scattered annihilation photons and hidden variables. *Nouvo Cimento B* **17**, 633–661.

Mandel, L. and Wolf, E. (1995). *Optical Coherence and Quantum Optics*. Cambridge University, Cambridge, U.K.

Peres, A. and Singer, P. (1960). On possible experimental tests for the paradox of Einstein, Podolsky and Rosen. *Nuovo Cimento* **15**, 907–915.

Pryce, M. H. L. and Ward, J. C. (1947). Angular correlation effects with annihilation radiation. *Nature* **160**, 435.

Snyder, H. S., Pasternack, S., and Hornbostel, J. (1948). Angular correlation of scattered annihilation radiation. *Phys. Rev.* **73**, 440–448.

Ward, J. C. (1949). *Some Properties of the Elementary Particles*. D. Phil Thesis, Oxford University, Oxford, U.K.

Wheeler, J. A. (1946). Polyelectrons. *Ann. NY Acad. Sci.* **48**, 219–238.

Wu, C. S. and Shaknov, I. (1950). The angular correlation of scattered annihilation radiation. *Phys. Rev.* **77**, 136.

18

Quantum Computing

18.1 Introduction

Quantum computing is a very active field producing many publications and varied perspectives. Here, we have chosen, once again, to follow Feynman's path given that his contributions are among the earliest in the field and the transparency of his approach is unmatched. Thus, the material presented here is primarily based on Feynman's paper entitled *Quantum Mechanical Computers* (Feynman, 1985, 1986). However, further material including additional concepts, not included in the Feynman paper, is also presented.

In the discussion that follows, the term *universal computer* is applied to traditional transistor-based computers using Boolean algebra executed by logical gates such as AND, OR, NOT, NAND, and NOR. The beauty of universal computers is that by via the use of mathematical equations, expressed in the logical language of the computer, it can accurately simulate scientific processes of interest. Their scope of applications is extremely wide. A disadvantage of the universal computer is that, for certain classes of calculations, they can consume more energy than desired, and they can be slower than desired. From a practical perspective, the issue of computational speed is very important.

There is a cost and efficiency motivation to replace universal computers that take too much time to perform computationally intensive calculations. This stimulates interest in new computer technologies.

Physical computers, such as optical computers, can be extremely fast to perform certain computational tasks that demand relatively long computational times in universal computers; however, the range of applications they offer is limited. An example of this class of computer, an interferometric computer, is described in the next section.

A quantum computer, as the name suggests, is a computer that can operate at the quantum level thus offering enormous improvements in energy consumption and size reductions (Bennett, 1982; Feynman, 1985, 1986). Significant size reductions and the use of photons should also lead to vast improvements in computational speed. The concept of a quantum computer goes beyond physical computers and seeks to offer an alternative universal computer. It does so by utilizing analogs to bits known as *qubits*.

These qubits can take a physical representation at the quantum level and allow the performance of Boolean logic operations. Here an introduction to this fascinating subject is provided while using the mathematical tools and quantum concepts already introduced in this book.

18.2 Interferometric Computer

The *N*-slit laser interferometer (NSLI), as an example of a physical or optical computer, is discussed by Duarte (2003). A simplified diagram of the NSLI is shown in Figure 18.1.

Interferograms recorded with the NSLI have been compared for numerous geometrical and wavelength parameters with interferograms calculated via the interferometric equation

$$|\langle x\,|\,s\rangle|^2 = \sum_{j=1}^{N}\Psi(r_j)\sum_{m=1}^{N}\Psi(r_m)e^{i(\Omega_m-\Omega_j)} \tag{18.1}$$

or

$$|\langle x\,|\,s\rangle|^2 = \sum_{j=1}^{N}\Psi(r_j)^2 + 2\sum_{j=1}^{N}\Psi(r_j)\left(\sum_{m=j+1}^{N}\Psi(r_m)\cos(\Omega_m-\Omega_j)\right) \tag{18.2}$$

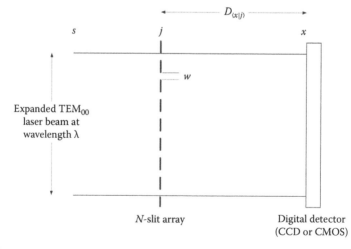

FIGURE 18.1
NSLI configuration. Critical parameters are the laser wavelength λ, the number of slits N, the dimension of the slits w, and the intra-interferometric distance $D_{(x|j)}$.

One such case is considered in Figure 18.2. The measured interferogram is displayed at the top of Figure 18.2 and the corresponding theoretical interferogram is included in the lower trace of Figure 18.2. In this regard, it should be mentioned that good agreement, between theory and experiment, exists from the near to the far field. Observed differences, especially at the baseline, are due to thermal noise in the digital detector that is used at room temperature. Further comparative aspects are discussed in Chapter 4.

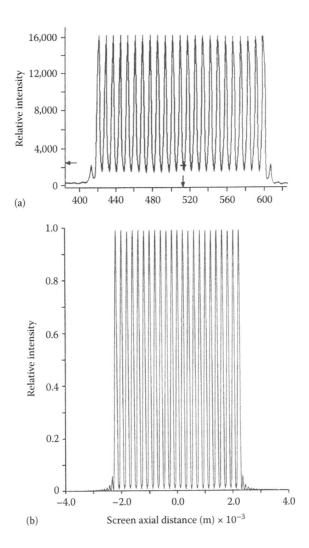

FIGURE 18.2
(a) Measured near-field interferogram using the NSLI for $N = 23$, $\lambda = 632.8$ nm, and $D_{(x|j)} = 1.5$ cm. Slits are 100 μm wide separated by 100 μm. (b) Calculated near-field interferogram for $N = 23$, $\lambda = 632.8$ nm, and $D_{(x|j)} = 1.5$ cm. Slits are 100 μm wide separated by 100 μm (Reproduced from Duarte, F.J., *Opt. Commun.* 103, 8, 1993, with permission from Elsevier.).

The interferometric calculations, using the interferometric equation to program a universal computer, require the following input information:

1. Slit dimensions: w
2. Standard of deviation of the slit dimensions: Δw
3. Interslit dimensions
4. Standard of deviation of interslit dimensions
5. Wavelength: λ
6. N-slit array, or grating, screen distance: $D_{\langle x|j\rangle}$
7. Number of slits: N

The Boolean algebra program, based on the interferometric equation, also gives options for the illumination profile and allows for multiple-stage calculations. That is, it allows for the propagation through several sequential N-slit arrays prior to arrival at x as considered in Chapter 4.

An interesting aspect of comparisons, between theory and experiment, is that for a given wavelength, set of slit dimensions, and distance from j to x, calculations in a conventional universal computer take longer as the number of slits N increases. In fact, the computational time $t(N)$ behaves in a nonlinear fashion as N increases. This is clearly illustrated in Figure 18.3 where $t = 0.96$ s for $N = 2$ and $t = 3111.2$ for $N = 1500$ s (Duarte, 1996). By contrast all of these calculations can be performed in the NSLI at a constant time of ~30 ms, which is a time mainly imposed by the integration time of the digital detector. Certainly, the generation and propagation of the interferogram is performed at speeds near the speed of light c.

In this regard, following the criteria outlined by Deutsch (1992), the NSLI can be classified as a physical, or interferometric, computer that can perform certain specific computations at times orders of magnitude below the computational time required by a universal computer. Among the computations that the *interferometric computer* can perform are

1. N-slit array interference calculations
2. Near- or far-field diffraction calculations
3. Beam divergence calculations
4. Wavelength calculations

For this limited set of tasks, the interferometric computer based on the NSLI outperforms, by orders of magnitude, universal computers. Hence, it can be classified as a very fast, albeit limited in scope, optical computer. The advantage of the universal computer remains its versatility and better signal to noise ratio. Also, in the universal computer, there is

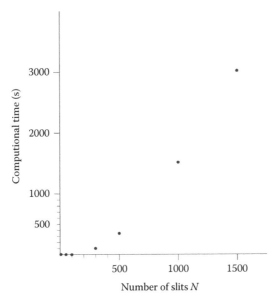

FIGURE 18.3
Computational time in a mainframe universal computer as a function of number of slits. The first four points are 0.96 s for $N = 2$, 1.14 s for $N = 10$, 7.03 s for $N = 50$, and 14.33 s for $N = 100$. For these calculations $\lambda = 632.8$ nm, $D_{(x|j)} = 75$ cm. Slits are 30 μm wide separated by 30 μm (Duarte, 1996).

access to intermediate results at all stages of the computation. This is not allowed in the NSLI where access is strictly limited to the input stage and the final stage of the computation. Attempts to acquire information about intermediate stages of the computation can destroy the final answer (see Chapter 11).

The interferometric computer is a macroscopic apparatus that allows the control of photon propagation via the probabilistic laws of quantum mechanics. The term quantum computer, however, applies to devices at the atomic scale that can be used to control the emission and/or propagation of quantum radiation obeying quantum logic rules.

18.3 Classical Logic Gates

Classical universal computers function based on logical operations performed by a series of gates such as the OR and AND gates and the negation gates NOT, NOR, and NAND.

Feynman, in his article, focuses in particular on the *primitive elements* NOT, AND, FAN OUT, and EXCHANGE. To this list he also adds the *reversible primitives* NOT, CONTROLLED NOT, and CONTROLED CONTROLLED NOT.

Figure 18.4 illustrates the NOT gate, Figure 18.5 the NAND, and Figure 18.6 the NOR gates with their corresponding transistor circuitry.

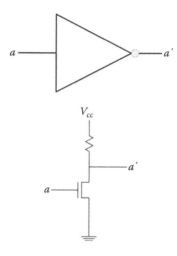

FIGURE 18.4
Symbol for NOT gate and transistor circuit of NOT gate using NMOS technology. NMOS technology refers to field effect transistors fabricated with n-type metal–oxide–semiconductors (MOS).

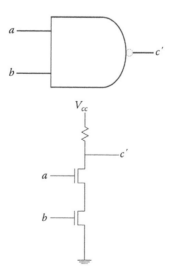

FIGURE 18.5
Symbol for NAND gate and transistor circuit of NAND gate using NMOS technology.

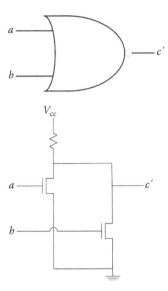

FIGURE 18.6
Symbol for NOR gate and transistor circuit of NOR gate using NMOS technology.

For an input a and an output a', the truth table for the NOT gate is

a	a'
0	1
1	0

For inputs a and b and output c', the truth table for the AND gate is

a	b	c'
0	0	0
0	1	0
1	0	0
1	1	1

For inputs a and b and output c', the truth table for the NAND gate is

a	b	c'
0	0	1
0	1	1
1	0	1
1	1	0

For inputs a and b and output c', the truth table for the OR gate is

a	b	c'
0	0	0
0	1	1
1	0	1
1	1	1

For inputs a and b and output c', the truth table for the NOR gate is

a	b	c'
0	0	1
0	1	0
1	0	0
1	1	0

Feynman pays particular attention to the CNOT gate. In this gate the value of the output b' is changed if and only if the value of $a = 1$, so that the truth table is

a	b	a'	b'
0	0	0	0
0	1	0	1
1	0	1	1
1	1	1	0

18.4 Qubits

In his paper, Feynman (1985, 1986) relates the concept of bit to the states $|1\rangle$ and $|0\rangle$ and goes on to explain that one bit can be represented by a single atom being in one of these two states. As we have learned before, any two states can be expressed as a linear combination forming a probability amplitude

$$|\psi\rangle = c_1|0\rangle + c_2|1\rangle \tag{18.3}$$

where c_1 and c_2 are *complex numbers*. States such as $|1\rangle$ and $|0\rangle$ are known as *qubit*, short for quantum bit, thus named in terminology introduced in 1995 (Schumacher, 1995). The normalized version of this probability amplitude becomes (see Chapter 17)

$$|\psi\rangle = \frac{1}{\sqrt{2}}\left(c_1\,|\,0\rangle + c_2\,|\,1\rangle\right) \tag{18.4}$$

The $|1\rangle$ and $|0\rangle$ states can also be entangled. In that case, as we saw in Chapter 17, the probability amplitude can be expressed via equations like

$$|\phi\rangle = \frac{1}{\sqrt{2}}\left(|\,0\rangle\,|\,1\rangle - |\,1\rangle\,|\,0\rangle\right) \tag{18.5}$$

For spin one-half particles, the $|1\rangle$ and $|0\rangle$ states can also be abbreviated as $|\uparrow\rangle$ and $|\downarrow\rangle$ so that the Pryce–Ward–Snyder (PWS) probability amplitude would read as

$$|\phi\rangle = \frac{1}{\sqrt{2}}\left(|\downarrow\rangle\,|\uparrow\rangle - |\uparrow\rangle\,|\downarrow\rangle\right) \tag{18.6}$$

which can also be written as

$$|\phi\rangle = \frac{1}{\sqrt{2}}\left(|\,0\rangle\,|\,1\rangle - |\,1\rangle\,|\,0\rangle\right) \tag{18.7}$$

or

$$|\phi\rangle = \frac{1}{\sqrt{2}}\left(|\,01\rangle - |\,10\rangle\right) \tag{18.8}$$

For more than two entangled states, we can have $|1010\rangle$ that is equivalent to $|\uparrow\downarrow\uparrow\downarrow\rangle$ and we can use the PWS probability amplitude to express linear combinations of these entangled states.

18.5 Quantum Logic

Feynman (1985) defines a 2 × 2 matrix that he calls A_a as

$$A_a = \begin{pmatrix} 0 & 1 \\ 1 & 0 \end{pmatrix} \tag{18.9}$$

and makes the observation that this matrix corresponds to the format of the NOT truth table. Previously, in Chapter 14, we had identified this matrix, which is also the σ_x Pauli matrix, with the matrix performing polarization rotation by $\pi/2$. Feynman then defines an operator matrix $\hat{a}$ that in atom a performs the operation $|1\rangle \rightarrow |0\rangle$ so that

$$\hat{a} = \begin{pmatrix} 0 & 1 \\ 0 & 0 \end{pmatrix}$$

(18.10)

However, if the initial state is $|0\rangle$, then $|0\rangle \rightarrow |0\rangle$. The transpose conjugate of this matrix is

$$\hat{a}^* = \begin{pmatrix} 0 & 0 \\ 1 & 0 \end{pmatrix}$$

(18.11)

so that

$$\hat{a}\hat{a}^* = \begin{pmatrix} 1 & 0 \\ 0 & 0 \end{pmatrix}$$

(18.12)

$$\hat{a}^*\hat{a} = \begin{pmatrix} 0 & 0 \\ 0 & 1 \end{pmatrix} = \begin{pmatrix} 1 & 0 \\ 0 & 1 \end{pmatrix} - \begin{pmatrix} 1 & 0 \\ 0 & 0 \end{pmatrix}$$

(18.13)

$$\hat{a}^*\hat{a} + \hat{a}\hat{a}^* = 1$$

(18.14)

Thus, the A_a matrix can be expressed as

$$A_a = \hat{a} + \hat{a}^*$$

(18.15)

and the CNOT matrix can be expanded into

$$A_{a,b} = \hat{a}^*\hat{a}(\hat{b} + \hat{b}^*) + \hat{a}\hat{a}^*$$

(18.16)

18.5.1 Pauli Matrices and Quantum Logic

In addition to the σ_x Pauli matrix

$$\sigma_x = \begin{pmatrix} 0 & 1 \\ 1 & 0 \end{pmatrix}$$

(18.17)

which materialized in the form of the A_a (Equation 18.9), the other two Pauli matrices

$$\sigma_y = \begin{pmatrix} 0 & -i \\ i & 0 \end{pmatrix}$$

(18.18)

and

$$\sigma_z = \begin{pmatrix} 1 & 0 \\ 0 & -1 \end{pmatrix} \qquad (18.19)$$

are also applicable to quantum computing. As we have already seen, the σ_x matrix rotates linear polarization by $\pi/2$ and can be applied to describe a quantum NOT gate.

The σ_y and σ_z can be applied for additional rotational operations in the Poincaré space.

18.5.2 Quantum Gates

Here, we briefly mention the experimental realization of the first quantum CNOT gate by Wineland and colleagues (Monroe et al., 1995) that applied a scheme proposed by Cirac and Zoller (1995). The experiment uses laser cooling of a single $^9Be^+$ ion. The transitions involve two $^2S_{1/2}$ hyperfine ground states, that is, four basis states. The control states are denoted by $|1\rangle$ and $|0\rangle$ type notation and are identified with $|n\rangle$. The spin one-half states are denoted by $|\uparrow\rangle$ and $|\downarrow\rangle$ type notation and are identified with $|S\rangle$.

The four basis states are $|0\rangle|\downarrow\rangle$, $|0\rangle|\uparrow\rangle$, $|1\rangle|\downarrow\rangle$, and $|1\rangle|\uparrow\rangle$. This register is achieved by applying two off-resonance tunable laser beams to the ion thus stimulating Raman transitions between the basis states.

Monroe et al. (1995) define the carrier frequency ω_0 as the frequency difference between the two $^2S_{1/2}$ hyperfine doublets and ω_x as the frequency difference from one hyperfine to the other ($\omega_x < \omega_0$). When the difference in frequency of the laser beams is approximately equal to ω_0, transitions are driven between $|S\rangle$ states while preserving $|n\rangle$. When the difference in frequency of the laser beams is approximately equal to ($\omega_0 - \omega_x$), transitions are driven between $|1\rangle|\downarrow\rangle$ and $|0\rangle|\uparrow\rangle$. When the laser frequency difference is ($\omega_0 + \omega_x$), transitions are driven between $|0\rangle|\downarrow\rangle$ and $|1\rangle|\uparrow\rangle$. Without further description of the experimental details, let us say that Moore et al. (1995) were able to observe the following relations between input and output states:

$|n\rangle_i$	$|S\rangle_i$	$|n\rangle_o$	$|S\rangle_o$
$|0\rangle$	$|\downarrow\rangle$	$|0\rangle$	$|\downarrow\rangle$
$|0\rangle$	$|\uparrow\rangle$	$|0\rangle$	$|\uparrow\rangle$
$|1\rangle$	$|\downarrow\rangle$	$|1\rangle$	$|\uparrow\rangle$
$|1\rangle$	$|\uparrow\rangle$	$|1\rangle$	$|\downarrow\rangle$

Comparison with the classical CNOT truth table

a	b	a'	b'
0	0	0	0
0	1	0	1
1	0	1	1
1	1	1	0

indicates a direct correlation, as intended.

A quantum NOT gate generated using single-photon emission in a macroscopic parametric amplification experiment is described by Pelliccia et al. (2003). These authors realized the quantum NOT operation via polarization measurements.

Beyond macroscopic realizations of the quantum NOT gate, it is not difficult to imagine half-wave plates at the atomic level or broadband prismatic polarization rotators at the atomic level.

Problems

18.1 Using Equations 18.10 and 18.11, show that $\hat{a}^*\hat{a} + \hat{a}\hat{a}^* = 1$.

18.2 Using Equations 18.10 and 18.11, show that $A_a = \hat{a} + \hat{a}^*$.

18.3 For the CNOT gate, define $\hat{b}$ and $\hat{b}^*$ and show that $A_{a,b} = \hat{a}^*\hat{a}(\hat{b} + \hat{b}^*) + \hat{a}\hat{a}^*$.

18.4 Discuss how a half-wave plate could be implemented at the atomic level to realize a quantum NOT gate.

18.5 Discuss how a broadband prismatic polarization rotator could be implemented at the atomic level to realize a quantum NOT gate.

References

Bennett, C. H. (1982). Thermodynamics of computation—A review. *Int. J. Theor. Phys.* **21**, 905–940.

Cirac, J. I. and Zoller, P. (1995). Quantum computations with cold trapped ions. *Phys. Rev. Lett.* **74**, 4091–4094.

Deutsch, D. (1992). Quantum computation. *Phys. World* **5**(6), 57–61.

Duarte, F. J. (1996). Generalized interference equation and optical processing. In *Proceedings of the International Conference on Lasers '95* (Corcoran, V. J. and Goldman, T. A., eds.). STS Press, McLean, VA, pp. 615–617.

Duarte, F. J. (2003). *Tunable Laser Optics*, Elsevier-Academic, New York.

Feynman, R. P. (1985). Quantum mechanical computers. *Opt. News* **11**(2), 11–20.

Feynman, R. P. (1986). Quantum mechanical computers. *Found. Phys.* **16**, 507–531.

Monroe, C., Meekhof, D. M., King, B. E., Itano, W. M., and Wineland, D. J. (1995). Demonstration of a fundamental quantum logic gate. *Phys. Rev. Lett.* **75**, 4714–4717.

Pelliccia, D., Schettini, V., Sciarrino, F., Sias, C., and De Martini, F. (2003). Contextual realization of the universal quantum cloning machine and of the universal-NOT gate by quantum-injected optical parametric amplification. *Phys. Rev. A* **68**, 042306.

Schumacher, B. (1995). Quantum coding. *Phys. Rev. Lett.* **51**, 2738–2747.

19

Quantum Cryptography and Teleportation

19.1 Introduction

Cryptography is a word derived from the Greek that is approximately translated as *hidden writing*. There are various forms and styles of cryptography; however, since ancient times the aim of cryptography has been the same: secure the transmission of information from an emitter to a receiver. A classical form of cryptography involves the sharing of a code between the emitter and the receiver. The emitter writes a message, using the shared code, and sends it to the intended receiver who uses the code to decipher the message. The integrity of the message is secured if and only if the code remains in the knowledge of the emitter and the intended receiver only. If the code is acquired, or broken, by a third party, then the message is no longer secured. An example of a simple classical numerical code is illustrated in Figure 19.1 and used to write the number π to 10 decimal places.

Albeit still in use, classical paper-based code systems have been largely replaced by various computerized mathematical methods that enable the electronic transmission of encrypted messages. The message is encrypted prior to transmission, transmitted via an unsecured channel in encrypted form, and decoded, once received. A widely used system of encryption, of this class, is known as *symmetric key algorithm*. This approach utilizes the same cryptographic key, or classical algorithm, for encryption and decryption (see Figure 19.2).

A different algorithmic approach allows the use of a public key and is known as *public key cryptography*. This approach uses a public key and a secured key, both of which are mathematically connected. One key encrypts the plaintext and the other unlocks the ciphertext (see Figure 19.3). The algorithmic methods just described enjoy compatibility with the vast array of existing computer networks and are thus widely used. There are many extensions and variations of these classical algorithmic methods.

Against this background, optical methods of communications offer alternatives that include some inherent advantages. In Chapter 11 we described a method of secure optical communications known as interferometric

FIGURE 19.1
Classical code representation for the truncated value of π to 10 decimal places. The key is included as a third item.

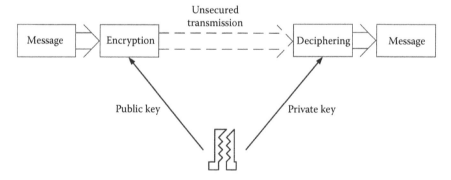

FIGURE 19.2
Classical symmetric key distribution. The same cryptographic key, or classical algorithm, is used for encryption and decryption.

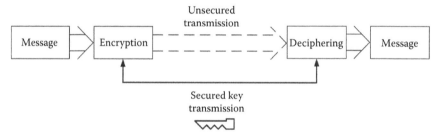

FIGURE 19.3
Classical public key cryptography. This approach uses a public key and a secured key, both of which are mathematically related.

communications that involve the transmission of *interferometric characters* at luminal speeds. This method does not require a key, although one might be added for added security, and is based on the fact that the interferometric character is catastrophically distorted or destroyed by attempts of interception. Thus, any attempt of intersection is immediately detected by the receiver.

An alternative method, which has received widespread attention in the open literature, is *quantum cryptography*. That method is described here in addition to a closely related discipline called *quantum teleportation*.

19.2 Quantum Cryptography

Quantum cryptography was first introduced by Weisner in 1983 and then by Bennett and Brassard (1984). A refined protocol was introduced by Bennett et al. (1992a). The Bennett and Brassard approach (BBA) relies on the straightforward polarization property of single photons. An alternative method, based on the quantum entanglement of pairs of photons, was introduced by Ekert (1991). Here, these two approaches to quantum cryptography are introduced with emphasis on the optics.

19.2.1 Bennett and Brassard Approach

In the BBA we have a single-photon emitter, often referred to in the literature as Alice, and a receiver, often referred to in the literature as Bob. There is also an adversary intruder, or eavesdropper, whose function is to optically intercept the transmission. This interceptor is often referred to in the literature as Eve.

Since our emphasis is the optics, we shall use mostly the terms emitter, receiver, and interceptor.

The BBA relies on the quantum polarization properties of single photons. In this regard, the polarization states of the photon are referred to as *base* states, and two bases are said to be conjugated in the sense that an attempt of measuring one base randomizes the other. Bennett et al. (1992a) refer to these conjugate bases as *canonical bases*. The canonical bases for polarized photons are

$$|H\rangle = |x\rangle \tag{19.1}$$

$$|V\rangle = |y\rangle \tag{19.2}$$

$$|L\rangle = \frac{1}{\sqrt{2}}\left(|x\rangle - i|y\rangle\right) \tag{19.3}$$

$$|R\rangle = \frac{1}{\sqrt{2}}\left(|x\rangle + i|y\rangle\right) \tag{19.4}$$

which correspond to horizontally ($|H\rangle$ or $\leftrightarrow$), vertically ($|V\rangle$ or $\updownarrow$), left-circularly ($|L\rangle$), and right-circularly ($|R\rangle$) polarized photons. The reader should recognize Equations 19.3 and 19.4 (see Chapter 17).

Bennett et al. (1992a) describe a quantum key distribution protocol that involves the emitter sending a random series of photons polarized in the canonical bases, the receiver choosing independently how to measure the polarizations (either rectilinearly [+] or circularly [O]), the receiver publicly announcing his measuring sequence (but not the results), the emitter publishing which of the receiver's bases were correct, then both parties agree to discard the data from incorrect measurements and null measurements, and finally the measurements are in bit form according to their polarization form. Following the style of Bennett et al. (1992a), an example of such protocol is given next with a summarized listing of each action:

1	$\leftrightarrow$	L	R	$\updownarrow$	L	$\updownarrow$	$\leftrightarrow$	L	$\updownarrow$	$\leftrightarrow$	R	$\updownarrow$	$\leftrightarrow$	L	$\updownarrow$
2	+	O	O	+	+	O	+	+	O	+	O	O	+	O	+
3	$\leftrightarrow$		R	$\updownarrow$		L	$\leftrightarrow$			$\leftrightarrow$		R		L	$\updownarrow$
4	+		O	+		O	+			+		O		O	+
5	•		•	•			•			•				•	•
6	$\leftrightarrow$		R	$\updownarrow$			$\leftrightarrow$			$\leftrightarrow$				L	$\updownarrow$
7	0		1	1			0			0				0	1

1. Emitter sends a random sequence of polarized photons: linear vertical ($\updownarrow$), linear horizontal ($\leftrightarrow$), right circular (R), and left circular (L).

2. Receiver measures the emitted photon's polarization using a series of random bases, which are either rectilinear (+) or circular (O).

3. Measurements result at the receiver, and some measurements are null.

4. Receiver reveals to emitter the bases he used for photon detection.

5. Emitter reveals to receiver which bases are correct.

6. The correct data are stored.

7. The data are interpreted in binary form using ($\leftrightarrow = L = 0$) and ($\updownarrow = R = 1$).

The series of qubits thus obtained is referred to as the *raw quantum transmission*. A diagram depicting this quantum transmission methodology is shown in Figure 19.4.

A schematic of an experimental apparatus based on the description of Bennett et al. (1992a) is given in Figure 19.5. In essence what we have here is a single-photon source followed by a polarizer that polarizes the light horizontally ($\leftrightarrow$). Next, the polarized photon goes through two electro-optics phase shifters (Pockels cells, see Chapter 15), which enable the generation of light

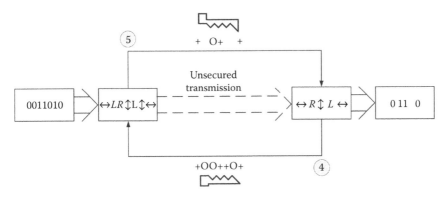

FIGURE 19.4
The Bennett–Brassard quantum cryptography method. The emitter is one of the two participants. The numbers 4 and 5 refer to the stages 4 and 5 in the Bennett–Brassard methodology
described in the text.

FIGURE 19.5
Top view of a generic experimental configuration applicable to the Bennett–Brassard protocol.
The emission from a single-photon source (SPS) is linearly polarized by a polarizer (P). Then
follows a polarization control system usually comprised of the EPC (see text). These are phase
shifters used to select ↔, ↕, L, or R polarization states. Following propagation the photon reaches
the RPC prior to being discriminated by a WP. For a description of the WP, refer to Chapter 15.

polarized in the ↔, ↕, L, and R states. These are the emitter's, or sender's,
Pockels cells (EPC). Next the polarized photon propagates in free space and
is received at the receiver's Pockels cell (RPC). The arriving photon then goes
through a calcite Wollaston prism (WP) that provides two different paths for
the ↔ and ↕ polarized photons onto their respective photomultiplier tubes.

Following the raw quantum transmission, the emitter and the receiver
continue comparing polarizations of transmitted photons. The main purpose of this exercise is to characterize the system for noise and to compare
its performance with statistical expectations for the effects of eavesdropping
or interception of polarized photons. The subject of signal interception is
rather extensive and will not be treated here; interested readers might refer
to the discussion offered by Bennett et al. (1992a). However, what is relevant,
from the optics perspective, is that the presence of errors in the transmission
is complicated by issues of detector noise, atmospheric turbulence, and the

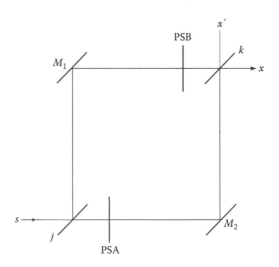

FIGURE 19.6

Mach–Zehnder interferometer configuration applied to the Bennett–Brassard protocol. PSA represents the PS alternative for the emitter (Alice), while PSB represents the PS alternative for the receiver (Bob). Detectors are positioned at x and x'.

inherent uncertainties related to single-photon emission and transmission. At this stage we should mention that Bennett et al. (1992b) are of the opinion that the Bennett and Brassard protocol is equivalent to methods based on quantum entanglement and Bell's theorem.

Finally, it should also be mentioned that albeit this discussion has centered on free-space propagation of polarized single photons, more recent alternative formats of the Bennett–Brassard protocol use fiber propagation, over distances beyond 100 km, and interferometric systems to prepare and detect the photons (see, e.g., Gobby et al., 2004).

An interferometric alternative applicable to the Bennett–Brassard protocol has been described by Bennett (1992). In this approach a Mach–Zehnder interferometer (see Chapter 10) is used to connect interferometrically the emitter (Alice) and the receiver (Bob) as illustrated in Figure 19.6. A phase shifter (PSA) allows the emitter to randomly induce $0, \pi/2, \pi, 3\pi/2$ phase shifts (PSs) in the $j - M_2 - k$ path while the receiver can apply random $0, \pi/2$ PSs in the $j - M_1 - k$ arm. Following transmission Alice and Bob agree publicly to keep the polarizations that only differ by 0 or π (Bennett, 1992).

19.2.2 Polarization Entanglement Approach

As mentioned in the previous section, an alternative approach, in quantum cryptography, to the straight Bennett and Brassard protocol involves the use of pairs of photons with entangled polarizations. Interest in this methodology was triggered by a paper by Ekert (1991) that described a cryptographic approach using spin one-half particles and Bell's theorem via the Bohm–EPR perspective (see Chapter 17). To make things simpler, our approach will consider directly the physics of entanglement as embodied by the Pryce–Ward–Snyder (PWS) probability amplitude equations of the form (see Chapter 17, Equations 17.29 through 17.32)

$$|s\rangle_- = \frac{1}{\sqrt{2}}\left(|x\rangle_1 |y\rangle_2 - |y\rangle_1 |x\rangle_2\right) \tag{19.5}$$

$$|s\rangle_+ = \frac{1}{\sqrt{2}}\left(|x\rangle_1 |y\rangle_2 + |y\rangle_1 |x\rangle_2\right) \tag{19.6}$$

$$|r\rangle_- = \frac{1}{\sqrt{2}}\left(|x\rangle_1 |x\rangle_2 - |y\rangle_1 |y\rangle_2\right) \tag{19.7}$$

$$|r\rangle_+ = \frac{1}{\sqrt{2}}\left(|x\rangle_1 |x\rangle_2 + |y\rangle_1 |y\rangle_2\right) \tag{19.8}$$

The equations apply to pairs of photons created at the same source and then propagate away in different optical paths. Although originally these optical paths were considered to be in directly opposite directions (Pryce and Ward, 1947; Snyder et al., 1948; Ward, 1949), obviously the same physics applies for divergent paths at any given angular separation. In the Ekert protocol the users, both receivers, use polarization analyzers randomly and independently.

Also, remember that in the PWS notation, x means horizontal polarization ($\leftrightarrow$) and y means vertical polarization ($\updownarrow$). Thus, the qubits $|x\rangle$ and $|0\rangle$ are equivalent and so are the qubits $|y\rangle$ and $|1\rangle$. Using commonly accepted photon cryptographic notation, Equations 19.5 and 19.6, for example, can be expressed as

$$|s\rangle_- = \frac{1}{\sqrt{2}}\left(|0\rangle_1 |1\rangle_2 - |1\rangle_1 |0\rangle_2\right) \tag{19.9}$$

$$|s\rangle_+ = \frac{1}{\sqrt{2}}\left(|0\rangle_1 |1\rangle_2 + |1\rangle_1 |0\rangle_2\right) \tag{19.10}$$

In this approach both participants share the entangled pair. Also, as suggested in Figure 19.7, the source of entangled photon pairs might be remotely located relative to both participants, which is a departure from the Bennett and Brassard method and the interferometric communications described in Chapter 11. In those methods, one of the participants, the emitter (or Alice), controls the emission process.

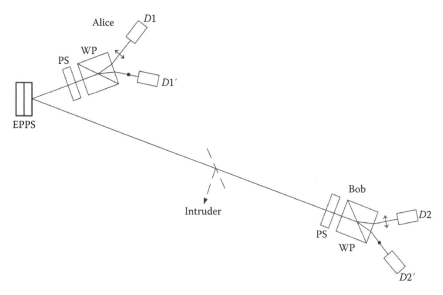

FIGURE 19.7
Top view of a generic experimental configuration applicable to cryptography using the entangled photon pair approach. In this configuration both participants might be separated from the source that could be remotely located in a satellite, for instance. Alternatively, one of the participants (Alice) could be next to the photon pair source, as depicted here. The emission is produced in an entangled photon pair source (EPPS) as available from parametric down conversion (see Appendix A). One photon goes to Alice where it is incident on an optical system consistent of PS optics followed by polarization selective optics such as a WP. Polarizations are either vertical (or perpendicular to the plane of incidence [•]) or horizontal (parallel to the plane of incidence [↔]).

A realization of the Ekert approach has been published by Naik et al. (2000). In their experiment these authors use spontaneous parametric down conversion to produce pairs of entangled photons propagating in divergent paths. The receiving optics includes a PS set composed of a randomly driven liquid crystal (LC) followed by a half-wave plate (HWP). The photon then proceeds to a polarizing beam splitter prior to arrival to a detector. An approximate depiction of this type of optical configuration is provided in Figure 19.7.

In their description, Naik et al. (2000) consider a probability amplitude of the form of Equation 19.8:

$$|s\rangle_+ = \frac{1}{\sqrt{2}}\left(|0\rangle_1\,|0\rangle_2 + |1\rangle_1\,|1\rangle_2\right) \qquad (19.11)$$

and each receiver proceeds to measure the polarizations using the bases

$$\left(|0\rangle_1 + e^{i\alpha}\,|1\rangle_1\right)$$

and

$$\left(|0\rangle_2 + e^{i\beta}\,|1\rangle_2\right)$$

In this process one receiver performs these measurements using four values of α and the other using four values of β (in increments of $\pi/4$ rads). As customary the receivers then publish the bases used but not the results of the measurements. These authors then use the detection probabilities, as a function of α and β, to perform Bell's inequality-type calculations (see Chapter 21) to test for the presence of an intruder (Naik et al., 2000).

A free-space experiment over a distance of 144 km was reported by Ursin et al. (2007).

In this configuration one of the participants (Alice) is at the source of emission and the receiver (Bob) at a remote location 144 km away. These authors used a photon source comprised by a laser-pumped spontaneous parametric down converter and receiver optical configurations including a phase shifter (HWP) and a polarizer beam splitter (PBS). The basic elements of this configuration are as depicted in Figure 19.7; however, this time Alice is next to the emitter and receives the photon pairs via fiber optics.

Ursin et al. (2007) consider a probability amplitude of the form of Equation 19.6

$$|s\rangle_{+} = \frac{1}{\sqrt{2}}\left(|0\rangle_1\,|1\rangle_2 + |1\rangle_1\,|0\rangle_2\right) \tag{19.12}$$

and proceeded to use four angular settings, for each receiver, to perform Bell's inequality-type calculations to test for the presence of an intruder. Their results yield Bell's parameter of $S = 2.508 \pm 0.0037$ (see Chapter 21) (Ursin et al., 2007).

19.3 Quantum Teleportation

Quantum entanglement of two linearly polarized photons is based on the PWS probability amplitudes (see Chapter 17; Duarte, 2012), which are often expressed as a simplified version of Equations 19.5 through 19.8:

$$|s\rangle_{-} = \frac{1}{\sqrt{2}}\left(|0_1 1_2\rangle - |1_1 0_2\rangle\right) \tag{19.13}$$

$$|s\rangle_{+} = \frac{1}{\sqrt{2}}\left(|0_1 1_2\rangle + |1_1 0_2\rangle\right) \tag{19.14}$$

$$|r\rangle_- = \frac{1}{\sqrt{2}}\left(|0_10_2\rangle - |1_11_2\rangle\right) \tag{19.15}$$

$$|r_+\rangle = \frac{1}{\sqrt{2}}\left(|0_10_2\rangle + |1_11_2\rangle\right) \tag{19.16}$$

Also, as shown in Chapter 17, we can add and subtract these equations to obtain

$$|0_11_2\rangle = \frac{1}{\sqrt{2}}\left(|s\rangle_+ + |s\rangle_-\right) \tag{19.17}$$

$$|1_10_2\rangle = \frac{1}{\sqrt{2}}\left(|s\rangle_+ - |s\rangle_-\right) \tag{19.18}$$

$$|0_10_2\rangle = \frac{1}{\sqrt{2}}\left(|r\rangle_+ + |r\rangle_-\right) \tag{19.19}$$

$$|1_11_2\rangle = \frac{1}{\sqrt{2}}\left(|r\rangle_+ - |r\rangle_-\right) \tag{19.20}$$

The physics derived from the quantum entanglement probability amplitude (Pryce and Ward, 1947; Snyder et al., 1948; Ward, 1949) was applied by Bennett et al. (1993) to the concept of teleportation. Quantum teleportation, as described by Bennett et al. (1993), consists in the disintegration of one quantum state, at the emitter's site, and in the subsequent reintegration of that quantum state at the receiver's site.

This concept is illustrated in more detail considering an example based on a description provided by Kim et al. (2001): assume that the emitter whishes to send the state

$$|\phi_1\rangle = \left(\alpha|0_1\rangle + \beta|1_1\rangle\right) \tag{19.21}$$

to the receiver. This, Equation 19.21, is the state that will be disassembled by the emitter and will be replicated remotely at the receiver's site. To do so, the emitter generates an entangled state of the form of any of the Equations 19.13 through 19.20; let's say

$$|s_{23}\rangle_+ = \frac{1}{\sqrt{2}}\left(|0_21_3\rangle + |1_20_3\rangle\right) \tag{19.22}$$

Then, the emitter creates a three-particle state:

$$|\phi_{123}\rangle = |\phi_1\rangle \otimes |s_{23}\rangle_+ = \frac{1}{\sqrt{2}}\left(\alpha|0_10_21_3\rangle + \alpha|0_11_20_3\rangle + \beta|1_10_21_3\rangle + \beta|1_11_20_3\rangle\right) \quad (19.23)$$

Using Equations 19.17 through 19.20 (with a slight modification in notation: $|s\rangle_+ = |s_{12}\rangle_+$, $|s\rangle_- = |s_{12}\rangle_-$, $|r\rangle_+ = |r_{12}\rangle_+$, $|r\rangle_- = |r_{12}\rangle_-$), Equation 19.23 can be expanded into

$$|\phi_{123}\rangle = \tfrac{1}{2}\left(|s_{12}\rangle_+|\phi_3\rangle_+ + |s_{12}\rangle_-|\phi_3\rangle_- + |r_{12}\rangle_+|\vartheta_3\rangle_+ + |r_{12}\rangle_-|\vartheta_3\rangle_-\right) \quad (19.24)$$

where

$$|\phi_3\rangle_+ = \left(\alpha|0_3\rangle + \beta|1_3\rangle\right) \quad (19.25)$$

$$|\phi_3\rangle_- = \left(\alpha|0_3\rangle - \beta|1_3\rangle\right) \quad (19.26)$$

$$|\vartheta_3\rangle_+ = \left(\alpha|1_3\rangle + \beta|0_3\rangle\right) \quad (19.27)$$

$$|\vartheta_3\rangle_- = \left(\alpha|1_3\rangle - \beta|0_3\rangle\right) \quad (19.28)$$

In this process, for example, if the emitter (Alice) measures $|s_{12}\rangle_-$, the qubit communicated to the receiver (Bob) is $|\phi_3\rangle_-$. According to Bennett et al. (1993), the recovery process at the receiver's end involves a multiplication of the form $\sigma|\phi_3\rangle$ where σ is either the identity matrix or one of the Pauli matrices (see Chapter 16):

$$I = \begin{pmatrix} 1 & 0 \\ 0 & 1 \end{pmatrix} \quad (19.29)$$

$$\sigma_x = \begin{pmatrix} 0 & 1 \\ 1 & 0 \end{pmatrix} \quad (19.30)$$

$$i\sigma_y = \begin{pmatrix} 0 & 1 \\ -1 & 0 \end{pmatrix} \quad (19.31)$$

$$\sigma_z = \begin{pmatrix} 1 & 0 \\ 0 & 1 \end{pmatrix} \quad (19.32)$$

Bob, the receiver, must apply one of these operators in order to convert $|\phi_3\rangle_-$ into the original $|\phi_1\rangle$. According to Bennett et al. (1993), an accurate teleportation can be performed by having Alice tell Bob the outcome of her measurement thus allowing Bob to apply the required rotation to transform the state of his particle into a replica of $|\phi_1\rangle$.

Alice is left *without any trace* of the original state $|\phi_1\rangle$.

Coming back to our example, if $|s_{12}\rangle_+$ is measured, the recovery transformation is just the identity and the receiver does not need to make a transformation since

$$I\,|\phi_3\rangle_+ = \left(\alpha\,|0_3\rangle + \beta\,|1_3\rangle\right) \tag{19.33}$$

if $|s_{12}\rangle_-$ is measured, the recovery transformation involves

$$\sigma_z\,|\phi_3\rangle_- = \sigma_z\left(\alpha\,|0_3\rangle - \beta\,|1_3\rangle\right) = \left(\alpha\,|0_3\rangle + \beta\,|1_3\rangle\right) \tag{19.34}$$

if $|r_{12}\rangle_+$ is measured, the recovery transformation involves

$$\sigma_x\,|\vartheta_3\rangle_+ = \sigma_x\left(\alpha\,|1_3\rangle + \beta\,|0_3\rangle\right) = \left(\alpha\,|0_3\rangle + \beta\,|1_3\rangle\right) \tag{19.35}$$

and, if $|r_{12}\rangle_-$ is measured, the recovery transformation involves

$$i\sigma_y\,|\vartheta_3\rangle_- = i\sigma_y\left(\alpha\,|1_3\rangle - \beta\,|0_3\rangle\right) = \left(\alpha\,|0_3\rangle + \beta\,|1_3\rangle\right) \tag{19.36}$$

In summary, this methodology for quantum teleportation can be summarized in the following sequence of steps:

1. The emitter creates $|s_{23}\rangle_+$.
2. The emitter creates the state to be sent in quantum teleportation, $|\phi_1\rangle$.
3. The emitter performs the operation $|\phi_{123}\rangle = |\phi_1\rangle \otimes |s_{23}\rangle_+$ thus disassembling the state to undergo teleportation, $|\phi_1\rangle$.
4. The emitter performs a measurement such as $|s_{12}\rangle_-$ thus creating $|\phi_3\rangle_-$.
5. The emitter sends $|s_{12}\rangle_-$ via a classical channel and $|\phi_3\rangle_-$ via a quantum channel.
6. The receiver transforms, in real time, $|\phi_3\rangle_-$ back into $|\phi_1\rangle$.

A simplified overview of this process is illustrated in Figure 19.8. Of the various experimental realizations of teleportation of quantum states, one experiment that should be mentioned is that performed, in free space between La Palma and Tenerife islands, over a distance of 143 km (Ma et al., 2012). Readers interested in the technical details involved photon pair generation, active polarization rotation, and the statistical processes involved in this type of experiments are encouraged to read (Ma et al., 2012).

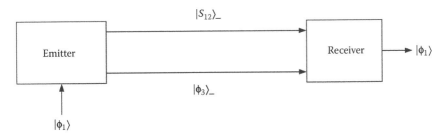

FIGURE 19.8
Simplified generic overview of quantum teleportation. At the emitter's site, an SPS creates $|\phi_1\rangle$, the state to be teleported. In addition, entangled photon pairs are produced by parametric down conversion (see Appendix A). This emission is used in the production, for example, of the $|s_{12}\rangle_-$ and $|\phi_3\rangle_-$ states by mixing the $|\phi_1\rangle$ state with the entangled photon pair; thus, $|\phi_1\rangle$ is disassembled and lost. Synchronized emission via free space enables the $|\phi_3\rangle_-$ state to reach the receiver where $|\phi_3\rangle_-$ undergoes the necessary unitary transformation to recreate $|\phi_1\rangle$.

Problems

19.1 Complete the Bennett table if the first three rows are

1	$\updownarrow$	$\leftrightarrow$	R	L	$\leftrightarrow$	$\updownarrow$	$\updownarrow$	L	R	$\leftrightarrow$	$\updownarrow$	$\leftrightarrow$	R	L	L
2	+	+	O	O	+	+	O	+	O	O	+	O	O	O	+
3	$\updownarrow$	$\leftrightarrow$	R	L	$\leftrightarrow$	$\updownarrow$		R		$\updownarrow$			R	L	

19.2 Show that $\sigma_x|0\rangle = |1\rangle$ and $\sigma_x|1\rangle = |0\rangle$.

19.3 Evaluate $i\sigma_y|0\rangle$ and $i\sigma_y|1\rangle$.

19.4 Perform the necessary recovery operations, in quantum teleportation, for $|\phi_1\rangle$, from $|r_{12}\rangle_+$ and $|r_{12}\rangle_-$ measurements, if the shared state is

$$|r_{23}\rangle_- = \frac{1}{\sqrt{2}}\left(|0_2 0_3\rangle - |1_2 1_3\rangle\right)$$

19.5 Perform the necessary recovery operations, in quantum teleportation, for $|\phi_1\rangle$, from $|s_{12}\rangle_+$ and $|s_{12}\rangle_-$ measurements, if the shared state is

$$|r_{23}\rangle_- = \frac{1}{\sqrt{2}}\left(|0_2 0_3\rangle - |1_2 1_3\rangle\right)$$

References

Bennett, C. H. (1992). Quantum cryptography using any two nonorthogonal states. *Phys. Rev. Lett.* **68**, 3121–3124.

Bennett, C. H., Bessette F., Brassard, G., Salvail, L., and Smolin, J. (1992a). Experimental quantum cryptography. *J. Cryptol.* **5**, 3–28.

Bennett, C. H. and Brassard, G. (1984). Quantum cryptography: Public key distribution and coin tossing. *Proceedings of IEEE International Conference on Computers Systems and Signal Processing*, Bangalore, India.

Bennett, C. H., Brassard, G., Crépeau, C., Jozsa, R., Peres, A., and Wootters, W. K. (1993). Teleporting an unknown quantum state via dual classical and Einstein-Podolsky-Rosen channels. *Phys. Rev. Lett.* **70**, 1895–1899.

Bennett, C. H., Brassard, G., and Mermin, N. D. (1992b). Quantum cryptography without Bells theorem. *Phys. Rev. Lett.* **68**, 557–559.

Duarte, F. J. (2012). The origin of quantum entanglement experiments based on polarization measurements. *Eur. Phys. J. H* **37**, 311–318.

Ekert, A. K. (1991). Quantum cryptography based on Bell's theorem. *Phys. Rev. Lett.* **67**, 661–663.

Gobby, C., Yuan, Z. L., and Shields, A. J. (2004). Quantum key distribution over 122 km of standard telecom fiber. *Appl. Phys. Lett.* **84**, 3762–3764.

Kim, Y. -H., Kulik, S. P., and Yanhua Shih, Y. (2001). Quantum teleportation of a polarization state with a complete Bell state measurement. *Phys. Rev. Lett.* **86**, 1370–1373.

Ma, X., Herbst, T., Scheidl, T., Wang, D., Kropatschek, S., Naylor, W., Mech, A. et al. (2012). Quantum teleportation over 143 kilometres using active feed-forward. *Nature* **489**, 269–273.

Naik, D. S., Peterson, C. G., White, A. G., Berglund, A. J., and Kwiat, P. G. (2000). Entangled state quantum cryptography: Eavesdropping on the Ekert protocol. *Phys. Rev. Lett.* **84**, 4733–4736.

Pryce, M. H. L. and Ward, J. C. (1947). Angular correlation effects with annihilation radiation. *Nature* **160**, 435.

Snyder, H. S., Pasternack, S., and Hornbostel, J. (1948). Angular correlation of scattered annihilation radiation. *Phys. Rev.* **73**, 440–448.

Ursin, R., Tiefenbacher, F., Schmitt-Manderbach, T., Weier, H., Scheidl, T., Lindenthal, M., Blauensteiner, B. et al. (2007). Free-space distribution of entanglement and single photons over 144 km. *Nat. Phys.* **3**, 481–486.

Ward, J. C. (1949). *Some Properties of the Elementary Particles*. D. Phil Thesis, Oxford University, Oxford, U.K.

20

Quantum Measurements

20.1 Introduction

Classically, the process of a measurement is relatively straightforward. The objects measured are macroscopic and they can be measured repeatedly without being disturbed in a nondestructive process. For example, using a meter rule, or a metric caliper, one can measure repeatedly the length of an object thus obtaining a series of measurements that can lead to an average dimension and a corresponding standard of deviation. In other words, the experimental physicist obtains the measurement and the error associated with that measurement ($x \pm \Delta x$). Thus, classically speaking, the measurement issue is settled.

All experimental measurements include an error. Measurements without an error, or without an uncertainty, are not physically possible.

Quantum mechanically speaking, the measurement problem is not as transparent as in the classical domain, since we are using relatively massive macroscopic classical instruments to obtain measurements on minute quantum objects such as photons and electrons. Therefore, the issue of uncertainties becomes even more important.

Going back to basics, we find that Dirac (1978) refers to the measurement process in a fairly abstract manner indicating that a succession of measurements should give identical results, thus implying that the measurement should be nondestructive. von Neumann (1932) introduced the reduction of the wave function hypothesis. Lamb (1989) is of the opinion that the Dirac and the von Neumann approaches are "essentially equivalent." And Lamb (1989) goes further to state that "neither Dirac nor von Neumann discusses his measurements in physical terms."

Furthermore, according to Lamb (1989), Pauli introduced a concept to use Stern–Gerlach type measurements to determine probability distributions, which was fine in principle but in practice "always destroyed the system of interest" (Lamb, 1989). Here, we should also add that one of van Kampen's theorems states explicitly that the measuring apparatus in quantum mechanics is *macroscopic* (van Kampen, 1988).

Thus, it is not surprising to read articles by noted physicists entitled "Against measurement" (Bell, 1990) where the concepts of Dirac, von Neumann, and van Kampen are criticized. In this chapter we attempt to describe practical approaches to the measurement of physical parameters associated with quantum entities such as the photons and ensembles of indistinguishable photons.

20.2 Interferometric Irreversible Measurements

The interferometric measurements described in Chapter 11 are macroscopic physical permanent records of interferometric photon distributions. As such, these recordings represent an inherently irreversible transformation of the event to be recorded. In other words, these measurements are destructive, and they illustrate what happens when a massive macroscopic classical system interacts with a quantum entity such as a photon or a photon interferometric distribution.

In Chapter 8 we saw that quantum mechanical probabilities are proportional to the photon intensity. Thus, when we measure an intensity distribution or an interferometric intensity distribution, we are recording the spatial information originally contained in the interferometric distribution, which is a probability distribution such as the generalized 1D N-slit interferometric distribution (Duarte, 1991, 1993):

$$|\langle x \,|\, s\rangle|^2 = \sum_{j=1}^{N} \Psi(r_j)^2 + 2\sum_{j=1}^{N} \Psi(r_j)\left(\sum_{m=j+1}^{N} \Psi(r_m)\cos(\Omega_m - \Omega_j)\right) \qquad (20.1)$$

where $\Psi(r_j)$ are wave functions (Dirac, 1978; Duarte, 2004), and the term in parentheses represents the phase that describes the exact geometry of the N-slit interferometer (Duarte, 1991, 1993). Again, the measured intensity is proportional to the probability $|\langle x \,|\, s\rangle|^2$, and it is this probability that gives origin to the spatial distribution of the observed intensity. This equation was originally derived for *single-photon propagation* (Duarte, 1993, 2004) albeit in practice it also applies to the propagation of an ensemble of indistinguishable photons, as in the case of narrow-linewidth laser emission.

In the interferometric measuring process, a photon within the interferometric distribution described by Equation 20.1 arrives at the interference plane or detection surface. The arrival of each individual photon, with an energy $E = h\nu$, is registered at the detector by the creation of a charge within the boundaries of the $\langle x|s\rangle\langle x|s\rangle^*$ distribution. For light associated with the emission of large numbers of indistinguishable photons, as in the case of narrow-linewidth laser emission, the energy becomes $E = Nh\nu$, and a cumulative charge distribution closely resembling $\langle x|s\rangle \langle x|s\rangle^*$ is registered at the detector (Duarte, 2004). Once the single photon, or the ensemble of indistinguishable photons, interacts with the detection surface, the process becomes *irreversible*, thus representing a destructive measurement. Nevertheless, this process is highly reproducible and repeatable with new photons prepared in the same state giving rise to the $\langle x|s\rangle \langle x|s\rangle^*$ distribution.

Further, it should be emphasized that, albeit irreversible and destructive, the interferometric measurement initially involves a quantum interaction between the photon and the detection surface. Once that first quantum interaction occurs, then the detection charge is registered at the interferometric plane. If the interferometric plane is comprised of a photographic plate, then the incident photon induces an image within the $\langle x|s\rangle \langle x|s\rangle^*$ distribution. If the interferometric plane is comprised of a photoelectric surface, the incident photon generates a charge, within the $\langle x|s\rangle \langle x|s\rangle^*$ distribution, that is amplified in a cascade process until it produces a classical manifestation.

In summary, macroscopic interferometric measurements involve the detection and recording of an intensity distribution, which is proportional to the calculated quantum probability, that is,

$$I(\lambda, N, \Omega) = K\langle x|s\rangle \langle x|s\rangle^* \tag{20.2}$$

where K is a constant with units of J s^{-1} m^{-2}. We also know, from comparisons from measurements and theory (Duarte, 1993), that this intensity distribution $I(\lambda, N, \Omega)$ closely reproduces the calculated probability distribution $\langle x|s\rangle \langle x|s\rangle^*$, which in turn originates from the multiplication of the probability amplitude series

$$\langle x|s\rangle = \sum_{j=1}^{N} \langle x|j\rangle \langle j|s\rangle \tag{20.3}$$

with its complex conjugate. Here, the probability amplitudes are wave functions of the form (Dirac, 1978)

$$\langle j|s\rangle = \Psi(r_{j,s})e^{-i\theta_j} \tag{20.4}$$

$$\langle x|j\rangle = \Psi(r_{x,j})e^{-i\phi_j} \tag{20.5}$$

that, albeit not measured directly, are essential to either predict or reproduce the correct measurable intensity distribution described in Equation 20.2. See Chapter 21 for further discussion on this topic.

20.2.1 Additional Irreversible Quantum Measurements

In addition to the interferometric measurements just described, other irreversible quantum measurements in optics include

1. The original quantum polarization entanglement measurements made by Wu and Shaknov (1950)
2. Quantum polarization measurements to test Bell's inequalities (see, e.g., Aspect et al., 1982)
3. Quantum cryptography measurements (see, e.g., Gobby et al., 2004)
4. Measurements of quantum teleported states (see, e.g., Ma et al., 2012)

In all these cases, quantum entities, namely, the photon, interact with massive macroscopic detectors in an irreversible, and destructive, manner.

20.3 Quantum Nondemolition Measurements

Lamb wrote several papers on the measurement problem in quantum mechanics. Here we refer to two of these papers. The first paper (Lamb, 1986) considers a sequence of quantum mechanical measurements on an isolated large macroscopic system, such as a gravity wave detector. The model used by Lamb (1986) consists of a 1D harmonic oscillator with momentum

$$p = m\frac{dx}{dt} \tag{20.6}$$

which is acted by an *unknown force F(t)* so that the Hamiltonian is

$$H = H_0 - F(t)x \tag{20.7}$$

where the *unperturbed* Hamiltonian is

$$H_0 = \frac{1}{2}\left(\frac{p^2}{m} + m\omega^2 x^2\right) \tag{20.8}$$

The wave function is assumed to evolve according to

$$-\frac{\hbar}{i}\frac{\partial\psi}{\partial t} = H\psi \tag{20.9}$$

Lamb (1986) goes on to explain that $\psi(x)$ is a linear combination of the eigenfunctions of H_0 of the form

$$\psi(x) = \sum c_n u_n(x) \tag{20.10}$$

and that according to what Lamb calls the Dirac–von Neumann hypothesis, a measurement would result in an eigenvalue of H_0:

$$E_{n_1} = \left(n_1 + \frac{1}{2} \right) \hbar \omega \qquad (20.11)$$

The value of $\psi(x)$ is changed to the n_1 th normalized eigenfunction $u_{n_1}(x)$. This wave function is then supposed to evolve according to time-dependent Schrödinger's equation until t_2, on the second measurement, and so on. This measuring strategy would yield a series of n values. Next, Lamb asks himself: "What can we learn about $F(t)$?"

   This description provided by Lamb (1986) can be used to illustrate the concept of *quantum nondemolition measurements* as related to the von Neumann (1932) approach: For a Hamiltonian defined as

$$H = H_0 - H_m \qquad (20.12)$$

and a wave function modified by a *weak measurement* according to

$$|\psi\rangle |A\rangle = \sum c_{\,n} |\psi_n\rangle |A_n\rangle \qquad (20.13)$$

the measurement is said to be nondemolition if

$$[H_0, H_m] = 0 \qquad (20.14)$$

In the second paper, Lamb (1989) is still interested in a large macroscopic system where "only the detector is treated quantum mechanically." He then discloses his idea of "making purely classical measurements on a quantum system" (Lamb, 1989). This approach appears to be consistent with van Kampen's fifth theorem that states that a measuring instrument in quantum mechanics "consists of a macroscopic system" (van Kampen, 1989).

20.4 Soft Polarization Measurements

More recently, researchers have reported to softly probe, nondestructively, propagating photons in *two-beam* interference experiments (Kocsis et al., 2011). In these experiments the emission of a single-photon source is divided into two components so that the adjacent, and initially parallel, Gaussian beams are eventually allowed to overlap and, hence, interfere.

   Initially, the two Gaussian beams are in a polarization state described by the usual

$$|\psi\rangle = \frac{1}{\sqrt{2}} \left(|x\rangle + |y\rangle \right) \qquad (20.15)$$

then a weak measurement is performed by using a macroscopic slab of calcite with its optic axis at $42°$ in the $x - z$ plane that rotates the polarization state to

$$| \psi' \rangle \langle \psi' | = \frac{1}{2} \begin{pmatrix} 1 & e^{-i\varphi} \\ e^{i\varphi} & 1 \end{pmatrix} \qquad (20.16)$$

The beams then propagate through a set of macroscopic optics including a quarter-wave plate, lenses, and a polarization beam splitter. After the polarization beam splitter, the two-beam interference pattern is observed. Kocsis et al. (2011) claim that this arrangement allows them to obtain information about the photon trajectory and still observe an interference spatial distribution associated with two-beam interference, although the interference pattern becomes slightly distorted. In other words, the photons are softly probed (using macroscopic classical elements) while producing a slightly altered interference pattern. The fact that the interference pattern *is altered* indicates that the soft probing conducted *does alter the interference* pattern, as would be expected from Feynman's teachings (Feynman et al., 1965).

20.5 Soft Intersection of Interferometric Characters

In Chapter 11 we saw that beam path proving via conventional high-surface quality beam splitters can be immediately detected due to the catastrophic collapse induced by the spatial disruption caused by the beam splitter. This is the case even if the classical beam splitter is highly transparent and only ~0.4 mm in thickness, since the observe phenomenon is a diffractive edge effect brought about by an abrupt change in the spatial distribution of the refractive index.

The question then becomes the following: Can we probe the intra-interferometric path $D_{\langle x|j \rangle}$ ever so gently as to avoid a violent disruption of the homogeneous refractive index serving as propagation medium between the N-slit array and the interference plane? The answer to this question is in the affirmative: Soft intersection of propagating N-slit interferometric characters, using spider web silk fibers, was demonstrated by Duarte et al. (2011) using an experimental configuration described in Figure 20.1. The fibers used in these experiments are ultrathin semitransparent and ultrathin transparent natural fibers. Two such fibers are fine human blond hair and spider web fibers. The fibers have a diameter of $d \approx 50$ µm in the case of blond human hair and $25 \leq d \leq 30$ µm in the case of transparent spider web fibers, which were collected in Western New York, near Lake Ontario (Duarte et al., 2011).

The interferometric character a $(N = 2)$, for an intra-interferometric distance of $D_{\langle x|j \rangle} = 7.235$ m, is illustrated in Figure 20.2. For the same

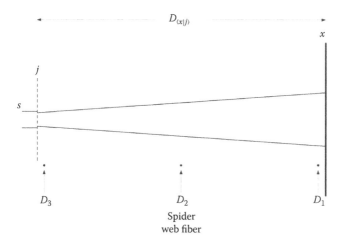

FIGURE 20.1

Top view schematics of the intra-interferometric path of the N-slit interferometer indicating (approximately) the three positions D_1, D_2, and D_3, in the $D_{\langle x|j \rangle}$ propagation path, where the spider web fiber is inserted perpendicular to the plane of incidence (i.e., orthogonal to the plane of the figure).

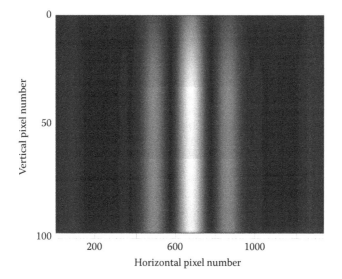

FIGURE 20.2

Interferometric distribution registered at x, for $D_{\langle x|j \rangle} = 7.235$ m, $\lambda = 632.8$ nm, and $N = 2$ (570 μm slits separated by 570 μm). This interferogram corresponds to the interferometric character a. This measurement was performed at a temperature of $T \approx 22°C$. Each pixel on the CCD screen is 20 μm wide. (Reproduced from Duarte, F.J. et al., *J. Opt.* 13, 035710, 2011, with permission from the Institute of Physics.)

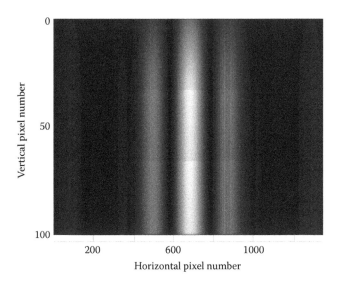

FIGURE 20.3
Photograph of the interferometric distribution registered at x, for $D_{\langle x|j\rangle} = 7.235$ m, $\lambda = 632.8$ nm, and $N = 2$ (570 μm slits separated by 570 μm) showing a superimposed diffractive pattern over the *outer right wing* of the interferogram. The superimposed diffractive distribution is caused by a spider web fiber deployed orthogonally to the propagation plane (i.e., perpendicular to the plane of Figure 20.1) at the distance of $D_1 = D_{\langle x|j\rangle} - 0.150$ m, or 15 cm from x. (Reproduced from Duarte, F.J. et al., *J. Opt.* 13, 035710, 2011, with permission from the Institute of Physics.)

interferometric character ($N = 2$) with the spider web fiber, positioned 15 cm from the interferometric plane at x (D_3), and displaced laterally, a diffraction pattern is superimposed over one of the outer wings of the interferogram (see Figure 20.3). The important and interesting effect here is that the propagating interferogram is physically intercepted but it is not destroyed. It is modified, or altered, in a suave orderly and controllable manner.

20.5.1 Comparison between Theoretical and Measured N-Slit Interferograms

Next, we describe in detail a series of additional soft probing experiments, conducted by Duarte et al. (2013), that involve the gentle and controlled insertion of the spider silk web fiber into the optical path $D_{\langle x|j\rangle}$ of the propagating interferometric character. The silk fiber is inserted into the intra-interferometric propagation path, under tension, perpendicular to the plane incidence at D_1, D_2, and D_3 (see Figure 20.1). That is, the spider web fiber is inserted orthogonal to the beam expansion defined by the multiple-prism expander. As illustrated in Figure 20.1, the N-slit array, or grating, is also deployed with the direction of the slits perpendicular to the plane of incidence.

The intra-interferometric distances of silk fiber insertion are (Duarte et al., 2013)

1. $D_1 = D_{\langle x|j \rangle} - 0.150$ m (15 cm from the detector).
2. $D_2 = D_{\langle x|j \rangle}/2$ (midway).
3. $D_3 = D_{\langle x|j \rangle} - 7.085$ m (15 cm from the grating).

while the overall intra-interferometric distance is maintained at $D_{\langle x|j \rangle} = 7.235$ m.

First, a control interferogram is generated from the illumination of $N = 3$ slits of the grating comprised with 570 μm slits separated by 570 μm, at $\lambda = 632.8$ nm, for an intra-interferometric distance of $D_{\langle x|j \rangle} = 7.235$ m, as illustrated in Figure 20.4. This interferometric character b is recorded at room temperature ($T \approx 22°C$), which becomes the standard measurement temperature.

An interferogram under identical propagation conditions, for $N = 3$, with the spider web silk fiber deployed orthogonally to the propagation plane, at D_1, is shown in Figure 20.5. The interferometric distribution thus obtained demonstrates a diffraction pattern superimposed over the outer right wing of the interferogram in Figure 20.5.

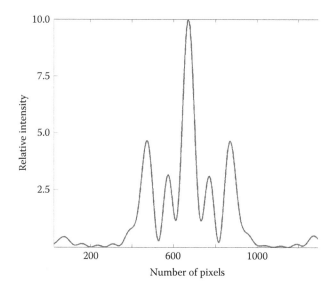

FIGURE 20.4
Control intensity profile of interferogram registered at x, for $D_{\langle x|j \rangle} = 7.235$ m, $\lambda = 632.8$ nm, and $N = 3$ (570 μm slits separated by 570 μm). The interferogram generated with $N = 3$ corresponds to the interferometric character b. This measurement was performed at a temperature of $T \approx 22°C$. Each pixel on the CCD screen is 20 μm wide. These parameters apply to all the measurements considered in the next set of figures. (Reproduced from Duarte, F.J. et al., *J. Mod. Opt.* 60, 136, 2013, with permission from Taylor and Francis.)

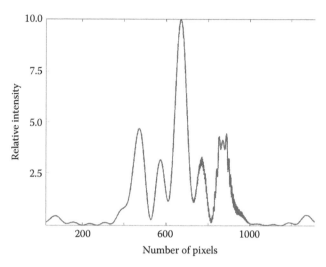

FIGURE 20.5
Measured intensity profile of interferogram registered at x, for $D_{\langle x|j\rangle} = 7.235$ m, $\lambda = 632.8$ nm, and $N = 3$ showing a superimposed diffractive intensity distribution over the *outer right wing* of the interferogram. The superimposed diffractive pattern is caused by a spider web fiber deployed orthogonally to the propagation plane (i.e., perpendicular to the plane of Figure 20.1) at the distance of $D_1 = D_{\langle x|j\rangle} - 0.150$ m or 15 cm from x. (Reproduced from Duarte, F.J. et al., *J. Mod. Opt.* 60, 136, 2013, with permission from Taylor and Francis.)

An interferogram under identical propagation conditions, for $N = 3$, but now with the spider web silk fiber deployed orthogonally to the propagation plane, at $D_2 = D_{\langle x|j\rangle}/2$, is shown in Figure 20.6.

An interferogram under identical propagation conditions, for $N = 3$, but now with the spider web silk fiber deployed orthogonally to the propagation plane, at D_3, is shown in Figure 20.7. For this experiment, the fiber is positioned 15 cm from the slits near the first (right) slit relative to the configuration of Figure 20.1. Duarte et al. (2013) note that a cooled detector and stabilized laser should improve detection conditions in this pre-interferometric regime.

A theoretical control interferogram, equivalent to the measured interferogram of Figure 20.4, is generated using the interferometric equation

$$| \langle x | s \rangle |^2 = \sum_{j=1}^{N} \Psi(r_j)^2 + 2 \sum_{j=1}^{N} \Psi(r_j) \left(\sum_{m=j+1}^{N} \Psi(r_m) \cos(\Omega_m - \Omega_j) \right)$$

and displayed in Figure 20.8. Reproduction, or prediction, of the interferograms with superimposed diffraction patterns is performed adopting an interferometric *cascade approach* (Duarte, 1993; Duarte et al., 2011). This cascade approach consists in using the interferometric equation (Equation 20.1) to create an interferometric distribution that becomes the illumination field

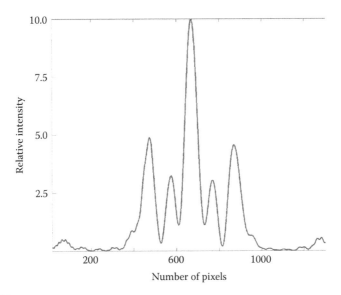

FIGURE 20.6
Measured intensity profile of interferogram registered at x, for $D_{\langle x|j\rangle} = 7.235$ m, $\lambda = 632.8$ nm, and $N = 3$ with a spider web fiber deployed orthogonally to the propagation plane at an intra-interferometric distance of $D_2 = D_{\langle x|j\rangle}/2$ m. (Reproduced from Duarte, F.J. et al., *J. Mod. Opt.* 60, 136, 2013, with permission from Taylor and Francis.)

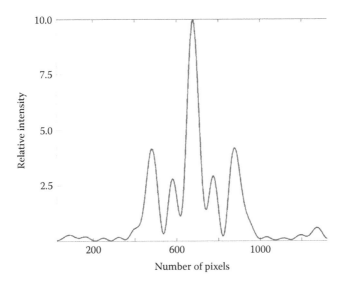

FIGURE 20.7
Measured intensity profile of interferogram registered at x, for $D_{\langle x|j\rangle} = 7.235$ m, $\lambda = 632.8$ nm, and $N = 3$ with a spider web fiber deployed orthogonally to the propagation plane at an intra-interferometric distance of $D_3 = D_{\langle x|j\rangle} - 7.085$ m (15 cm from the slits). (Reproduced from Duarte, F.J. et al., *J. Mod. Opt.* 60, 136, 2013, with permission from Taylor and Francis.)

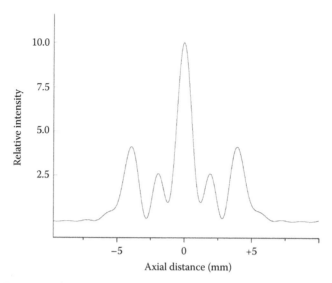

FIGURE 20.8

Calculated control interferogram at x, using Equation 20.1, for $D_{\langle x|j\rangle} = 7.235$ m, $\lambda = 632.8$ nm, and $N = 3$. This calculated interferogram corresponds to the measured interferogram displayed in Figure 20.4. (Reproduced from Duarte, F.J. et al., *J. Mod. Opt.* 60, 136, 2013, with permission from Taylor and Francis.)

of the next N-slit array (Duarte, 1993). Using this approach, the spider web silk fiber is represented by two wide slits separated by the diameter of the fiber. Thus, we can predict or replicate the diffraction effect induced by the spider web fiber at various intra-interferometric distances.

The calculations representing insertion of a 25 μm fiber in the intra-interferometric path $D_{\langle x|j\rangle}$ at D_1, D_2, and D_3 are displayed in Figures 20.9 through 20.11, respectively. At D_1 the superposition of the diffraction signal over the interferometric distribution is beautifully predicted as illustrated in Figure 20.9. At D_2 the silk fiber induces only a minor effect as shown in Figure 20.10. Furthermore, at D_3, the silk fiber produces almost no disturbance when placed between slits and tends to only very slightly modify the whole interferometric distribution when positioned at the center of one of the slits (Figure 20.11). Either in the case of the beautiful superimposed diffraction distribution (Figure 20.9) or in the other two more subtle interactions, the theoretical interferograms nicely reproduce the corresponding measured interferograms. Previously, it was demonstrated that insertion of a conventional thin and highly transparent beam splitter into the intra-interferometric path led, as expected (see Chapter 11), immediately to a *catastrophic collapse* of the interferometric character or signal (Duarte, 2002, 2005; Duarte et al., 2010). However, the experiments described by Duarte et al. (2013) demonstrate a remarkable suave and controlled way to alter the propagating interferograms in a soft and nondestructive manner. Thus, we have transitioned from a regime of *total signal collapse*, using classical beam splitters, to

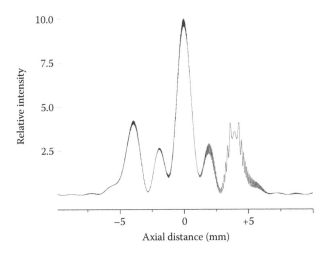

FIGURE 20.9
Calculated interferogram at x, for $D_{\langle x|j\rangle} = 7.235$ m, $\lambda = 632.8$ nm, and $N = 3$ with a 30 μm diameter fiber deployed orthogonally to the propagation plane at a distance of 15 cm from x, or $D_1 = D_{\langle x|j\rangle} - 0.150$ m. The fiber is positioned 4 mm from the center. This calculated interferogram corresponds to the measured interferogram displayed in Figure 20.5. (Reproduced from Duarte, F.J. et al., *J. Mod. Opt.* 60, 136, 2013, with permission from Taylor and Francis.)

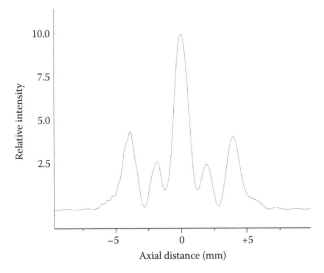

FIGURE 20.10
Calculated interferogram at x, for $D_{\langle x|j\rangle} = 7.235$ m, $\lambda = 632.8$ nm, and $N = 3$ with a 30 μm web fiber deployed orthogonally to the propagation plane at an intra-interferometric distance of $D_2 = D_{\langle x|j\rangle}/2$ m (i.e., 3.6175 m). The fiber is positioned 2.227 mm from the center. This calculated interferogram corresponds to the measured interferogram displayed in Figure 20.6. (Reproduced from Duarte, F.J. et al., *J. Mod. Opt.* 60, 136, 2013, with permission from Taylor and Francis.)

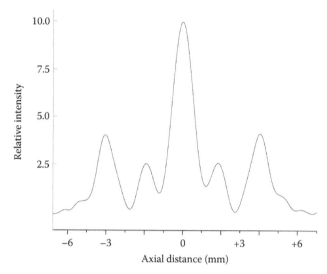

FIGURE 20.11

Calculated interferogram at x, for $D_{\langle x|j\rangle} = 7.235$ m, $\lambda = 632.8$ nm, and $N = 3$ with a 30 μm web fiber deployed orthogonally to the propagation plane at an intra-interferometric distance of $D_3 = D_{\langle x|j\rangle} - 7.085$ m. The fiber is positioned at the center right slit. This calculated interferogram corresponds to the measured interferogram displayed in Figure 20.7. (Reproduced from Duarte, F.J. et al., *J. Mod. Opt.* 60, 136, 2013, with permission from Taylor and Francis.)

a suave regime of *nondestructive* gentle detection with the use of spider silk threads. This appears to be allowed by the unusual geometry of the N-slit interferometer where the dimensions and separation of the slits (570 μm) are relatively enormous to the dimensions of the fiber diameter (25–30 μm).

20.5.2 Soft Interferometric Probing

As indicated earlier, Equation 20.1 is a quantum expression that was originally derived for single-photon propagation that also describes accurately the interferometric propagation generated by ensembles of indistinguishable photons as available from narrow-linewidth lasers. Thus, as previously observed (Duarte, 2002, 2005), and according to Feynman's teachings, we have expected and observed the *catastrophic collapse* of interferometric characters at any macroscopic attempt to extract information.

These attempts involved the insertion of thin highly transparent beam splitters. In this regard, the orderly nondestructive diffractive effects reported by Duarte et al. (2011, 2013), following soft interrogation via the insertion of a microscopic spider web fiber, are extremely interesting given that the information contained in the interferometric character is largely preserved albeit the presence of the fiber is nicely detected. Even more interesting are the results obtained at $D_2 = D_{\langle x|j\rangle}/2$ and D_3. Probing with the fiber at the later position causes a nearly indistinguishable effect. These results demonstrate the capability of interacting with the propagating interferograms nonde-structively causing only slight alterations to the information relative to the pristine interferograms.

20.5.3 Mechanics of Soft Interferometric Probing

According to W. E. Lamb, quantum mechanics can be extended to measure-ments on a "rather large and otherwise macroscopic system" (Lamb, 1989). He goes on to explain that the detector is treated quantum mechanically, in other words the process of the measurement that in our case refers directly to the interferometric probability distribution (Equation 20.1) arising from either the propagation of single photons or an ensemble of indistinguishable photons (Duarte, 2004).

Albeit the measured interferograms are neatly reproduced using the cas-cade interferometric approach, where the interferometric distribution at one plane becomes the input for the next plane, as described by Duarte (1993), next we use an alternative approach to describe the physics of the fiber inter-section with the propagating interferogram (Duarte et al., 2013).

In a straightforward interferometric propagation, as outlined in Figure 20.1, and in the absence of the probing fibers, the interferograms generated are the corresponding control interferogram as shown in Figure 20.4 (mea-sured) and Figure 20.8 (calculated), and the probability amplitudes from the grating to the interferometric plane are simply given by $\langle x|j\rangle$; however, with the insertion of the spider web fiber, the probability amplitudes are altered, so that

$$\langle x|j\rangle \rightarrow \langle x|j'\rangle\langle j'|j\rangle \tag{20.17}$$

Indeed, as soon as the spider web fiber is introduced, the original probability amplitude

$$\langle x|s\rangle = \sum_{j=1}^{N} \langle x|j\rangle\langle j|s\rangle$$

is replaced by a probability amplitude relevant to the experimental configu-ration of Figure 20.1, that is,

$$\langle x|s\rangle' = \sum_{j'=1}^{N} \sum_{j=1}^{N} \langle x|j'\rangle\langle j'|j\rangle\langle j|s\rangle \tag{20.18}$$

where the $\langle j'|j\rangle$ term represents the probability amplitude of transmission via the fiber's plane. In reality, this is an undetermined spatially unsymmetric transmission that results in the alteration of the original interferometric pattern.

Under those circumstances, and using the wave function notation of Chapter 4, the probability amplitude has the form of

$$\langle x \,|\, s \rangle' = \sum_{j'=1}^{N} \Psi(r_{j'}) e^{-i\Lambda_{j'}} \sum_{j=1}^{N} \Psi(r_j) e^{-i\Omega_j} \tag{20.19}$$

An alternative and complementary description is to think of $\langle j' | j \rangle$ as representing the propagation from the grating to the new plane established by two narrowly separated large slits. The width of the separation corresponds to the diameter of the spider web fiber. Toward the sides, away from the fiber,

$$\langle j' | j \rangle \approx 1 \tag{20.20}$$

and Equation 20.17 reduces to

$$\langle x \,|\, s \rangle' \approx \sum_{j=1}^{N} \langle x \,|\, j \rangle \langle j \,|\, s \rangle \tag{20.21}$$

so that the interferometric probability distribution is very close to Equation 20.1 or

$$|\langle x \,|\, s \rangle'|^2 \approx \sum_{j=1}^{N} \Psi(r_j)^2 + 2 \sum_{j=1}^{N} \Psi(r_j) \left(\sum_{m=j+1}^{N} \Psi(r_m) \cos(\Omega_m - \Omega_j) \right) \tag{20.22}$$

as shown in Figures 20.10 and 20.11.

However, in the strong interferometric regime, and immediately around the spider web fiber, $\langle j' | j \rangle$ alters the overall probability amplitude in a subtle but measurable way thus leading to the beautiful effect illustrated in Figure 20.5 (measured) and Figure 20.9 (calculated). In that case, the probability amplitude given in Equation 20.18 applies.

From a dimensional perspective, the experiments described here document the unusual opportunity to softly prove, nondestructively, the intra-interferometric propagation using natural fibers with diameters 11–22 times smaller than the width of the grating slits.

20.5.4 Discussion

The soft probing technique described here illustrates that an interferogram can be delicately contacted in a nondestructive manner by a microscopic semitransparent fiber.

However, a straightforward extension of this concept to a measurement technique is not obvious.

Illumination of the *N*-slit array by a single photon means that a single photon illuminates the *N*-slit array and that new probability amplitudes are generated at each slit.

In this case, insertion of a detecting microfiber, or a detecting nanofiber, should result in the absorption of the photon and the immediate collapse of the interferogram.

Illumination of the *N*-slit array by an ensemble of indistinguishable photons means that each of these photons generates a new set of probability amplitudes. Insertion of a detecting microfiber, or a detecting nanofiber, should result in the absorption of a fraction of the photons. The photons that are not absorbed should yield a modified interferogram, as previously seen. In neither case this technique can be used to determine the trajectory of a particular photon or "which hole" a particular photon went through.

Once the idea that a single photon illuminates the whole *N*-slit array, and that each photon in an ensemble of indistinguishable photons illuminates the whole *N*-slit array, is accepted, the questions of "which hole" or "which slit" the photon went through do not apply, and it does not arise.

Problems

20.1 In addition to the list provided in Section 20.2.1, provide three further examples, from the open literature of photon-based, irreversible quantum measurements.

20.2 Rewrite the probability amplitude given in Equation 20.15 in density matrix notation.
That is, as $|\psi\rangle\langle\psi|$.

20.3 Rewrite the density matrix given in Equation 20.16 in the usual probability amplitude notation. That is, simply as $|\psi'\rangle$.

20.4 For $N = 2$, expand Equation 20.18 and multiply it by its complex conjugate to obtain the corresponding probability.

References

Aspect, A., Dalibard, J., and Roger, G. (1982). Experimental test of Bell's inequalities using time-varying analyzers. *Phys. Rev. Lett.* **49**, 1804–1807.

Bell, J. S. (1990). Against measurement. *Phys. World* **3**(8), 33–39.

Dirac, P. A. M. (1978). *The Principles of Quantum Mechanics*, 4th edn. Oxford University, Oxford, U.K.

Duarte, F. J. (1991). Dispersive dye lasers. In *High Power Dye Lasers* (Duarte, F. J., ed.). Springer, Berlin, Germany, Chapter 2.

Duarte, F. J. (1993). On a generalized interference equation and interferometric measurements. *Opt. Commun.* **103**, 8–14.

Duarte, F. J. (2002). Secure interferometric communications in free space. *Opt. Commun.* **205**, 313–319.

Duarte, F. J. (2004). Comment on 'Reflection, refraction and multislit interference. *Eur. J. Phys.* **25**, L57–L58.

Duarte, F. J. (2005). Secure interferometric communications in free space: Enhanced sensitivity for propagation in the metre range. *J. Opt. A: Pure Appl. Opt.* **7**, 73–75.

Duarte, F. J., Taylor, T. S., Black, A. M., and Olivares, I. E. (2013). Diffractive patterns superimposed over propagating *N*-slit interferograms. *J. Mod. Opt.* **60**, 136–140.

Duarte, F. J., Taylor, T. S., Black, A. M., Davenport, W. E., and Varmette, P. G. (2011). *N*-slit interferometer for secure free-space optical communications: 527 m intra interferometric path length. *J. Opt.* **13**, 035710.

Duarte, F. J., Taylor, T. S., Clark, A. B., and Davenport, W. E. (2010). The *N*-slit interferometer: An extended configuration. *J. Opt.* **12**, 015705.

Feynman, R. P., Leighton, R. B., and Sands, M. (1965). *The Feynman Lectures on Physics*, Vol. III, Addison-Wesley, Reading, MA.

Gobby, C., Yuan, Z. L., and Shields, A. J. (2004). Quantum key distribution over 122 km of standard telecom fiber. *Appl. Phys. Lett.* **84**, 3762–3764.

Kocsis, S., Braverman, B., Ravets, S., Stevens, M. J., Mirin, R. P., and Shalm, L. K. (2011). Observing the average trajectories of single photons in a two-slit interferometer. *Science* **332**, 1170–1173.

Lamb, W. E. (1986). Theory of quantum mechanical measurements. In *Proceedings of the 2nd International Symposium Foundations of Quantum Mechanics*, Physical Society of Japan, Tokyo, Japan, pp. 185–192.

Lamb, W. E. (1989). Classical measurements on a quantum mechanical system. *Nucl. Phys. B* **6**, 197–201.

Ma, X., Herbst, T., Scheidl, T., Wang, D., Kropatschek, S., Naylor, W., Mech, A. et al. (2012). Quantum teleportation over 143 kilometres using active feed-forward. *Nature* **489**, 269–273.

van Kampen, N. G. (1988). Ten theorems about quantum mechanical measurements. *Physica A* **153**, 97–113.

von Neumann, J. (1932). *Mathematical Foundations of Quantum Mechanics*, Springer, Berlin, Germany.

Wu, C. S. and Shaknov, I. (1950). The angular correlation of scattered annihilation radiation. *Phys. Rev.* **77**, 13.

21

Interpretational Issues in Quantum Mechanics

21.1 Introduction

In 1935 Einstein, Podolsky, and Rosen (EPR) (1935) wrote a famous paper entitled *Can quantum mechanical description of physical reality considered complete?* This is a cleverly crafted document that argues that in quantum mechanics *"when the momentum of a particle is known, its coordinate has no physical reality"* (Einstein et al., 1935).

Subsequently, the authors go on to conclude that *"the quantum mechanical description of reality given by the wave function is not complete"* (Einstein et al., 1935).

In a reply also entitled *Can quantum mechanical description of physical reality considered complete?* Bohr (1935) uses the Heisenberg uncertainty principle and the principle of *complementarity* to refute the argument of Einstein et al. (1935). Although the matter was considered by many quantum physicists to have been resolved, Bohr's reply, in six plus pages, was somewhat nontransparent and convoluted. Thus, it was not surprising that clever critics, such as Bell (1990), persisted in their criticism.

In this chapter, we examine the EPR argument and the opinion toward interpretational matters from various noted quantum physicists.

21.2 EPR

The central argument in Einstein et al. (1935), from now on also referred as EPR, begins with a definition of reality: *"if... we can predict with certainty the value of a physical quantity, then there exist an element of physical reality corresponding to this physical quantity"* (Einstein et al., 1935).

Here the EPR argument is examined as presented in the paper by Einstein et al. (1935) with a slight change in notation (using P rather than A to describe

the momentum operator). EPR begin their discussion introducing an eigenfunction ψ modified by an operator P so that

$$\psi' \equiv P\psi = a\psi \tag{21.1}$$

where a is said to be a number. Then, EPR indicate that P has the value a
when the particle is in a state ψ. Also, EPR define ψ as

$$\psi = e^{(2\pi i/h)px} \tag{21.2}$$

and introduce the operator

$$P = \frac{h}{2\pi i}\frac{\partial}{\partial x} \tag{21.3}$$

so that

$$\psi' = \frac{h}{2\pi i}\frac{\partial \psi}{\partial x} = p\psi \tag{21.4}$$

Thus, EPR say that in the state defined by Equation 21.2, the momentum has
a certain value P and this momentum is real according to their definition of
reality.

Next, EPR argue that if Equation 21.1 does not hold, then P does not have a
particular value. EPR further argue that this is the case for the coordinate of
the particle; in other words,

$$q\psi = x\psi \neq a\psi \tag{21.5}$$

At this stage EPR say that according to quantum mechanics there is only a
relative probability that a measurement of the coordinate will yield a result
between a and b

$$P(a,b) = \int_a^b \bar{\psi}\psi dx = \int_a^b dx = (b-a) \tag{21.6}$$

Finally, EPR argue that this probability is independent of a and depends only
on the difference $(b - a)$ so that "all values of the coordinate are equally probable" (Einstein et al., 1935). Hence their conclusion "*when the momentum of a
particle is known, its coordinate has no physical reality*" (Einstein et al., 1935). This
is the essence of the EPR argument.

21.3 Bohm Polarization Projection of the EPR Argument

In 1957, Bohm and Aharanov wrote a paper entitled *"Discussion of experimental proof for the paradox of Einstein, Rosen, and Podolsky"* (Bohm and Aharanov, 1957). In this paper Bohm and Aharanov introduce the probability amplitudes for the polarizations of entangled photons traveling in opposite directions (see Chapter 17) originally introduced by Pryce and Ward (1947), Ward (1949), and Snyder et al. (1948). However, Bohm and Aharanov cited only Snyder et al. (1948). As previously described, in Chapter 17, these probability amplitudes are of the form

$$|s\rangle = \frac{1}{\sqrt{2}} (|x\rangle_1 |y\rangle_2 + |y\rangle_1 |x\rangle_2) \tag{21.7}$$

and

$$|s\rangle = \frac{1}{\sqrt{2}} (|x\rangle_1 |y\rangle_2 - |y\rangle_1 |x\rangle_2) \tag{21.8}$$

Clearly in this paper Bohm and Aharanov propose to test the EPR paradox using the polarization properties of correlated photons. As such they refer to the experiment conducted by Wu and Shaknov (1950) that, we know, was based on the polarization correlation theory of Pryce and Ward (1947) (Duarte, 2012a). In this regard, these authors write: "we have essentially the same puzzling kind of correlations in the properties of distant particles, in which the property of anyone photon that is definite is determined by a measurement of a far away photon. Thus, the paradox of EPR can equally well be tested by polarization properties of pairs of photons" (Bohm and Aharanov, 1957).

21.4 Bell's Inequalities

Following the paper of Bohm and Aharanov (1957), J. S. Bell published a famous paper in 1964 entitled *On the Einstein Podolsky Rosen Paradox* (Bell, 1964). In this paper Bell showed that local theories are incompatible with the predictions of quantum mechanics.

An important and profound result indeed.

Bell did so by deriving a set of inequalities and then showing that the quantum predictions were outside the boundaries of these inequalities. Here, we follow Bell (1964) in the introduction of his famous inequalities.

Bell defines a hidden variable λ having a probability density $\rho(\lambda)$ such that

$$\int \rho(\lambda)d\lambda = 1 \tag{21.9}$$

then the correlation between observables $A(a, \lambda)$ and $B(a, \lambda)$ is given by Bell as

$$P(a,b) = \int A(a,\lambda)B(b,\lambda)\rho(\lambda)d\lambda \tag{21.10}$$

and since $|A(a, \lambda)| = 1$

$$|P(a,b) - P(a,b')| \leq \int |[B(b,\lambda) - B(b',\lambda)]| \rho(\lambda)d\lambda \tag{21.11}$$

$$|P(a',b) + P(a',b')| \leq \int |[B(b,\lambda) + B(b',\lambda)]| \rho(\lambda)d\lambda \tag{21.12}$$

$$|P(a,b) - P(a,b')| + |P(a',b) + P(a',b')| \leq$$

$$\int [|B(b,\lambda) - B(b',\lambda)| + |B(b,\lambda) + B(b',\lambda)|]\rho(\lambda)d\lambda \tag{21.13}$$

given that $|B(b, \lambda)| = \pm 1$, $|B(b', \lambda)| = \pm 1$, and using Equation 21.9, we have *Bell's inequality* (Bell, 1964)

$$|P(a,b) - P(a,b')| + |P(a',b) + P(a',b')| \leq 2 \tag{21.14}$$

Now, using the PWS probability amplitude, given in Equation 21.8, to calculate the corresponding quantum probabilities for the $|P(a,b) - P(a,b')| + |P(a',b) + P(a',b')|$ expression, yields numerical values *greater than* 2. Hence, Bell's inequality is violated.

Now consider the PWS probability amplitude, that is, Equation 21.8, for photon 2 emerging through polarizer 2 and photon 1 emerging through polarizer 1, as illustrated in Figure 21.1:

$$|s\rangle = \frac{1}{\sqrt{2}}(|x\rangle_1 |y\rangle_2 - |y\rangle_1 |x\rangle_2) \tag{21.15}$$

$$|s\rangle = \frac{1}{\sqrt{2}}(\sin\theta_1 \cos\theta_2 - \cos\theta_1 \sin\theta_2) \tag{21.16}$$

$$|s\rangle = \frac{1}{\sqrt{2}}\sin(\theta_1 - \theta_2) \tag{21.17}$$

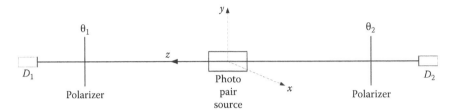

FIGURE 21.1
Generic two-photon entanglement experimental configuration. θ_1 and θ_2 refer to the orientation of the polarizers. D_1 and D_2 designate the respective detectors.

(see Chapter 17, Equations 17.48 through 17.54). The alternative situation corresponds to photon 1 not being allowed passage through polarizer 1 (due to a rotation of $\pi/2$ in θ_1) so that

$$|s\rangle = \frac{1}{\sqrt{2}}\cos(\theta_1 - \theta_2) \qquad (21.18)$$

Using the notation of Mandel and Wolf (1995) (+) to denote transmission through a polarizer and (–) to denote absorption, then the probability alternatives can be described by

$$p(+\theta_1, +\theta_2) = p(-\theta_1, -\theta_2) = \frac{1}{2}\sin^2(\theta_1 - \theta_2) \qquad (21.19)$$

$$p(+\theta_1, -\theta_2) = p(-\theta_1, +\theta_2) = \frac{1}{2}\cos^2(\theta_1 - \theta_2) \qquad (21.20)$$

so that, using the appropriate geometrical identity, the overall probability $P(\theta_1, \theta_2)$, is (Mandel and Wolf, 1995)

$$P(\theta_1, \theta_2) = p(+\theta_1, +\theta_2) + p(-\theta_1, -\theta_2) - p(+\theta_1, -\theta_2) - p(-\theta_1, +\theta_2) = -\cos 2(\theta_1 - \theta_2)$$

$$P(\theta_1, \theta_2) = -\cos 2(\theta_1 - \theta_2) \qquad (21.21)$$

21.4.1 Example

Using a set of angles defined by $\theta_1 = 0, \theta_2 = \pi/3, \theta_1' = \pi/6, \theta_2' = 0$, Equation 21.21 applied to the left-hand side of Bell's inequality yields

$$|0.5 - (-1.0)| + |(-0.5) + (-0.5)| = 2.5$$

which means

$$| P(\theta_1, \theta_2) - P(\theta_1, \theta_2') | + | P(\theta_1', \theta_2) + P(\theta_1', \theta_2') | = 2.5$$

thus violating Bell's inequality.

21.4.2 Discussion

Implicit in this section has been the assumption that the detectors are ideal detectors functioning at ~100% efficiency. That was done to keep the presentation simple and with the benefit of knowing that overwhelmingly published experimental results of optical configurations measuring polarization correlations of counter propagating entangled photons do confirm violation of Bell's inequalities. Thus, the pragmatic conclusion is to confirm the validity of quantum mechanics and to dismiss local hidden variable theories.

   Remarkably, years after his insightful contribution Bell became a sharp critic of orthodox quantum mechanics. His criticisms were aimed at the foundations of quantum mechanics and toward what he called the *"why bother"* attitude among distinguished practitioners of quantum mechanics such as Dirac: "It seems to me that it is among the most sure-footed of quantum physicist, those who have it *in their bones*, that one finds the greatest impatience with the idea that the 'foundations of quantum mechanics' might need some attention" (Bell, 1990). In particular, Bell was critical of the concept of *measurement* as used in quantum mechanics (Bell, 1990). In the next section attention is brought up on issues of interpretation.

21.5 Some Prominent Quantum Physicists on Issues of Interpretation

Quantum mechanics is a wondrous branch of physics. As such, this section should begin by stating the thoughts John Clive Ward on quantum mechanics (Ward, 2004): *"The inner mysteries of quantum mechanics require a willingness to extend one's mental processes into a strange world of phantom possibilities, endlessly branching into more and more abstruse chains of coupled logical networks, endlessly extending themselves forward and even backwards in time."* John Ward was one of those prominent quantum physicists who never expressed any doubts on the correctness or effectiveness of quantum mechanics. Nor did he *bother* on issues of interpretation (Ward, 2004; Duarte, 2012b).

   The interpretation of quantum mechanics has been the subject of many publications and books (Bell, 1988; Selleri, 1988; Wallace, 1996). In this section, a pragmatic perspective on this topic is presented that is mainly derived from the Feynman school of thought: *"unless a thing can be defined by measurement, it has no place in the theory"* and *"already in classical mechanics there was indeterminability from a practical point of view"* (Feynman et al., 1965). In this regard, Feynman was keenly aware of the crucial role of the Heisenberg uncertainty principle in the formulation of quantum mechanics: *"the uncertainty principle protects quantum mechanics… quantum mechanics maintains its perilous but still correct existence"* (Feynman et al., 1965).

Dirac was famously not impressed by discussions on the interpretation of quantum mechanics and in one of his last papers he wrote: *"The interpretation of quantum mechanics has been dealt with by many authors, and I do not want to discuss it here. I want to deal with more fundamental things"* (Dirac, 1987). Nevertheless, in a visit to Sydney, in August 1975, Dirac did refer to the interpretation of quantum mechanics and did refer to the Bohr–Einstein controversy. In this regard he went on to say that according to standard atomic physics, "Bohr was right." However, he left open the possibility for some kind of improved quantum mechanics of the future. But he warned that an improved version would be "at the expense of some other basic idea" (Duarte, 2012b).

More recently Willis Lamb, yet another noted quantum physicist, assigned the interpretational problems of quantum mechanics to *"historical misunderstandings"* (Lamb, 2001). In his broad critique, Lamb included the EPR argument in a list associated to these misunderstandings and went as far as describing Bell's inequalities as "unnecessary" (Lamb, 2001).

As a footnote to this discussion, I will bring into focus a letter I wrote in regard to the Dirac description of interference (Duarte, 1998). For a while many people criticized the Dirac description of interference due to the sentence: *"Each photon then interferes only with itself"* (Dirac, 1978). This sentence is the conclusion of an argument that begins considering a beam of monochromatic light, which means a population of indistinguishable photons, "with a large number of photons" (Dirac, 1978). The beam is made to split into two components of equal intensity and then made to interfere (as in a Mach–Zehnder interferometer). Dirac then explains that each photon goes partially into each component, and hence *"each photon then interferes with itself"* (Dirac, 1978). That is, each photon has a probability amplitude to follow either path, and it is the addition of these probability amplitudes, multiplied with its corresponding complex conjugate, that accurately and correctly describe the observed interference patterns (Duarte, 1998, 2003).

In other words, Dirac writes about a single beam of monochromatic light with a large number of photons that is equivalent to a single narrow-linewidth high-power laser beam. In other words, it is equivalent to a laser beam comprised of a population of indistinguishable photons. The key principle here is that photon interference is a phenomenon that involves either single photons or populations of *indistinguishable photons*. In this regard, Dirac outlined the principles of laser interference back in 1947 (Duarte, 2003), and thus he should be considered the father of *laser optics* and *quantum optics*.

Finally, Dirac's statement does not exclude interference between two different laser sources as long as the emission from these two sources is indistinguishable. In other words, interference from two separate narrow-linewidth lasers at the same central wavelength will register with sharp high-contrast interferograms with high visibility approaching unity ($\mathcal{V} \approx 1$).

Going back to Dirac's observation about a possible future improved quantum mechanics, that might well be possible. However, from our perspective a possible future improvement will have to wait for a yet undiscovered principle.

21.6 Heisenberg's Uncertainty Principle and EPR

"When the momentum of a particle is known, its coordinate has no physical reality" (Einstein et al., 1935) is an essential component of the EPR argument. As mentioned previously, Bohr (1935) used both the principle of complementarity and the Heisenberg uncertainty principle in response to the EPR argument. However, Bohr's did not develop a transparent response based on the uncertainty principle.

Here, a direct approach to this issue is taken making explicit use of the uncertainty principle itself (Duarte, 2012b). The Heisenberg uncertainty principle is given by (Dirac, 1978)

$$\Delta x \Delta p \approx h \tag{21.22}$$

and one of its alternative forms is (Feynman et al., 1965)

$$\Delta x \approx \frac{h}{\Delta p} \tag{21.23}$$

Now, if we measure the momentum p of a particle, we can only measure

$$p \pm \Delta p \tag{21.24}$$

An *absolutely exact measurement of momentum* p with $\Delta p = 0$ is *physically impossible*.

The presence of uncertainties and errors in measurements has been known to physicists since the dawn physics and optics (Newton, 1687, 1704). The EPR sentence *"when the momentum of a particle is known, its coordinate has no physical reality"* (Einstein et al., 1935) implies an idealized exact measurement of momentum p with $\Delta p = 0$. Once a real physical measurement of momentum is made with a nonzero estimate of the error in the measurement, that is, $p \pm \Delta p$, then the coordinate can be determined according to the uncertainty principle

$$\Delta x \approx \frac{h}{\Delta p}$$

and the *"all values"* spread in the coordinate, as feared by Einstein et al. (1935), is not allowed. Once the *"all values"* spread in the coordinate is found physically untenable, the claim of *"no physical reality"* is neutralized. Hence, the EPR conclusion that *"the quantum mechanical description of physical reality ... is not complete"* can be negated.

In summary, the EPR claim of an *"all values"* spread in the coordinate depends on an idealized absolute and exact measurement of p with $\Delta p = 0$. Since this is physically impossible, the claim of *"no physical reality"* can be negated.

While writing this book we came across a paragraph in Dirac's book that very succinctly reinforces the argument made here. The paragraph in question follows a discussion on the uncertainty principle: "it is evident physically that a state for which all values of q are equally probable, or one for which all values of p are equally probable, cannot be attained in practice" (Dirac, 1978). Notice that the book we are using is a revised printing of the 1947 edition. This means that observant physicists like Feynman, Ward, and Lamb might have been very much aware of the existence of this little-known Dirac dictum.

21.7 van Kampen's Quantum Theorems

A useful pragmatic perspective on quantum mechanics is provided by van Kampen (1988). He argues that the difficulties in quantum mechanics only surface when one starts philosophizing: "this philosophizing has given rise to a number of 'interpretations', in which ψ is endowed with more physical significance than is needed" (van Kampen, 1988). Next, some of the van Kampen theorems are outlined and explained in some detail:

Theorem I states that *"quantum mechanics works."* This perspective was already eloquently exposed by Feynman in his lectures when he refers to quantum mechanics as "a great triumph of twentieth century physics" (Feynman et al., 1965). The transistor and the laser are both testimonies that quantum mechanics works. This concept is self-explanatory and one of the central motivations for this book.

Theorem II states that quantum mechanics *"is concerned with macroscopic phenomena... not perturbed by observation."* Here, van Kampen (1989) argues that what is observed is $|\psi|^2$, multiplied by N, which is the number of either electron or photons in a beam. He then states that $|\psi|^2$, for a single photon or electron, occurs in the calculation but is not observed experimentally. This perspective is restrictive given recent single photon experiments. Thus, this theorem should be extended to include the word *microscopic*, along the lines *quantum mechanics is concerned with macroscopic, or microscopic, phenomena unperturbed by observation. Theorem III* states that both ψ and $|\psi|^2$ are mathematical tools and more specifically that $|\psi|^2$ is not observed directly. This perspective agrees with our experience of measuring interferometric intensity distribution that is proportional to the probability distributions; in other words, $I = K|\psi|^2$, where K is a constant.

Theorem IV reemphasizes the role of ψ simply as a computational tool.

Theorem V states that the measuring apparatus, in quantum experiments, *"consists of a macroscopic system prepared in a metastable state."* Here, van Kampen (1988) argues that the transition from the metastable into the stable macrostate provides the energy needed to make the microscopic phenomena macroscopically visible. He further argues that this is the reason why the measuring process is irreversible. This posture is compatible with our previous observation of the measurement process (see Chapter 20) in which a quantum microscopic event irreversibly interacts with the detector and triggers a cascade process of amplification in the detector that makes the initial quantum event macroscopically visible.

21.8 On Probabilities and Probability Amplitudes

When a light beam propagates in free space, its spatial intensity distribution is proportional to its probability distribution, that is,

$$I(P) = K P(\lambda, N, \Omega) \tag{21.25}$$

where K is a constant of proportionality with units of J s^{-1} m^{-2}. This intensity distribution $I(P)$ is what is usually measured or recorded by the macroscopic detector and the corresponding classical apparatus connected to the detector. For the 1D case, the probability distribution can have the form of

$$P(\lambda, N, \Omega) = \sum_{i=1}^{N} \vartheta(r_j)^2 + 2\sum_{i=1}^{N} \vartheta(r_j) \left(\sum_{m=i+1}^{N} \vartheta(r_m)\cos(\Omega_m - \Omega_j) \right) \tag{21.26}$$

This equation applies either to the propagation of a single photon or to the propagation of an ensemble of indistinguishable photons. If the undetected propagating interferogram, described by Equation 21.25, is allowed to illuminate a new interferometric plane (j), composed of N slits, a whole new array of probability amplitudes is generated (see Figure 21.2). Once this interaction occurs, the original $I(P)$ distribution gives origin to a whole new *array* of probability amplitudes represented by $\langle x|j\rangle\langle j|s\rangle$, where s is the source and x the new interferometric plane.

The new series of probability amplitudes is strongly influenced by the geometry of the interferometric array and is described by the Dirac principle

$$\langle x|s\rangle = \sum_{j=1}^{N} \langle x|j\rangle\langle j|s\rangle$$

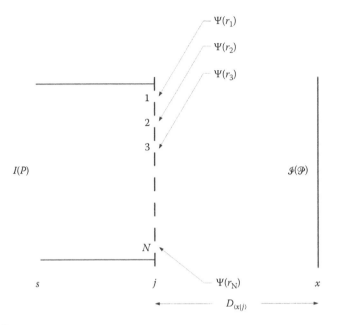

FIGURE 21.2
Propagating interferometric intensity distribution $I(P)$ interacts with N-slit array, thus producing a new interferometric intensity distribution described by $\mathcal{I}(\mathcal{P})$. Wave function $\Psi(r_1)$ is associated with slit $j = 1$, $\Psi(r_2)$ is associated with slit $j = 2$, $\Psi(r_3)$ is associated with slit $j = 3$, and so on.

where the probability amplitudes are (see Chapter 4)

$$\langle j \,|\, s \rangle = \Psi(r_{\langle j|s \rangle})e^{-i\theta_j} \tag{21.27}$$

$$\langle x \,|\, j \rangle = \Psi(r_{\langle x|j \rangle})e^{-i\phi_j} \tag{21.28}$$

so that

$$\langle x \,|\, s \rangle = \sum_{j=1}^{N} \Psi(r_j)e^{-i\Omega_j} \tag{21.29}$$

Here, the amplitudes can have a Gaussian form and

$$\Psi(r_j) = \Psi(r_{\langle x|j \rangle})\Psi(r_{\langle j|s \rangle}) \tag{21.30}$$

while the new phase term becomes

$$\Omega_j = (\theta_j + \phi_j) \tag{21.31}$$

Multiplication of Equation 21.29 with its complex conjugate gives rise to a *new measurable* probability distribution

$$|\langle x | s \rangle|^2 = \sum_{j=1}^{N} \Psi(r_j)^2 + 2 \sum_{j=1}^{N} \Psi(r_j) \left(\sum_{m=j+1}^{N} \Psi(r_m) \cos(\Omega_m - \Omega_j) \right) \quad (21.32)$$

at any interferometric plane x at a distance $D_{\langle x|j \rangle}$ from the interferometric array j. Thus, the new probability distribution becomes

$$\mathcal{P}(\lambda, N, \Omega) = |\langle x | s \rangle|^2 \quad (21.33)$$

and the new intensity distribution is given by

$$\mathcal{I}(\mathcal{P}) = K \mathcal{P}(\lambda, N, \Omega) \quad (21.34)$$

In summary, a measurable interferometric distribution $\mathcal{I}(\mathcal{P})$ interacts with a new interferometric array (j), in an irreversible process, thus creating a whole new series of probability amplitudes. This new series of probability amplitudes are represented by a corresponding series of wave functions

$$\Psi(r_1), \Psi(r_2), \Psi(r_3), ... \Psi(r_N)$$

Immediately following passage of the N-slit array, and due to diffraction, these wave functions become entangled and give rise to a new measurable interferometric intensity distribution as described by Equations 21.32 through 21.34.

21.9 Comment on the Interpretational Issue

There is a plethora of interpretations of quantum mechanics. Among the most prominent interpretations one finds are the Copenhagen interpretation and the many-worlds interpretation. Albeit this is an interesting area of discussion, from the pragmatic perspective of physics, interpretations are not necessary to solve problems or predict the result of an experiment.

Our approach to quantum mechanics is to use the Dirac principles, as presented by Dirac (1978) and elucidated by Feynman (Feynman et al., 1965), while being perfectly aware that it is the experiment that has the final say.

In this regard, we very humbly assume that nature is far more subtle than our philosophical abilities and accept, as Ward and others have done, the physics of entanglement and the intricacy of the machinery of interference, as *nature's way*. In other words, we postulate that the most efficient and practical interpretation of quantum mechanics is... *no interpretation at all*.

Problems

21.1 Show that Equation 21.14 follows from Equation 21.13.

21.2 For the following set of polarizer angles, $\theta_1 = 0, \theta_2 = \pi/6, \theta_1' = \pi/3, \theta_2' = 0$, calculate Bell's inequality.

21.3 How would the argument in Section 21.6 change if the general form of the uncertainty principle, as given in Chapter 3, were used rather than $\Delta x \Delta p \approx h$? Would the conclusion remain unchanged?

21.4 Refer to Equation 21.25 or 21.34: comment on the possible components that might comprise the overall constant K.

References

Bell, J. S. (1964). On the Einstein-Podolsky-Rosen paradox. *Physics* **1**, 195–200.

Bell, J. S. (1990). Against measurement. *Phys. World* **3**(8), 33–39.

Bell, J. S. (1988). *Speakable and Unspeakable in Quantum Mechanics*, Cambridge University, Cambridge, U.K.

Bohm, D. and Aharonov, Y. (1957). Discussion of experimental proof for the paradox of Einstein, Rosen, and Podolsky. *Phys. Rev.* **108**, 1070–1076.

Bohr, N. (1935). Can quantum-mechanical description of physical reality be considered complete? *Phys. Rev.* **48**, 696–701.

Dirac, P. A. M. (1978). *The Principles of Quantum Mechanics*, 4th edn. Oxford, London, U.K.

Dirac, P. A. M. (1987). The inadequacies of quantum field theories. In *Paul Adrien Maurice Dirac* (Kursunoglu, B. N. and Wigner, E. P., eds.). Cambridge University, Cambridge, U.K., Chapter 15.

Duarte F. J. (1998). Interference of two independent sources. *Am. J. Phys.* **66**, 662–663.

Duarte, F. J. (2003). *Tunable Laser Optics*, Elsevier-Academic, New York.

Duarte, F. J. (2012a). The origin of quantum entanglement experiments based on polarization measurements. *Eur. Phys. J. H* **37**, 311–318.

Duarte, F. J. (2012b). *Laser Physicist*, Optics Journal, New York.

Einstein, A., Podolsky, B., and Rosen, N. (1935). Can quantum mechanical description of physical reality be considered complete? *Phys. Rev.* **47**, 777–780.

Feynman, R. P., Leighton, R. B., and Sands, M. (1965). *The Feynman Lectures on Physics*, Vol. III, Addison-Wesley, Reading, MA.

Lamb, W. E. (2001). Super classical quantum mechanics: the best interpretation of non- relativistic quantum mechanics. *Am. J. Phys.* **69**, 413–421.

Mandel, L. and Wolf, E. (1995). *Optical Coherence and Quantum Optics*, Cambridge University, Cambridge, U.K.

Newton, I. (1687). *Principia Mathematica*, Royal Society, London, U.K.

Newton, I. (1704). *Opticks*, Royal Society, London, U.K.

Pryce, M. H. L. and Ward, J. C. (1947). Angular correlation effects with annihilation radiation. *Nature* **160**, 435.

Selleri, F. (1988). *Quantum Mechanics Versus Local Realism*, Plenum, New York.

Snyder, H. S., Pasternack, S., and Hornbostel, J. (1948). Angular correlation of scattered annihilation radiation. *Phys. Rev.* **73**, 440–448.

van Kampen, N. G. (1988). Ten theorems about quantum mechanical measurements. *Physica A* **153**, 97–113.

Wallace, R. P. (1996). *Paradox Lost*, Springer, Berlin, Germany.

Ward, J. C. (1949). *Some Properties of the Elementary Particles*. D. Phil Thesis, Oxford University, Oxford.

Ward, J. C. (2004). *Memoirs of a Theoretical Physicists*, Optics Journal, New York.

Wu, C. S. and Shaknov, I. (1950). The angular correlation of scattered annihilation radiation. *Phys. Rev.* **77**, 136.

Appendix A: Survey of Laser Emission Characteristics

A.1 Introduction

Lasers and laser emission characteristics are central to quantum optics. Here, we provide, in tabular form, a brief survey of laser emission characteristics. This exposition gives emphasis to wavelength coverage and includes published data compiled in a review by Duarte (2003), in addition to newly selected results from the open literature. The vast majority of the lasers included in this survey can be classified as macroscopic emission sources of coherent radiation. An alternative description, for these lasers, would be macroscopic sources of quantum emission.

The aim of this appendix is to provide an expeditious first reference to the various types of lasers available and their respective wavelength coverage. References therein should direct the reader to further details and more specialized information.

A.2 Gas Lasers

The emission of high-pulsed energies, or high-average powers, requires the rapid removal of heat. Gas lasers are well suited for the efficient removal of excess heat. In this section, representative gas lasers from the subclasses of molecular and atomic transition laser are included. Some of the molecular lasers can be tuned.

A.2.1 Pulsed Molecular Gas Lasers

Table A.1 lists the molecular transition and wavelength of some representative ultraviolet molecular lasers including excimer lasers and the nitrogen laser.

TABLE A.1

Ultraviolet and Visible Molecular Pulsed Gas Lasers

Laser	Transition	λ (nm)	prf[a] (Hz)	Bandwidth (GHz)	Ref.[b]
KrF	$B^2\Sigma_{1/2}^+ - X^2\Sigma_{1/2}^+$	248	200	10500[c]	Loree et al. (1978)
XeCl		308	1000	204	McKee (1985)
XeF		351	200	187	Yang et al. (1988)
N_2	$C^3\Pi_u - B^3\Pi_g$	337.1	100	203	Woodward et al. (1973)
HgBr	$B^2\Sigma_{1/2}^+ - X^2\Sigma_{1/2}^+$	502	100	918	Shay et al. (1981)

Source: Duarte, F.J., *Tunable Laser Optics*, Elsevier-Academic, New York, 2003.
[a] prf: pulse repetition frequency (typical).
[b] References provide bandwidth only.
[c] Tuning range.

Table A.2 lists a few tunable narrow-linewidth excimer lasers. MPLG refers to multiple-prism Littrow grating configuration, while GIG refers to the grazing-incidence grating configuration (see Appendix B). A review on tunable excimer lasers is given by Sze and Harris (1995).

Table A.3 lists pulsed narrow-linewidth tunable CO_2 lasers. HMPGIG refers to a hybrid multiple-prism GIG configuration.

TABLE A.2

Narrow-Linewidth Tunable Pulsed Excimer Laser Oscillators

Laser	Oscillator	λ (nm)	$\Delta\nu$	Pulse Energy	Ref.
KrF	GI	248	9 GHz	15 µJ	Caro et al. (1982)
XeCl	GI	308	1.5 GHz	~1 mJ	Sze et al. (1986)
XeCl	GI	308	1 GHz	4 mJ	Sugii et al. (1987)
XeCl	3 etalons	308	150 MHz	2–5 µJ	Pacala et al. (1984)

Source: Duarte, F.J., *Tunable Laser Optics*, Elsevier-Academic, New York, 2003.

TABLE A.3

Narrow-Linewidth Tunable Pulsed CO_2 Lasers

Laser	Oscillator	λ (nm)	$\Delta\nu$ (MHz)	Pulse Energy (mJ)	Ref.
CO_2	GIG[a]	10,591	700	230	Bobrovskii et al. (1987)
CO_2	MPLG	10,591	140	200	Duarte (1985)
CO_2	HMPGIG	10,591	107[b]	85	Duarte (1985)

Source: Duarte, F.J., *Tunable Laser Optics*, Elsevier-Academic, New York, 2003.
[a] Open-cavity configuration (see Appendix B).
[b] This laser linewidth corresponds to near single-longitudinal-mode oscillation.

A.2.2 Pulsed Atomic Metal Vapor Lasers

Pulsed atomic metal vapor lasers include the copper vapor laser (CVL) that is a high-prf device particularly well suited for industrial applications and the excitation of high-power tunable dye lasers (Webb, 1991). Table A.4 includes transition assignments and emission wavelengths for the copper and gold metal vapor lasers.

A.2.3 Continuous Wave Gas Lasers

This category includes a number of important lasers such as the He–Ne laser and the Ar ion laser. Due to their cavity geometry, long active lengths, and relatively narrow apertures, these lasers emit single-transverse-mode (TEM$_{00}$) beams with Gaussian or near-Gaussian profiles. These features are quite advantageous for a number of applications including interferometry and metrology. Also, since emission originates from distinct atomic or ionic transitions, their intrinsic emission linewidth can be relatively narrow (~1 GHz, for some He–Ne lasers, for instance). Table A.5 lists visible transitions available from the He–Ne laser, while Table A.6 includes principal visible transition for ionic lasers.

TABLE A.4

Atomic Pulsed Gas Lasers

Laser	Transition	λ (nm)	prf[a] (kHz)	Bandwidth (GHz)	Ref.[b]
Cu	$^2P_{3/2} - {}^2D_{5/2}$	510.5	2–30	~7	Tenenbaum et al. (1980)
	$^2P_{1/2} - {}^2D_{3/2}$	578.2	2–30	~11	Tenenbaum et al. (1980)
Au	$^2P_{1/2} - {}^2D_{3/2}$	627.8	5–20		

Source: Duarte, F.J., *Tunable Laser Optics*, Elsevier-Academic, New York, 2003.
[a] prf: pulse repetition frequency.
[b] References relate to bandwidth only.

TABLE A.5

Visible Transitions of CW He–Ne Lasers

Transition[a]	λ (nm)[b]
$3s_2 - 2p_{10}$	543.30
$3s_2 - 2p_8$	593.93
$3s_2 - 2p_7$	604.61
$3s_2 - 2p_6$	611.80
$3s_2 - 2p_4$	632.82

Source: Duarte, F.J., *Tunable Laser Optics*, Elsevier-Academic, New York, 2003.
[a] Transition assignment from Willett (1974).
[b] Wavelength values from Beck et al. (1976).

TABLE A.6

Principal Visible Transitions of Ionized CW Gas Lasers

Active Medium	Transition[a]	λ (nm)[b]
Ar⁺	$4p\,^2S^0_{1/2} - 4s\,^2P_{1/2}$	457.93
	$4p\,^2P^0_{3/2} - 4s\,^2P_{1/2}$	476.49
	$4p\,^2D^0_{5/2} - 4s\,^2P_{3/2}$	487.99
	$4p\,^2D^0_{3/2} - 4s\,^2P_{1/2}$	496.51
	$4p'\,^2F^0_{5/2} - 3d\,^2D_{3/2}$	501.72
	$4p\,^4D^0_{5/2} - 4s\,^2P_{3/2}$	514.53
Kr⁺	$4p\,^4P^0_{3/2} - 5s\,^4P_{3/2}$	520.83
	$5p\,^4P^0_{5/2} - 5s\,^4P_{3/2}$	530.87
	$5p\,^4D^0_{5/2} - 5s\,^2P_{3/2}$	568.19
	$5p\,^4P^0_{5/2} - 5s\,^2P_{3/2}$	647.09
Zn⁺	$4f\,^2F^0_{7/2} - 4d\,^2D_{5/2}$	492.40
	$4s^2\,^2D_{3/2} - 4p\,^2P^0_{1/2}$	589.44
	$4s^2\,^2D_{3/2} - 4p\,^2P^0_{1/2}$	610.25
Cd⁺	$5s^2\,^2D_{5/2} - 5p\,^2P^0_{3/2}$	441.56
	$4f\,^2F^0_{5/2} - 5d\,^2D_{3/2}$	533.75
	$6g\,^2G_{7/2} - 4f\,^2F^0_{5/2}$	635.48
	$6g\,^2G_{9/2} - 4f\,^2F^0_{7/2}$	636.00
I⁺	$6p'\,^3D_2 - 6s'\,^3D^0_2$	540.73
	$6p'\,^3F_2 - 6s'\,^3D_2$	567.81
	$6p'\,^3D_2 - 6s'\,^3D^0_1$	576.07
	$6p'\,^3D_1 - 6s'\,^3D^0_2$	612.75

Source: Duarte, F.J., *Tunable Laser Optics*, Elsevier-Academic,
          New York, 2003.
[a] Transition assignment from Willett (1974).
[b] Wavelength values from Beck et al. (1976).

A.3 Dye Lasers

Dye lasers are inherently tunable and well suited for the generation of high-average powers (Duarte, 1991). These lasers can be divided in two main categories: pulsed dye lasers and CW dye lasers. These lasers, using numerous dyes, span the electromagnetic spectrum from the near ultraviolet to the near infrared. More specifically, the wavelength coverage available from dye lasers is approximately 320–1000 nm (Duarte, 2003). Some individual dyes, such as rhodamine 6G, can provide tuning ranges in excess of 50 nm.

A.3.1 Pulsed Dye Lasers

Laser-pumped dye lasers provide either very large pulsed energies or high-average powers as indicated in Table A.7. Narrow-linewidth pulsed dye-laser oscillators are highlighted in Table A.8. Oscillator–amplifier systems can produce tunable narrow-linewidth emission at average powers in the kilowatt regime (Bass et al., 1992).

Flashlamp-pumped dye lasers can emit hundreds of Joules per pulse and high-average powers as indicated in Table A.9. The performance of a ruggedized narrow-linewidth flashlamp-pumped dye-laser oscillators is included in Table A.8.

TABLE A.7

High-Power Laser-Pumped Dye Lasers

Excitation Laser	Pulse Energy	prf	Average Power	η (%)[a]	Tuning Range (nm)	Ref.
XeCl	800 J[b]	Low		27	475[c]	Tang et al. (1987)
XeCl	200 mJ	250 Hz	50 W	20	401.5[c]	Tallman and Tennant (1991)
CVL	190 mJ	13.2 kHz	2.5 kW	50	550–650	Bass et al. (1992)

Source: Duarte, F.J., *Tunable Laser Optics*, Elsevier-Academic, New York, 2003.
[a] Efficiency.
[b] Pulse length quoted at 500 ns.
[c] Central wavelength.

TABLE A.8

Narrow-Linewidth Liquid Dye Lasers

Excitation Source	Cavity	$\Delta \nu$ (MHz)	Δt (ns)	Tuning Range (nm)	η (%)
CVL[a,b]	MPLG[c,d]	60[e]	30	560–600	5
CVL[f]	HMPGIG[g]	400[e]	12	565–603	4–5
Nd:YAG[h,i]	HMPGIG[g]	650	5	425–457	~9[j]
Flashlamp[k]	MPLG[c]	300	70	565–605	

[a] Copper vapor laser.
[b] From Bernhardt and Rasmussen (1981).
[c] Multiple-prism Littrow grating configuration.
[d] Includes an intracavity etalon.
[e] Single-longitudinal-mode emission.
[f] From Duarte and Piper (1984).
[g] Hybrid multiple-prism GIG configuration.
[h] Uses the third harmonics of the fundamental (3ν).
[i] From Dupre (1987).
[j] This efficiency includes amplification via two stages.
[k] From Duarte et al. (1991); output laser energy is ~3 mJ per pulse.

TABLE A.9

High-Energy Flashlamp-Pumped Dye Lasers

Excitation	Pulse Duration (μs)	Pulse Energy (J)	η (%)	Ref.
Linear[a,b]	7	40	0.4	Fort and Moulin (1987)
Transverse[b,c]	5	140[d]	1.8	Klimek et al. (1992)
Coaxial[b]	10	400	0.8	Baltakov et al. (1974)

Source: Duarte, F.J., *Tunable Laser Optics*, Elsevier-Academic, New York, 2003.
[a] Employs 12 flashlamps.
[b] Uses rhodamine 6G dye.
[c] Employs 16 flashlamps in a transverse configuration.
[d] Yields an average power of 1.4 kW at a prf of 10 Hz.

TABLE A.10

Narrow-Linewidth Solid-State Organic Dye-Laser Oscillators

Cavity	Matrix	$\Delta \nu$ (MHz)	Δt (ns)	Tuning Range (nm)	Pulse Energy (mJ)
MPLG[a,b]	MPMMA[c]	350[d]	3	550–603	~0.1[e]
MPLG[f,g]	HEMA:MMA[h]	650	105	564–602	~0.4

[a] Laser-pumped dye laser; optimized cavity. LPDL,[a,b] FLPDL.[e,f]
[b] From Duarte (1999).
[c] MPMMA doped with rhodamine 6 G.
[d] Linewidth corresponds to single-longitudinal-mode oscillation.
[e] Power per pulse ~33 kW.
[f] Flashlamp-pumped dye laser.
[g] From Duarte et al. (1998).
[h] HEMA: MMA doped with rhodamine 6 G.

Narrow-linewidth solid-state dye-laser oscillators, in the short- and long-pulse regimes, are listed in Table A.10. These organic lasers use highly homogeneous rhodamine 6G dye-doped polymer matrices as gain medium (Duarte, 1994).

A.3.2 Continuous Wave Dye Lasers

High-power CW dye lasers are listed in Table A.11. Femtosecond pulse dye lasers utilizing prismatic pulse compression are listed in Table A.12. A thorough review on femtosecond dye lasers is given by Diels (1990). Frequency-stabilized CW dye lasers can be found elsewhere (Hollberg, 1990; Duarte, 2003).

TABLE A.11

High-Power CW Dye Lasers

Cavity	Spectral Range (nm)	Linewidth	Power	η (%)	Ref.
Linear[a]	560–650	SLM[b]	33 W[c]	17	Baving et al. (1982)
Ring[a]	407–887[d]	SLM	5.6 W[e]	23.3	Johnston and Duarte (2002)

Source: Duarte, F.J., *Tunable Laser Optics*, Elsevier-Academic, New York, 2003.
[a] Under Ar⁺ laser excitation.
[b] Linewidth values can be in the few MHz range.
[c] Using rhodamine 6G at 0.94 mM.
[d] Using 11 dyes.
[e] Using rhodamine 6G.

TABLE A.12

Femtosecond Pulse Dye Lasers

Pulse Compressor	Δt (fs)	Ref.
Single prism	60	Dietel et al. (1983)
Double prism	50	Kafka and Baer (1987)
Double prism	18	Osvay et al. (2005)
Four prisms	29	Kubota et al. (1988)
Four prisms plus grating pair	6	Fork et al. (1987)

A.4 Solid-State Lasers

The solid-state laser field is a vast field that includes traditional crystalline materials and fiber gain media. These lasers can emit both in the pulse and the CW regime. In this section, some of the most well-known gain media are surveyed with emphases on spectral characteristics. An authoritative review on this class of lasers is given by Barnes (1995a).

A.4.1 Ionic Solid-State Lasers

These lasers emit at fixed wavelengths in the infrared. They are often used with frequency multiplication techniques to reach the visible spectrum. Table A.13 includes transitions and corresponding wavelengths of various lasers in this class.

TABLE A.13

Emission Wavelengths from Ionic
Solid-State Lasers

Ion	Transition	λ (nm)
Yb^{3+}	$^2F_{5/2} - {}^2F_{7/2}$	1015
Nd^{3+}	$^4F_{3/2} - {}^4I_{11/2}$	1064
Er^{3+}	$^4I_{13/2} - {}^4I_{15/2}$	1540
Tm^{3+}	$^3H_4 - {}^3H_6$	2013
Ho^{3+}	$^5I_7 - {}^5I_8$	2097

Source: Duarte, F.J., *Tunable Laser Optics,*
Elsevier-Academic, New York, 2003.

A.4.2 Transition Metal Solid-State Lasers

This class of lasers include widely tunable lasers such as the Ti:sapphire and the alexandrite lasers. Table A.14 lists transitions and tuning ranges corresponding to these lasers.

Broadly tunable solid-state lasers have produced some of the shortest pulses recorded directly from a laser system (see Table A.15). These lasers

TABLE A.14

Transition Metal Solid-State Lasers

Ion: Host Crystal	Transition	λ (nm)	Ref.[a]
$Cr^{3+}:Al_2O_3$	$^2E(\bar{E})-{}^4A_2$	694.3	Maiman (1960)
$Cr^{3+}:Be_3Al_2(SiO_3)^6$	$^4T_2-{}^4A_2$	695–835	Shand and Walling (1982)
$Cr^{3+}:BeAl_2O_4$	$^4T_2-{}^4A_2$	701–818	Walling et al. (1980)
$Ti^{3+}:Al_2O_3$	$^2T_2-{}^2E$	660–986	Moulton (1986)

Source: Duarte, F.J., *Tunable Laser Optics,* Elsevier-Academic, New York, 2003.
[a]   References relate to the wavelength range only.

TABLE A.15

Ultrashort-Pulse Solid-State Lasers

Laser	Post Laser	Δt (fs)	λ (nm)	Energy	Ref.
$Ti^{3+}:Al_2O_3$	Non-collinear OPA[a]	6.9		20 μJ	Travella et al. (2010)
$Ti^{3+}:Al_2O_3$	Chirped mirror compressor	3.64[b]	810[c]		Demmler et al. (2011)

[a]   Optics includes a 2-prism stretcher, a fiber stretcher, and two fiber preamplifiers.
[b]   1.3 cycle at 810 nm.
[c]   Central wavelength.

have improved on the previous performance established by dye lasers. A reference work on ultrashort-pulse lasers in general is given by Diels and Rudolph (2006).

A.4.3 Diode-Laser-Pumped Fiber Lasers

Emission wavelengths and output powers for diode-laser-pumped Yb-doped fiber lasers are listed in Table A.16. Wavelength ranges of tunable narrow-linewidth fiber lasers are listed in Table A.17.

A.4.4 Optical Parametric Oscillators

Although the optical parametric oscillator (OPO) does not involve the process of population inversion in its excitation mechanism, it is included, nevertheless, since it is a source of spatially, and spectrally, coherent emission, which is inherently tunable. For detailed review articles on this subject, the reader is referred to Barnes (1995b) and Orr et al. (2009). Spectral characteristics of several well-known OPOs are given in Table A.18. Among their many applications, these sources are used in the generation of correlated photon pairs for quantum entanglement (see Chapter 19).

TABLE A.16

Diode-Laser-Pumped Yb-Doped Fiber Lasers

Cavity	λ (nm)	$\Delta\nu$	CW Power	η (%)	Ref.
Linear	~1120	BB[a]	110 W[b]	58	Dominic et al. (1999)
HTGI[c]	1032–1124	2.5 GHz	10 W[d]	68	Auerbach et al. (2002)
	~1100	BB[a]	1.36 kW[e]	86	Jeong et al. (2004)

Source: Duarte, F.J., *Tunable Laser Optics*, Elsevier-Academic, New York, 2003.
[a] BB: broadband.
[b] Excitation wavelength at ~915 nm.
[c] Hybrid-telescope GIG configuration in a ring cavity.
[d] Excitation wavelength at 980 nm.
[e] Excitation wavelength at 975 nm.

TABLE A.17

Tunable Fiber Lasers

Tuning Technique	Tuning Range (nm)	Linewidth (MHz)	CW Power (mW)	Ref.
Bragg grating (Er^{3+})	1510–1580	100	0.5	Chen et al. (2003)
Diffraction grating (Tm^{3+})	2275–2415	210	6.0	Yeh et al. (2007)

TABLE A.18

Pulsed OPOs

Crystal	λ_p (nm)	Tuning Range (μm)	Ref.
KTP	532	0.61–4.0	Orr et al. (1995)
BBO	355	0.41–3.0	Orr et al. (1995)
LBO	355	0.41–2.47	Schröder et al. (1994)
LiNbO₃	532	0.61–4.4	Orr et al. (1995)
AgGaS₂	1064	1.4–4.0	Fan et al. (1984)

Source: Duarte, F.J., *Tunable Laser Optics,* Elsevier-Academic, New York, 2003.

A.5 Semiconductor Lasers

The area of semiconductor lasers is a rapidly evolving field, which can be classified into high-power lasers, tunable external cavity lasers, tunable miniature lasers, tunable infrared lasers, and femtosecond lasers. Semiconductor lasers work via direct electrical excitation, are compact, are inherently tunable, and can be very stable. Six tables are included in this section, each highlighting output emission characteristics in a different area of semiconductor laser technology. The tables themselves are self-explanatory. Reviews on tunable external cavity tunable semiconductor lasers are given by Zorabedian (1995) and Duarte (2009), among others (Tables A.19 through A.24).

A.6 Free Electron Lasers

An important source of widely tunable coherent radiation, not included in this survey, is the free electron laser (FEL). FELs are fairly large high-power devices that require accelerator technology. In his review of FELs,

TABLE A.19

Wavelength Coverage of Semiconductor Laser Materials

Semiconductor Type	Wavelength Range (nm)
II–VI (GaN)	395–410
III–V (AlGaInP/GaAs)	610–690
III–V (AlGaAs)	815–825
III–V (InGaAsP/InP)	1255–1335
III–V (InGaAs/InP)	1530–1570

Source: Duarte, F.J., Broadly tunable dispersive external-cavity semiconductor lasers, in *Tunable Laser Applications,* Duarte, F. J. ed., CRC, New York, Chapter 5, 2009.

TABLE A.20

Power Performance of CW Diode-Laser Arrays

λ (nm)	Output Power (W)	Ref.
791	10	Srinivasan et al. (1999)
~915	45	Dominic et al. (1999)
940	200	Morita et al. (2012)

Source: Duarte, F.J., *Tunable Laser Optics*, Elsevier-Academic, New York, 2003.

TABLE A.21

External Cavity Tunable Semiconductor Lasers

Semiconductor	Cavity	Δν	Tuning Range (nm)	Power (mW)	Ref.
InGaAsP/InP[a]	LG[b]	10 kHz	55 @ 1500		Wyatt and Devlin (1983)
InGaAsP/InP	LG[b,c]	31 kHz	1285–1320	1	Shan et al. (1991)
	LG[b]	20 kHz	15 @ 1260		Favre et al. (1986)
InGaAsP/InP[a]	MPLG[c]	100 kHz	1255–1335		Zorabedian (1992)
GaAlAs[a]	LG[b,c]	1.5 MHz	815–825	5	Fleming and Mooradian (1981)
GaAlAs[a]	LG[b,c]	200 kHz	32 @ 850	1	De Labachelerie (1985)
GaAlAs[a]	2 Etalons[c]	32 MHz	10 @ 875		Voumard (1977)
GaAlAs[a]	Etalon[c]	4 kHz			Harrison and Mooradian (1989)
GaAlAs[a]	GIG[d]	10 kHz	20 @ 780		Harvey and Myatt (1991)
	GIG[d]	15 MHz	30 @ 820	30	Gavrilovic et al. (1992)

Source: Duarte, F.J., Broadly tunable dispersive external-cavity semiconductor lasers, in *Tunable Laser Applications*, Duarte, F.J., ed., CRC, New York, Chapter 5, 2009.

[a] AR coating of the internal facet adjacent to the frequency-selective optics.
[b] Littrow grating.
[c] Closed cavity configuration (see Appendix B).
[d] Open cavity configuration (see Appendix B).

Benson (1995) reports on various devices covering the electromagnetic spectrum from 2 to 2500 μm. More recent developments have seen the extension of the FEL emission well into the extreme ultraviolet (see, e.g., Allaria et al., 2012).

Duarte (2003) refers to additional miscellaneous type lasers such as chemical lasers, far-infrared lasers, and nuclear pumped lasers.

TABLE A.22

MEMS[a] Tunable Semiconductor Lasers

Semiconductor	Cavity	Tuning Range (nm)	$\Delta\nu$	Output Power (mW)	Ref.
InGaAsP/InP	GIG[b]	1531.5–1557.9	2 MHz	20	Berger et al. (2001)
InGaAsP/InP	GIG[b,c]	42 @ 1550	2 MHz	70	Berger and Anthon (2003)
GaAs/AlGaAs[d]	Mirror[e]	1533–1555	SM[f]	0.9	Kner et al. (2002)
	GIG[g]	1530–1570	50 kHz		Zhang et al. (2012)

Source: Duarte, F.J., *Tunable Laser Optics*, Elsevier-Academic, New York, 2003.
[a] MEMS: microelectromechanical systems.
[b] Uses a silicon MEMS driven GIG cavity.
[c] Open cavity configuration (see Appendix B).
[d] VCEL.
[e] Tuning is achieved by displacing a voltage driven micromirror, which is supported by a cantilever.
[f] Single mode.
[g] Open cavity GIG plus etalon.

TABLE A.23

Tunable External Cavity Quantum Cascade Lasers

Stages in Cascade	Cavity	Tuning Range (μm)	$\Delta\nu$	Output Power	Ref.
20	Littrow	8.2–10.4	SM[a]		Maulini et al. (2006)
74	Littrow	7.6–11.4	~3.59 GHz	15 mW	Hugi et al. (2009)
	DFB[b]		5 GHz[a]	65 μW	Lu et al. (2013)

[a] Single mode.
[b] Distributed feedback: primary emissions around 9 and 10.22 μm.

TABLE A.24

Ultrashort-Pulse External Cavity Semiconductor Lasers

Laser	Cavity	Mode Locking Technique	Δt (fs)	λ (nm)	Ref.
InGaAsP[a]	Étalon[b]	Active	580	1300	Corzine et al. (1988)
AlGaAs	4 prisms[c]	Hybrid MQW SA[d]	200	~838	Delfyett et al. (1992)
AlGaAs[a]	6 prisms[b]	Active, SA[d]	650	805	Pang et al. (1992)
MQW[a]	Grating pair[c,e]	Passive	260	~846	Salvatore et al. (1993)

Source: Duarte, F.J., Broadly tunable dispersive external-cavity semiconductor lasers, in *Tunable Laser Applications*, Duarte, F.J. (ed.), CRC, New York, Chapter 5, 2009.
[a] AR coating of the internal facet next to the frequency-selective optics.
[b] Closed cavity configuration.
[c] Open cavity configuration.
[d] Saturable absorber.
[e] Uses a Littrow grating for tuning.

References

Allaria, E., Appio, R., Badano, L., Barletta, W. A., Bassanese, S., Biedron, S. G., Borga, A. et al. (2012). Highly coherent and stable pulses from the FERMI seeded free-electron laser in the extreme ultraviolet. *Nat. Photon.* **6**, 699–704.

Baltakov, F. N., Barikhin, B. A., and Sukhanov, L. V. (1974). 400-J pulsed laser using a solution of rhodamine-6G in ethanol. *JETP Lett.* **19**, 174–175.

Barnes (1995a). Transition metal solid-state lasers. In *Tunable Lasers Handbook* (Duarte, F. J., ed.). Academic Press, New York, Chapter 6.

Barnes (1995b). Optical parametric oscillators. In *Tunable Lasers Handbook* (Duarte, F. J., ed.). Academic Press, New York, Chapter 7.

Bass, I. L., Bonanno, R. E., Hackel, R. H., and Hammond, P. R. (1992). High-average power dye laser at Lawrence Livermore National Laboratory. *Appl. Opt.* **31**, 6993–7006.

Baving, H. J., Muuss, H., and Skolaut, W., (1982). CW dye laser operation at 200W pump power. *Appl. Phys. B.* **29**, 19–21.

Beck, R., Englisch, W., and Gürs, K. (1976). *Table of Laser Lines in Gases and Vapors*, Springer-Verlag, Berlin, Germany.

Benson, S. V. (1995). Tunable free-electron lasers. In *Tunable Lasers Handbook* (Duarte, F. J., ed.). Academic Press, New York, Chapter 9.

Berger, J. D. and Anthon, D. (2003). Tunable MEMS devices for optical networks. *Opt. Photon. News* **14**(3), 43–49.

Berger, J. D., Zhang, Y., Grade, J. D., Howard, L., Hrynia, S., Jerman, H., Fennema, A., Tselikov, A., and Anthon, D. (2001). External cavity diode lasers tuned with silicon MEMS. *IEEE LEOS Newslett.*, **15** (5), 9–10.

Bernhardt, A. F. and Rasmussen, P. (1981). Design criteria and operating characteristics of a single-mode pulsed dye laser. *Appl. Phys. B* **26**, 141–146.

Bobrovskii, A. N., Branitskii, A. V., Zurin, M. V., Koshevnikov, A. V., Mishchenko, V. A., and Myl'nikov, G. D. (1987). Continuously tunable TEA CO_2 laser. *Sov. J. Quant. Electron.* **17**, 1157–1159.

Caro, R. G., Gower, M. C., and Webb, C. E. (1982). A simple tunable KrF laser system with narrow bandwidth and diffraction-limited divergence. *J. Phys. D: Appl. Phys.* **15**, 767–773.

Chen, H., Babin, F., Leblanc, M., and Schinn, G. W. (2003). Widely tunable single-frequency Erbium-doped fiber lasers. *IEEE Photon. Tech. Lett.* **15**, 185–187.

Corzine, S. W., Bowers, J. E., Przybylek, G., Koren, U., Miller, B. I., and Soccolich, C. E. (1988). Actively mode-locked GaInAsP laser with subpicosecond output. *Appl. Phys. Lett.* **52**, 348–350.

DeLabachelerie, M. and Cerez, P. (1985). An 850 nm semiconductor laser tunable over a 30 nm range. *Opt. Commun.* **55**, 174–178.

Delfyett, P. J., Florez, L., Stoffel, N., Gmitter, T., Andreadakis, N., Alphonse, G., and Ceislik, W. (1992). 200 fs optical pulse generation and intracavity pulse evolution in a hybrid mode-locked semiconductor diode-laser/amplifier system, *Opt. Lett.* **17**, 670–672.

Demmler, S., Rothhardt, J., Heidt, A. M., Hartung, A., Rohwer, E. G., Bartelt, H., Limpert, J., and Tünnermann A. (2011). Generation of high quality, 1.3 cycle pulses by active phase control of an octave spanning supercontinuum. *Opt. Exp.* **19**, 20151–20158.

Diels, J. -C. (1990). Femtosecond dye lasers. In *Dye Laser Principles* (Duarte, F. J. and Hillman, L. W., eds.). Academic Press, New York, Chapter 3.

Diels, J. -C. and Rudolph, W. (2006). *Ultrashort Laser Pulse Phenomena*, 2nd edn. Academic Press, New York.

Dietel, W., Fontaine, J. J., and Diels, J. -C. (1983). Intracavity pulse compression with glass: a new method of generating pulses shorter than 60 fs. *Opt. Lett.* **8**, 4–6.

Dominic, V., MacCormack, S., Waarts, R., Sanders, S., Bicknese, S., Dohle, R., Wolak, E., Yeh, P. S., and Zucker, E. (1999). 110W fibre laser. *Electron. Lett.* **35**, 1158–1160.

Duarte, F. J. (1985). Multiple-prism Littrow and grazing-incidence pulsed CO_2 lasers. *Appl. Opt.* **24**, 1244–1245.

Duarte, F. J. (1991). Dispersive dye lasers. In *High Power Dye Lasers* (Duarte, F. J., ed.). Springer-Verlag, Berlin, Germany, Chapter 2.

Duarte, F. J. (1994). Solid-state multiple-prism grating dye laser oscillators. *Appl. Opt.* **33**, 3857–3860.

Duarte, F. J. (1999). Multiple-prism grating solid-state dye laser oscillator: Optimized architecture. *Appl. Opt.* **38**, 6347–6349.

Duarte, F. J. (2003). *Tunable Laser Optics.* Elsevier-Academic, New York.

Duarte, F. J. (2009). Broadly tunable dispersive external-cavity semiconductor lasers. In *Tunable Laser Applications* (Duarte, F. J. ed.). CRC Press, New York, Chapter 5.

Duarte, F. J., Davenport, W. E., Ehrlich, J. J., and Taylor, T. S. (1991). Ruggedized narrow-linewidth dispersive dye laser oscillator. *Opt. Commun.* **84**, 310–316.

Duarte, F. J. and Piper, J. A. (1984). Narrow-linewidth, high prf copper laser-pumped dye laser oscillators. *Appl. Opt.* **23**, 1391–1394.

Duarte, F. J., Taylor, T. S., Costela, A., Garcia-Moreno, I., and Sastre, R. (1998). Long-pulse narrow-linewidth dispersive solid-state dye-laser oscillator. *Appl. Opt.* **37**, 3987–3989.

Dupre, P. (1987). Quasiunimodal tunable pulsed dye laser at 440 nm: Theoretical development for using quad prism beam expander and one or two gratings in a pulsed dye laser oscillator cavity. *Appl. Opt.* **26**, 860–871.

Fan, Y. X., Eckardt, R. C., Byer, R. L., Route, R. K., and Feigelson, R. S. (1984). $AgGaS_2$ infrared parametric oscillator. *Appl. Phys. Lett.* **45**, 313–315.

Favre, F., LeGuen, D., Simon, J. C., and Landousies, B. (1986). External-cavity semiconductor laser with 15 nm continued tuning range. *Electron. Lett.* **22**, 795–796.

Fleming, M. W. and Mooradian, A. (1981). Spectral characteristics of external-cavity controlled semiconductor lasers. *IEEE J. Quant. Electron.* **QE-17**, 44–59.

Fork, R. L., Brito Cruz, C. H., Becker, P. C., and Shank, C. V. (1987). Compression of optical pulses to six femtoseconds by using cubic phase compression. *Opt. Lett.* **12**, 483–485.

Fort, J. and Moulin, C. (1987). High-power high-energy linear flashlamp-pumped dye Laser. *Appl. Opt.* **26**, 1246–1249.

Gavrilovic, P., Chelnokov, A. V., O'Neill, M. S., and Beyea, D. M. (1992). Narrow- linewidth operation of broad-stripe single quantum well laser diodes in a grazing incidence external cavity. *Appl. Phys. Lett.* **60**, 2977–2979.

Harrison, J. and Mooradian, A. (1989). Linewidth and offset frequency locking of external cavity GaAlAs lasers. *IEEE J. Quant. Electron.* **QE-25**, 1152–1155.

Harvey, K. C. and Myatt, C. J. (1991). External-cavity diode laser using a grazing- incidence diffraction grating. *Opt. Lett.* **16**, 910–912.

Hollberg, L. (1990). CW dye lasers. In *Dye Laser Principles* (Duarte, F. J. and Hillman, L. W., eds.). Academic Press, New York, Chapter 5.

Hugi, A., Terazzi, R., Bonetti, Y., Wittmann, A., Fischer, M., Beck, M., Faist, J., and Gini, E. (2009). External cavity quantum cascade laser tunable from 7.6 to 11.4 μm. *Appl. Phys. Letts.* **95**, 061103.

Jeong, Y., Sahu, J. K., Payne, D. N., and Nilsson, J. (2004). Ytterbium-doped large-core fiber laser with 1.36 kW continuous-wave output power. *Opt. Exp.* **12**, 6086–6092.

Johnston, T. F. and Duarte, F. J. (2002). Lasers, dye. In *Encyclopedia of Physical Science and Technology*, 3rd edn. (Meyers, R. A., ed.). Academic Press, New York, pp. 315–359.

Kafka, J. D. and Baer, T. (1992). Prism-pair delay lines in optical pulse compression. *Opt. Lett.* **12**, 401–403.

Klimek, D. E., Aldag, H. R., and Russell, J. (1992). In *Conference on Lasers and Electro-Optics*, Optical Society of America, Washington, DC, p. 332.

Kner, P., Sun, D., Boucart, J., Floyd, P., Nabiev, R., Davis, D., Yuen, W., Jansen, M., and Chang-Hasnain, C. J. (2002). VCSELS. *Opt. Photon. News* **13**(3), 44–47.

Kubota, H., Kurokawa K., and Nakazawa, M. (1988). 29-fs pulse generation from a linear-cavity synchronously pumped dye laser. *Opt. Lett.* **13**, 749–751.

Loree, T. R. Butterfield, K. B., and Barker, D. L. (1978). Spectral tuning of ArF and KrF discharge lasers. *Appl. Phys. Lett.* **32**, 171–173.

Lu, Q. Y., Bandyopadhyay, N., Slivken, S., Bai, Y., and Razeghi, M. (2013). High performance terahertz quantum cascade laser sources based on intracavity difference frequency generation. *Opt. Exp.* **21**, 968–973.

Maiman, T. H. (1960). Stimulated optical radiation in ruby. *Nature* **187**, 493–494.

Maulini, R., Mohan, A., Giovannini, M., Faist, J., Gini, E. (2006). External cavity quantum-cascade laser tunable from 8.2 to 10.4 μm using a gain element with a heterogeneous cascade. *Appl. Phys. Lett.* **88**, 201113.

McKee, T. J. (1985). Spectral-narrowing techniques for excimer lasers oscillators. *Can. J. Phys.* **63**, 214–219.

Morita, T., Kageyama, N., Torii, K., Nagakura, T., Takauji, M., Maeda, J., Miyamoto, M., Miyajima, H., and Yoshida, H. (2012). Developments of high-power 9xx-nm single emitter laser diodes and laser diode bars. In *Photonics Society Summer Topical Meeting Series, IEEE.* pp. 33–34.

Moulton, P. F. (1986). Spectroscopic and laser characteristics of Ti:Al$_2$O$_3$. *J. Opt. Soc. Am. B* **3**, 125–132.

Orr, B. J., He, Y., and White, R. T. (2009). Spectroscopic applications of pulsed tunable optical parametric oscillators. In *Tunable Laser Applications*, 2nd edn. (Duarte, F. J., ed.). CRC Press, New York, Chapter 2.

Orr, B. J., Johnson, M. J., and Haub, J. G. (1995). Spectroscopic applications of pulsed tunable optical parametric oscillators. In *Tunable Laser Applications* (Duarte, F. J., ed.). Marcel Dekker, New York, Chapter 2.

Osvay, K., Kovács, A. P., Kurdi, G., Heiner, Z., Divall, M., Klebniczki, J., and Ferincz, I. E. (2005). Measurements of non-compensated angular dispersion and the subsequent temporal lengthening of femtosecond pulses in a CPA laser. *Opt. Commun.* **248**, 201–209.

Pacala, T. J., McDermid, I. S., and Laudenslager, J. B. (1984). Single-longitudinal-mode operation of an XeCl laser. *Appl. Phys. Lett.* **45**, 507–509.

Pang, L. Y., Fujimoto, J. G., and Kintzer, E. S. (1992). Ultrashort-pulse generation from high-power diode arrays by using intracavity optical nonlinearities. *Opt. Lett.* **17**, 1599–1601.

Salvatore, R. A., Schrans, T., and Yariv, A. (1993). Wavelength tunable source of subpicosecond pulses from CW passively mode-locked two-section multiple-quantum-well laser. *IEEE Photo. Tech. Lett.* **5**, 756–758.

Schröder, T., Boller, K. -J., Fix, A., and Wallenstein R. (1994). Spectral properties and numerical modelling of a critically phase-matched nanosecond LiB_3O_5 optical parametric oscillator. *Appl. Phys. B.* **58**, 425–438.

Shan, X., Siddiqui, A. S., Simeonidou, D., and Ferreira, M. (1991). Rebroadening of spectral linewidth with shorter wavelength detuning away from the gain curve peak in external avity semiconductor lasers sources. In *Conference on Lasers and Electro- Optics*. Optical Society of America, Washington, DC, pp. 258–259.

Shand, M. L. and Walling, J. C. (1982). A tunable emerald laser. *IEEE J. Quant. Electron.* **QE-18**, 1829–1830.

Shay, T., Hanson, F., Gookin, D., and Schimitschek, E. J. (1981). Line narrowing and enhanced efficiency of an HgBr laser by injection locking. *Appl. Phys. Lett.* **39**, 783–785.

Srinivasan, B., Tafoya, J., and Jain, R. K. (1999). High power Watt-level CW operation of diode-pumped 2.7 mm fiber lasers using efficient cross-relaxation and energy transfer mechanisms. *Opt. Exp.* **4**, 490–495.

Sugii, M., Ando, M., and Sasaki, K. (1987). Simple long-pulse XeCl laser with narrow-line output. *IEEE J. Quant. Electron.* **QE-23**, 1458–1460.

Sze R. C. and Harris, D. G. (1995). Tunable excimer lasers. In *Tunable Lasers Handbook* (Duarte, F. J., ed.). Academic Press, New York, Chapter 3.

Sze, R. C., Kurnit, N. A., Watkins, D. E., and Bigio, I. J. (1986). Narrow band tuning with small long-pulse excimer lasers. In *Proceedings of the International Conference on Lasers '85* (Wang, C. P., ed.). STS Press, McLean, VA, pp. 133–144.

Tallman, C. and Tennant, R. (1991). Large-scale excimer-laser-pumped dye lasers. In *High Power Dye Lasers* (Duarte, F. J., ed.). Springer-Verlag, Berlin, Germany, Chapter 4.

Tang, K. Y., O'Keefe, T., Treacy, B., Rottler, L., and White, C. (1987). Kilojoule output XeCl dye laser: Optimization and analysis. In *Proceedings: Dye Laser/Laser Dye Technical Exchange Meeting, 1987* (Bentley, J. H., ed.). U. S. Army Missile Command, Redstone Arsenal, AL, pp. 490–502.

Tavella, F., Willner, A., Rothhardt, J., Hädrich, S., Seise, E., Düsterer, S., Tschentscher, T. et al. (2010). Fiber-amplifier pumped high average power few-cycle pulse non-collinear OPCPA. *Opt. Exp.* **18**, 4689–4694.

Tenenbaum, J., Smilanski, I., Gabay, S., Levin, L. A., Erez, G., and Lavi. S. (1980). Structure of 510.6 and 578.2 nm copper laser lines. *Opt. Commum.* **32**, 473–477.

Voumard, C. (1977). External-cavity-controlled 32-MHz narrow-band CW GaAlAs-diode lasers. *Opt. Lett.* **1**, 61–63.

Walling, J. C., Peterson, O. G., Jensen, H. P., Morris, R. C., and O'Dell, E. W. (1980). Tunable alexandrite lasers. *IEEE J. Quant. Electron.* **QE-16**, 1302–1315.

Webb, C. E. (1991). High-power dye lasers pumped by copper vapor lasers. In *High Power Dye Lasers*, (Duarte, F. J., ed.). Springer-Verlag, Berlin, Germany, Chapter 5.

Willett, C. S. (1974). *An introduction to Gas Lasers: Population Inversion Mechanisms*. Pergamon, New York.

Woodward, B. W., Ehlers, V. J., and Lineberger, W. C. (1973). A reliable, repetitively pulsed, high-power nitrogen laser. *Rev. Sci. Instrum.* **44**, 882–887.

Wyatt, R. and Devlin, W. J. (1983). 10 kHz linewidth 1.5 μm InGaAsP external cavity laser with 55 nm tuning range. *Electron. Lett.* **19**, 110–112.

Yang, T. T., Burde, D. H., Merry, G. A., Harris, D. G., Pugh, L. A. Tillotson, J. H., Turner, C. E., and Copeland, D. A. (1988). Spectra of electron-beam pumped XeF lasers. *Appl. Opt.* **27**, 49–57.

Yeh, C-H., Huang, T-T., Chien, H-C., Ko, C-H., and Chi, S. (2007). Tunable S-band erbium-doped triple-ring laser with single-longitudinal-mode operation, *Opt. Exp.* **15**, 382–386.

Zhang, D., Zhao, J., Yang, O., Liu, W., Fu, Y., Li, C., Luo, M., Hu, S. Q., and Wang, L. (2012). Compact MEMS external cavity tunable laser with ultra-narrow linewidth for coherent detection. *Opt. Exp.* **20**, 19670–19682.

Zorabedian, P. (1992). Characteristics of a grating-external-cavity semiconductor laser containing intracavity prism beam expanders. *J. Lightwave Technol.* **10**, 330–335.

Zorabedian, P. (1995). Tunable external-cavity semiconductor lasers. In *Tunable Lasers Handbook* (Duarte, F. J., ed.). Academic Press, New York, Chapter 8.

Appendix B: Brief Survey of Laser Resonators and Laser Cavities

B.1 Introduction

A laser is composed of a gain medium, a mechanism to excite that medium, and an optical resonator and/or optical cavity (Duarte, 2003). The terms optical cavity, optical oscillator, and optical resonator are approximately equivalent, but usage of these terms indicates that an *optical cavity* applies equally to an optical oscillator or an optical resonator. The term *optical resonator* tends to be associated with *unstable resonators*, while *optical oscillator* tends to be identified with narrow-linewidth oscillators. In this regard, narrow-linewidth multiple-prism grating oscillators can behave as unstable resonators depending on the lensing properties of the gain medium (Duarte et al., 1997).

The initial laser radiation from a gain medium can be multiple transverse modes, and each of these modes can include a multitude of longitudinal modes (Duarte, 2003). Such class of radiation constitutes broadband laser radiation. In this regard, it is the function of the laser cavity, laser resonator, or laser oscillator to restrict the laser emission to a single transverse mode, that is, TEM_{00}, and to a single longitudinal mode. In other words, for radiation of a given temporal characteristic, it is the geometry and the optical configuration of the cavity, resonator, or oscillator that determine the degree of spatial coherence and spectral coherence of the laser emission.

In this appendix we provide a brief survey of some widely used laser cavity configurations.

B.2 Resonators and Oscillators

Here, we briefly introduce broadband resonators, tunable resonators, and narrow-linewidth oscillators. For comprehensive reviews on this subject, the reader should refer to Duarte (1990a, 1991, 1995a, 2003).

B.2.1 Broadband Resonators

The most elemental of resonators is that composed of two flat reflective surfaces aligned perpendicular to the optical axis of the cavity as illustrated in Figure B.1. In this flat-mirror resonator, one of the mirrors is usually ~100% reflective, at the wavelength or wavelengths of interest, and the other mirror, known as the output coupler, is partially reflective. However, in high-gain lasers, such as dye lasers, simply the windows of the gain region can act as partial reflectors and sustain broadband laser action. This very rudimentary cavity class is outlined in Figure B.2. Here it is important to observe that this is not "mirror-less" laser action since each window is acting as a mirror. Also, by broadband laser emission, we mean powerful laser emission, which can be a few nm wide and even up to 10 nm wide as reported by Schäfer et al. (1966).

In this class of broadband resonator, the optimum reflectivity for the output coupler is often determined empirically. For a low-gain laser medium, this reflectivity can approach 99%, while, for a high-gain laser medium, the reflectivity can be as low as 10%. In Figure B.1 the gain region is depicted with its output windows at an angle relative to the optical axis. If the angle of incidence of the laser emission, on the windows, is the Brewster angle, then the emission will be highly linearly polarized. For the case illustrated in Figure B.1, the laser emission will be polarized parallel to the plane of incidence. On the

FIGURE B.1
Generic broadband flat-mirror resonator. M_1 is a total reflector and M_2 is a partial reflector also known as output coupler. The diameter of the laser beam is determined by the dimensions of the aperture next to M_2.

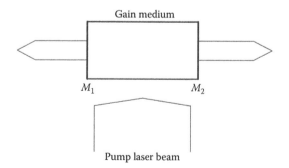

FIGURE B.2
Optically pumped high-gain broadband resonator with flat partially reflecting surfaces, which perform the function of mirrors as M_1 and M_2. The emission is bidirectional, and the dimensions of the output beam depend on the focusing characteristics of the pump laser beam.

other hand, the configuration illustrated in Figure B.2 could allow the emission of unpolarized laser radiation (Duarte, 1990b).

The transverse-mode structure in these resonators is determined by the emission wavelength and the ratio of the cavity aperture to the cavity length. In this regard, the beam profile can be calculated using the interferometric equation:

$$
|\langle x\,|\,s\rangle|^2 = \sum_{j=1}^{N} \Psi(r_j)^2 + 2\sum_{j=1}^{N} \Psi(r_j)\left(\sum_{m=j+1}^{N} \Psi(r_m)\cos(\Omega_m - \Omega_j) \right) \qquad \text{(B.1)}
$$

where the width of the intracavity aperture is represented by the dimensions of subslits and the number N of these subslits. The calculation also includes the wavelength λ and the intra-interferometric distance $D_{\langle x|j\rangle}$ (see Chapter 9 [Duarte, 1991, 1993, 2003]).

Equation B.1 tells us that short cavities with wide apertures lead to the emission of beams with a large number of transverse modes, while long cavities with narrow apertures yield TEM_{00} emission. This equation explains why lasers such as He–Ne, Ar$^+$, and Kr$^+$ have exquisite beam quality.

An alternative to the flat-mirror approach is to use a pair of optically matched curved mirrors.

B.2.2 Frequency-Selective Resonators

These resonators are available in a variety of configurations. The simplest frequency-selective resonator, or frequency-selective cavity, incorporates a diffraction grating as a tuning element as illustrated in Figure B.3. This class of cavity can also incorporate other intracavity frequency-selective elements, such as etalons (Figure B.3b). Prisms can also be used as intracavity tuning elements as shown in Figure B.4. Prisms can be used in conjunction with a diffraction grating, in Littrow configuration, or a simple flat mirror (Duarte, 1990a).

B.2.3 Unstable Resonators

An additional class of broadband laser resonators is the *unstable resonators*. These resonators depart from the flat-mirror design and incorporate curved mirrors as depicted in Figure B.5. These mirror configurations are adopted from the field of reflective telescopes. A widely used design is a variation of the Newtonian telescope known as the Cassegrain telescope. In this configuration the two mirrors have a high reflectivity.

Unstable resonators allow the use of large-gain medium volumes, which are suitable for high-power laser generation. Albeit some designs emit a hollow beam in the near field, the far field exhibits beam profiles with a single-transverse-mode structure. Unstable resonators are particularly well suited for high-power lasers and for the amplification stage of MOFO configurations

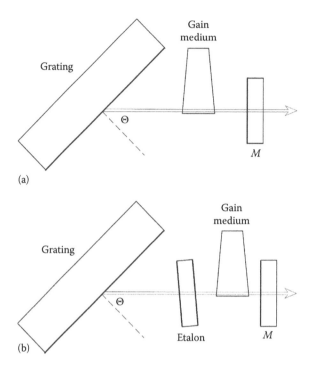

FIGURE B.3
(a) Frequency-selective cavity tuned by a diffraction grating in Littrow configuration. This is a *closed cavity* configuration in which the laser output is coupled via the partially reflective surface of M. (b) Further linewidth narrowing can be provided via an intracavity etalon.

as illustrated in Figure B.6 (Duarte and Conrad, 1987; Duarte 1990a). Unstable resonators can be evaluated using parameters from *ABCD* matrices as indicated in Appendix C. For a detailed treatment on the subject of unstable resonators, the reader should refer to Siegman (1986).

B.2.4 Narrow-Linewidth Oscillators

Narrow-linewidth tunable laser oscillators can be configured using either grazing-incidence grating (GIG) designs (Figure B.7) or multiple-prism grating configurations as illustrated in Figures B.8 and B.9. The GIG cavity depicted in Figure B.7 is *open cavity* as the laser output is coupled via the reflection losses of the grating.

Multiple-prism grating architectures can be of the hybrid multiple-prism preexpanded near-grazing-incidence (HMPGI) grating class (Figure B.8) or the multiple-prism Littrow (MPL) grating class (Figure B.9). Both these types of oscillators are inherently compatible with *closed cavity* configurations, which produce lower spectral noise, and offer better protection against optical coupling with extracavity elements (Duarte 1990a, 2003). In addition to narrow-linewidth emission, these cavities are inherently tunable.

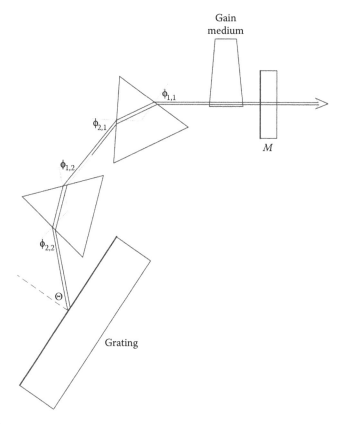

FIGURE B.4
Frequency-selective cavity tuned by a double-prism grating assembly. Wavelength selectivity is achieved by rotation of the grating. In some broadband configurations, the grating is replaced by a tuning mirror. This is a *closed cavity* configuration in which the laser output is coupled via the output coupler mirror.

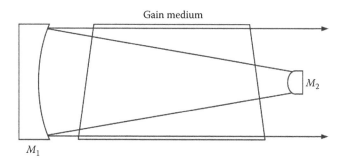

FIGURE B.5
Generic unstable resonator configuration (see Appendix C).

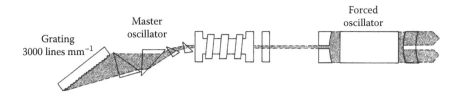

FIGURE B.6
MOFO laser system. The amplifier stage (FO) is configured with an unstable resonator cavity. (Reproduced from Duarte, F.J. and Conrad, R.W., *Appl. Opt.* 26, 2567, 1987, with permission from the Optical Society of America.)

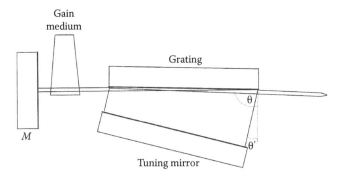

FIGURE B.7
Tunable narrow-linewidth cavity incorporating a GIG configuration. This is an *open cavity* configuration in which the laser output is coupled via the reflection losses of the grating.

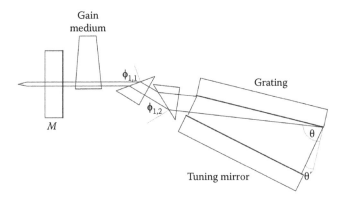

FIGURE B.8
Tunable narrow-linewidth cavity incorporating a double-prism beam expander and a grating in a near-grazing-incidence configuration. In the literature these cavities are referred to as hybrid multiple-prism preexpanded near-grazing-incidence (HMPGI) grating cavities (Duarte and Piper, 1981, 1984). This is a *closed cavity* configuration in which the laser output is coupled via the output coupler mirror.

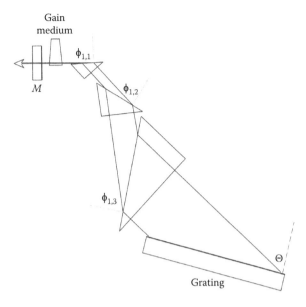

FIGURE B.9
Tunable narrow-linewidth cavity incorporating an MPL grating configuration (Duarte and Piper, 1981, 1984). This is a *closed cavity* configuration in which the laser output is coupled via the output coupler mirror.

These dispersive cavities, or narrow-linewidth tunable laser oscillators, allow the designer to obtain laser beams that are

1. Spatially coherent: that is, single-transverse (TEM_{00})-mode emission with a Gaussian beam profile (Duarte, 1995b, 1999).
2. Spectrally coherent: that is, narrow-linewidth laser emission that is single longitudinal mode. For optimized oscillators this emission can be near the limit allowed by Heisenberg's uncertainty principle:

$$\Delta v \, \Delta t \approx 1 \tag{B.2}$$

For example, for $\Delta v \approx 350$ MHz, and $\Delta t \approx 3$ ns (Duarte, 1999), the product above becomes $\Delta v \, \Delta t \approx 1.05$.

Further details can be found in Chapter 9 and review articles are given by Duarte (1990a, 1995a, 2003). Also, these oscillator architectures are applicable to lasers in the gas, the liquid, and the solid state (Duarte, 2003). The performance of tunable narrow-linewidth oscillators as described, for various gain media, is described in Appendix A.

B.3 CW Tunable Cavities

CW tunable laser cavities include linear (Figure B.10) and ring laser resonators (Figure B.11) developed for dye lasers and later applied, in modified
form, to the generation of ultrashort pulses (Diels, 1990; Diels and Rudolph,
2006). A generic unidirectional ring resonator deployed in an 8-shape
configuration is illustrated in Figure B.11 (Hollberg, 1990).

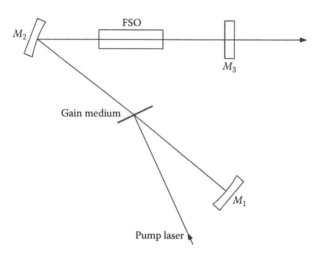

FIGURE B.10
Generic linear CW laser cavity. FSO represents frequency-selective optics.

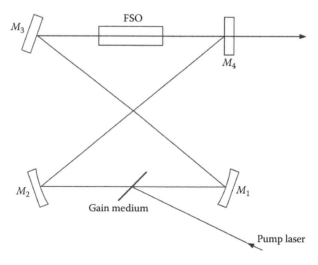

FIGURE B.11
Unidirectional CW ring laser cavity. FSO represents frequency selective optics.

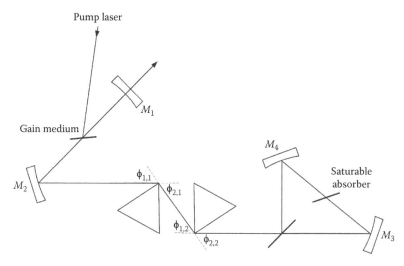

FIGURE B.12
Generic linear laser cavity including a double-prism compressor.

B.3.1 Femtosecond Laser Cavities

A linear cavity incorporating an *antiresonat ring* is depicted in Figure B.12. At the ring, a *collision* between two counterpropagating pulses occurs at the saturable absorber; thus, collision causes the two pulses to interfere. This interference induces a transient grating that shortens the emission pulse. This effect is known as colliding-pulse-mode locking (CPM) (Fork et al., 1981).

The prisms in this cavity are deployed to provide negative dispersion and thus reduce the intracavity dispersion (see Chapter 6). For gain media with an intrinsically large tuning range (such as laser dyes or Ti:sapphire), minimization of the intracavity dispersion allows the emission of radiation with a wide spectral width, or continuum. For a well-designed cavity, wide spectral emission is associated with very short pulses (see Equation B.2). Thus, specially designed prismatic arrays introduce *negative dispersion* thus reducing the overall intracavity dispersion, which leads to *pulse compression*. Details of pulse compression are given in Chapter 6. Although originally developed for dye lasers, these cavities are widely used with a variety of gain media.

References

Diels, J.-C. (1990). Femtosecond dye lasers. In *Dye Laser Principles* (Duarte, F. J. and Hillman, L. W., eds.). Academic Press, New York, Chapter 3.

Diels, J.-C. and Rudolph, W. (2006). *Ultrashort Laser Pulse Phenomena*, 2nd edn. Academic Press, New York.

Duarte, F. J. (1990a). Narrow-linewidth pulsed dye laser oscillators. In *Dye Laser Principles* (Duarte, F. J. and Hillman, L. W., eds.). Academic Press, New York, Chapter 4.

Duarte, F. J. (1990b). Technology of pulsed dye lasers. In *Dye Laser Principles* (Duarte, F. J. and Hillman, L. W., eds.) Academic Press, New York, Chapter 6.

Duarte, F. J. (1991). Dispersive dye lasers. In *High Power Dye Lasers* (Duarte, F. J., ed.). Springer-Verlag, Berlin, Germany, Chapter 2.

Duarte, F. J. (1993). On a generalized interference equation and interferometric measurements. *Opt. Commun.* **103**, 8–14.

Duarte, F. J. (1995a). Narrow linewidth laser oscillators and intracavity dispersion. In *Tunable Lasers Handbook* (Duarte, F. J., ed.). Academic Press, New York, Chapter 5.

Duarte, F. J. (1995b). Solid-state dispersive dye laser oscillator: Very compact cavity. *Opt. Commun.* **117**, 480–484.

Duarte, F. J. (1999). Multiple-prism grating solid-state dye laser oscillator: Optimized architecture. *Appl. Opt.* **38**, 6347–6349.

Duarte, F. J. (2003). *Tunable Laser Optics*. Elsevier-Academic, New York.

Duarte, F. J. and Conrad, R. W. (1987). Diffraction-limited single-longitudinal-mode multiple-prism flashlamp-pumped dye laser oscillator: Linewidth analysis and injection of amplifier system. *Appl. Opt.* **26**, 2567–2571.

Duarte, F. J., Costela, A., Garcia-Moreno, I., Sastre, R., Ehrlich, J. J., and Taylor, T. S. (1997). Dispersive solid-state dye laser oscillators. *Opt. Quant. Electron.* **29**, 461–472.

Duarte, F. J. and Piper, J. A. (1981). A prism preexpanded grazing incidence pulsed dye laser. *Appl. Opt.* **20**, 2113–2116.

Duarte, F. J. and Piper, J. A. (1984). Narrow-linewidth, high prf copper laser-pumped dye laser oscillators. *Appl. Opt.* **23**, 1391–1394.

Fork, R. L., Greene, B. I., and Shank, C. V. (1981). Generation of optical pulses shorter than 0.1 psec by colliding pulse mode locking, *Appl. Phys. Lett.* **38**, 671–672.

Hollberg, L. (1990). CW dye lasers. In *Dye Laser Principles* (Duarte, F. J. and Hillman, L. W., eds.). Academic Press, New York, Chapter 5.

Schäfer, F. P., Schmidt, W., and Volze, J. (1966). Organic dye solution laser. *Appl. Phys. Lett.* **9**, 306–309.

Siegman, A. (1986). *Lasers*. University Science Books, Mill Valley, CA.

Appendix C: Ray Transfer Matrices

C.1 Introduction

A practical method to characterize and design laser optics systems is the use of beam propagation matrices also known as *ray transfer matrices*. This method applies to the propagation of laser beams with a Gaussian profile. In other words, it applies to the propagation of single-transverse-mode beams or *spatially coherent* beams, which is a crucial characteristic to the concept of laser emission. In this appendix the basic principles of propagation matrices are outlined and a survey of matrices for various widely applicable optical elements is given. This appendix follows the style of a review on the subject by Duarte (2003). For early references on the subject, the reader is referred to Brouwer (1964), Kogelnik (1979), Siegman (1986), and Wollnik (1987).

C.2 ABCD Propagation Matrices

The basic idea with propagation matrices is that one vector, at a given plane, is related to a second vector, at a different plane, via a linear transformation. This transformation is represented by a propagation matrix. This concept is applicable to the characterization of the deviation of a ray, or beam, of light through either free space or any linear optical media. The rays of light are assumed to be a paraxial ray that propagates in proximity and almost parallel to the optical axis (Kogelnik, 1979).

Consider the propagation of a paraxial ray of light from an original plane to a secondary plane, in free space, as depicted in Figure C.1. Here it is noted that, in moving from the original plane to the secondary plane, the ray of light experiences a linear deviation in the x direction and a small angular spread, that is,

$$x_2 = x_1 + l\theta_1 \tag{C.1}$$

$$\theta_2 = \theta_1 \tag{C.2}$$

FIGURE C.1
Geometry for propagation through distance l in vacuum free space. The x direction is perpendicular to the direction of l (i.e., z-axis).

which in matrix form becomes

$$\begin{pmatrix} x_2 \\ \theta_2 \end{pmatrix} = \begin{pmatrix} 1 & l \\ 0 & 0 \end{pmatrix}\begin{pmatrix} x_1 \\ \theta_1 \end{pmatrix}$$ (C.3)

the resulting 2×2 matrix is known as a *ray transfer matrix*. Here, it should be noted that some authors use derivatives instead of the angular quantities, that is, $dx_1/dz = \theta_1$ and $dx_2/dz = \theta_2$.

For a thin lens the geometry of propagation is illustrated in Figure C.2. In this case, there is no displacement in the x direction and the ray is concentrated, or focused, toward the optical axis so that

$$x_2 = x_1$$ (C.4)

$$\theta_2 = -\left(\frac{1}{f}\right)x_1 + \theta_1$$ (C.5)

which can be expressed as

$$\begin{pmatrix} x_2 \\ \theta_2 \end{pmatrix} = \begin{pmatrix} 1 & 0 \\ 1/f & 1 \end{pmatrix}\begin{pmatrix} x_1 \\ \theta_1 \end{pmatrix}$$ (C.6)

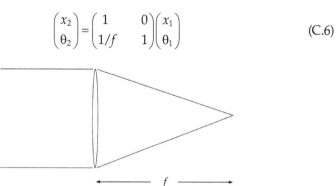

FIGURE C.2
Geometry depicting a thin convex lens with a focal length f.

In more general terms, the X_2 vector is related to the X_1 vector by a transfer matrix T known as the $ABCD$ matrix so that

$$X_2 = T\,X_1 \tag{C.7}$$

where

$$T = \begin{pmatrix} A & B \\ C & D \end{pmatrix} \tag{C.8}$$

A is a ratio of spatial dimensions
B is an optical length
C is the reciprocal of an optical length.
D is the reciprocal of A

Consideration of various imaging systems leads to the conclusion that the spatial ratio represented by A is a beam magnification factor (M), while D is the reciprocal of such magnification (M^{-1}). These observations are very useful to verify the physical validity of newly derived matrices.

C.2.1 Properties of ABCD Matrices

$ABCD$ matrices can be cascaded, via matrix multiplication, to produce a single overall matrix describing the propagation properties of an optical system. For example, if a linear optical system is composed of N optical elements deployed from left to right, as depicted in Figure C.3, then the overall

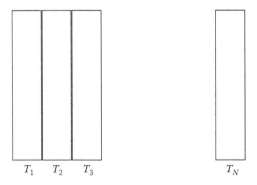

FIGURE C.3
N optical elements in series. (From Duarte, F.J., *Tunable Laser Optics*, Elsevier-Academic, New York, 2003.)

transfer matrix is given by the multiplication of the individual matrices *in the reverse order*, that is (Duarte, 2003),

$$\prod_{m=1}^{N} T_m = T_N \dots T_3 T_2 T_1 \tag{C.9}$$

It is easy to see that the complexity in the form of these product matrices can increase rather rapidly. Thus, it is always useful to remember any resulting matrix must have the dimensions of Equation C.8 and a determinant equal to unity, that is,

$$AD - BC = 1 \tag{C.10}$$

C.2.2 Survey of ABCD Matrices

In Table C.1 a number of representative and widely used optical components are included in ray transfer matrix form. This is done without derivation and using the published literature as reference.

C.2.3 Example: The Astronomical Refractive Telescope

The astronomical refractive telescope (Figure C.4) is composed of an input lens with focal length f_1, an intra-lens distance L, and an output lens with focal lens f_2. Following Equation C.9 the matrix multiplication proceeds as

$$\begin{pmatrix} 1 & 0 \\ -1/f_2 & 1 \end{pmatrix} \begin{pmatrix} 1 & L \\ 0 & 1 \end{pmatrix} \begin{pmatrix} 1 & 0 \\ -1/f_1 & 1 \end{pmatrix} \tag{C.11}$$

For a well-adjusted telescope, where

$$L = f_2 + f_1 \tag{C.12}$$

the transfer matrix becomes

$$\begin{pmatrix} A & B \\ C & D \end{pmatrix} = \begin{pmatrix} -f_2/f_1 & f_2 + f_1 \\ 0 & -f_1/f_2 \end{pmatrix} \tag{C.13}$$

which is the matrix given in Table C.1. Defining

$$-M = -\left(\frac{f_2}{f_1}\right) \tag{C.14}$$

TABLE C.1

ABCD Ray Transfer Matrices

Optical Element or System	*ABCD* Matrix	Ref.
Distance l in free space	$\begin{pmatrix} 1 & l \\ 0 & 1 \end{pmatrix}$	Kogelnik (1979)
Distance l in a medium with refractive index n	$\begin{pmatrix} 1 & l/n \\ 0 & 1 \end{pmatrix}$	Kogelnik (1979)
Slab of material with refractive index n	$\begin{pmatrix} 1 & (l/n)(\cos\phi/\cos\psi)^2 \\ 0 & 1 \end{pmatrix}$	Duarte (1991)
	ϕ is the angle of incidence. ψ is the angle of refraction.	
Thin convex (positive) lens of focal length f	$\begin{pmatrix} 1 & 0 \\ -1/f & 1 \end{pmatrix}$	Kogelnik (1979)
Thin concave (negative) lens	$\begin{pmatrix} 1 & 0 \\ 1/\lvert f\rvert & 1 \end{pmatrix}$	Siegman (1986)
Galilean telescope	$\begin{pmatrix} f_2/\lvert f_1\rvert & f_2-\lvert f_1\rvert \\ 0 & \lvert f_1\rvert/f_2 \end{pmatrix}$	Siegman (1986)
Astronomical telescope	$\begin{pmatrix} -f_2/f_1 & f_2+f_1 \\ 0 & -f_1/f_2 \end{pmatrix}$	Siegman (1986)
Curved mirror	$\begin{pmatrix} 1 & 0 \\ -2/R & 1 \end{pmatrix}$	Siegman (1986)
Flat mirror	$\begin{pmatrix} 1 & 0 \\ 0 & 1 \end{pmatrix}$	Duarte (2003)
Double pass in Cassegrain telescope	$\begin{pmatrix} M & (M+1)L/M \\ 0 & 1/M \end{pmatrix}$	Siegman (1986)
Flat grating	$\begin{pmatrix} \cos\Theta/\cos\Phi & 0 \\ 0 & \cos\Phi/\cos\Theta \end{pmatrix}$	Siegman (1986)
	Θ is the angle of incidence. Φ is the angle of diffraction.	
Flat grating in Littrow configuration ($\Theta = \Phi$)	$\begin{pmatrix} 1 & 0 \\ 0 & 1 \end{pmatrix}$	Duarte (1991)

(continued)

TABLE C.1 (continued)

ABCD Ray Transfer Matrices

Optical Element or System	*ABCD* Matrix	Ref.
Single right-angle prism	$\begin{pmatrix} \cos\psi/\cos\phi & (l/n)\cos\phi/\cos\psi \\ 0 & \cos\phi/\cos\psi \end{pmatrix}$	Duarte (1989)
	ϕ is the angle of incidence. ψ is the angle of refraction.	
Multiple-prism beam expander	$\begin{pmatrix} M_1M_2 & B \\ 0 & (M_1M_2)^{-1} \end{pmatrix}$	Duarte (1991)
Multiple-prism beam expander (return pass)	$\begin{pmatrix} (M_1M_2)^{-1} & B \\ 0 & M_1M_2 \end{pmatrix}$	Duarte (1991)

Source: Duarte, F.J., *Tunable Laser Optics*, Elsevier-Academic, New York, 2003.

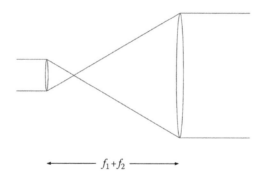

$$f_1 + f_2$$

FIGURE C.4
Schematics of an astronomical refractive telescope comprised by two convex lenses.

this matrix can be restated as

$$\begin{pmatrix} A & B \\ C & D \end{pmatrix} = \begin{pmatrix} -M & L \\ 0 & -1/M \end{pmatrix} \tag{C.15}$$

For this matrix it can be easily verified that the condition

$$|AD - BC| = 1$$

holds.

C.2.4 Multiple-Prism Beam Expanders

For a generalized multiple-prism array, as illustrated in Figure C.5, the ray transfer matrix is given by (Duarte, 1989, 1991)

$$\begin{pmatrix} A & B \\ C & D \end{pmatrix} = \begin{pmatrix} M_1 M_2 & B \\ 0 & (M_1 M_2)^{-1} \end{pmatrix} \tag{C.16}$$

where

$$M_1 = \prod_{m=1}^{r} k_{1,m} \tag{C.17}$$

$$M_2 = \prod_{m=1}^{r} k_{2,m} \tag{C.18}$$

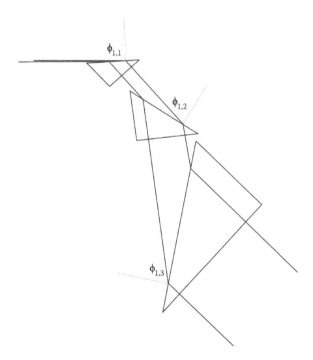

FIGURE C.5
Multiple-prism beam expander.

and

$$B = M_1 M_2 \sum_{m=1}^{r-1} L_m \left(\prod_{j=1}^{m} k_{1,j} \prod_{j=1}^{m} k_{2,j} \right)^{-2} + \left(\frac{M_1}{M_2} \right) \sum_{m=1}^{r} \left(\frac{l_m}{n_m} \right) \left(\prod_{j=1}^{m} k_{1,j} \right)^{-2} \left(\prod_{j=m}^{r} k_{2,j} \right)^{2}$$

(C.19)

For a straightforward multiple-prism beam expander with orthogonal beam exit, $\cos \psi_{2,j} = 0$ and $k_{2,j} = 1$, so that the equations reduce to

$$\begin{pmatrix} A & B \\ C & D \end{pmatrix} = \begin{pmatrix} M_1 & B \\ 0 & (M_1)^{-1} \end{pmatrix}$$

(C.20)

where

$$B = M_1 \sum_{m=1}^{r-1} L_m \left(\prod_{j=1}^{m} k_{1,j} \right)^{-2} + M_1 \sum_{m=1}^{r} \left(\frac{l_m}{n_m} \right) \left(\prod_{j=1}^{m} k_{1,j} \right)^{-2}$$

(C.21)

For a single prism these equations reduce further to

$$M_1 = k_{1,1}$$

(C.22)

$$B = l(nk_{1,1})^{-1}$$

(C.23)

Thus, the matrix for a single prism can be expressed as

$$\begin{pmatrix} A & B \\ C & D \end{pmatrix} = \begin{pmatrix} k_{1,1} & l(nk_{1,1})^{-1} \\ 0 & k_{1,1}^{-1} \end{pmatrix}$$

(C.24)

which is an alternative version of the single-prism matrix given in Table C.1 where

$$k_{1,1} = \frac{\cos \psi_{1,1}}{\cos \phi_{1,1}}$$

(C.25)

C.2.5 Telescopes in Series

For some applications it is necessary to propagate TEM_{00} laser beams through optical systems including telescopes in series. For a series comprised of a telescope followed by a free-space distance, followed by a second telescope, and so on, the single-pass cumulative matrix is given by (Duarte, 2003)

$$A = M^r$$

(C.26)

$$B = \sum_{m=1}^{r} M^{r-2m+2} L_m + M^{r-2m+1} B_{T_m} \qquad \text{(C.27)}$$

$$C = 0 \qquad \text{(C.28)}$$

$$D = M^{-r} \qquad \text{(C.29)}$$

where
r is the total number of telescopes
B_{T_m} is the B term of the mth telescope

This result applies to a series of well-adjusted Galilean or astronomical telescopes or a series of prismatic telescopes.

C.3 Applications

ABCD can be used to characterize laser resonators and narrow-linewidth oscillators (Duarte, 2003). For instance, the single return-pass beam divergence of a dispersive laser cavity can be expressed as (Duarte et al., 1997)

$$\Delta\theta = \frac{\lambda}{\pi w}\left[1 + \left(\frac{L_R}{B}\right)^2 + \left(\frac{AL_R}{B}\right)^2\right]^{1/2} \qquad \text{(C.30)}$$

$$L_R = \left(\frac{\pi w^2}{\lambda}\right) \qquad \text{(C.31)}$$

where A and B are the corresponding return-pass elements of the ray transfer matrix as calculated by Duarte (2003). The characterization can also be extended to a multiple return-pass analysis (Duarte, 2003). Besides providing the parameters of interest to perform beam divergence calculations, the *ABCD* matrix also provides information about the stability of a cavity or resonator. In this regard, the condition for lasing in the unstable regime is determined by the inequality

$$\left|\frac{(A+D)}{2}\right| > 1 \qquad \text{(C.32)}$$

Siegman (1986).

The description of optical systems using 3×3, 4×4, and 6×6 matrices has been considered by several authors (Brouwer, 1964; Siegman, 1986; Wollnik, 1987). Multiple-prism expanders have been described in 4×4 matrices by Duarte (1992, 2003).

_____

References

Brouwer, W. (1964). *Matrix Methods in Optical Design*. Benjamin, New York.

Duarte, F. J. (1989). Ray transfer matrix analysis of multiple-prism dye laser oscillators. *Opt. Quant. Electron.* **21**, 47–54.

Duarte, F. J. (1991). Dispersive dye lasers. In *High Power Dye Lasers* (Duarte, F. J., ed.). Springer-Verlag, Berlin, Germany, Chapter 2.

Duarte, F. J. (1992). Multiple-prism dispersion and 4×4 ray transfer matrices. *Opt. Quant. Electron.* **24**, 49–53.

Duarte, F. J. (2003). *Tunable Laser Optics*. Elsevier-Academic, New York.

Duarte, F. J., Costela, A., Garcia-Moreno, I., Sastre, R., Ehrlich, J. J., and Taylor, T. S. (1997). Dispersive solid-state dye laser oscillators. *Opt. Quant. Electron.* **29**, 461–472.

Kogelnik, H. (1979). Propagation of laser beams. In *Applied Optics and Optics Engineering*, (Shannon, R. R. and Wyant., J. C., eds.). Academic Press, New York, pp. 155–190.

Siegman, A. (1986). *Lasers*. University Science Books, Mill Valley, CA.

Wollnik, H. (1987). *Optics of Charged Particles*. Academic Press, New York.

Appendix D: Multiple-Prism Dispersion Series

D.1 Multiple-Prism Dispersion Series

In Chapter 5, the generalized multiple-prism dispersion equation, applicable to multiple-prism arrays of any geometry, configuration, or materials, is given as (Duarte, 2009)

$$\nabla_\lambda \phi_{2,m} = \pm \mathcal{H}_{2,m} \nabla_\lambda n_m \pm \left(k_{1,m} k_{2,m} \right)^{-1} \left(\mathcal{H}_{1,m} \nabla_\lambda n_m (\pm) \nabla_\lambda \phi_{2,(m-1)} \right) \tag{D.1}$$

For positive refraction only, this equation becomes (Duarte and Piper, 1982, 1983)

$$\nabla_\lambda \phi_{2,m} = \mathcal{H}_{2,m} \nabla_\lambda n_m + \left(k_{1,m} k_{2,m} \right)^{-1} \left(\mathcal{H}_{1,m} \nabla_\lambda n_m \pm \nabla_\lambda \phi_{2,(m-1)} \right) \tag{D.2}$$

where the $\pm$ sign refers to either a positive (+) or compensating configuration (–).

In Chapter 5, Equation D.2 is expressed in a series format directly applicable to the geometry at hand. Duarte and Piper (1982) also provide further examples of simple special cases leading to explicit engineering-type equations. For instance, for increasing values of m, for the *very special case* of r identical prisms deployed at the same angle of incidence (i.e., $\phi_{1,1} = \phi_{1,2} = \dots = \phi_{1,m}$ and $\psi_{1,1} = \psi_{1,2} = \dots = \psi_{1,m}$) and orthogonal beam exit (i.e., $\phi_{2,1} = \phi_{2,2} = \dots = \phi_{2,m} = 0$ and $\psi_{2,1} = \psi_{2,2} = \dots = \psi_{2,m} = 0$), Equation D.2 reduces to a simple power series (Duarte and Piper, 1982; Duarte, 1990):

$$\nabla_\lambda \phi_{2,r} = \tan \psi_{1,1} \nabla n_1 \left(1 \pm k_{1,1}^{-1} \pm k_{1,1}^{-2} \pm k_{1,1}^{-3} \pm \dots \pm k_{1,1}^{-(r-1)} \right) \tag{D.3}$$

Also, as shown in Chapter 5, for orthogonal beam exit, Equation D.2 reduces to the explicit series:

$$\nabla_\lambda \phi_{2,r} = \sum_{m=1}^{r} (\pm 1) \mathcal{H}_{1,m} \left(\prod_{j=m}^{r} k_{1,j} \right)^{-1} \nabla_\lambda n_m \tag{D.4}$$

which was disclosed in print, in its double-pass version by Duarte (1985). This simple explicit equation obviously can be expressed in its long-hand version (Duarte, 2012):

$$\nabla_\lambda \phi_{2,r} = \pm \mathcal{H}_{1,1}(k_{1,1}k_{1,2}...k_{1,r})^{-1}\nabla_\lambda n_1 \pm \mathcal{H}_{1,2}(k_{1,2}...k_{1,r})^{-1}\nabla_\lambda n_2 \pm \cdots \pm \mathcal{H}_{1,r}(k_{1,r})^{-1}\nabla_\lambda n_r$$

$$(D.5)$$

These examples are included here to illustrate that the generalized dispersion Equation D.1 leads directly to easy-to-use explicit results that might appear as "new" to some.

References

Duarte, F. J. (1985). Note on achromatic multiple-prism beam expanders. *Opt. Commun.* **53**, 259–262.
Duarte, F. J. (1990). Narrow linewidth pulsed dye laser oscillators. In *Dye Laser Principles* (Duarte, F. J. and Hillman, L. W., eds.). Academic Press, New York, Chapter 4.
Duarte, F. J. (2009). Generalized multiple-prism dispersion theory for laser pulse compression: higher order phase derivatives. *Appl. Phys. B* **96**, 809–814.
Duarte, F. J. (2012). Tunable organic dye lasers: Physics and technology of high-performance liquid and solid-state narrow-linewidth oscillators. *Prog. Quant. Electron.* **36**, 29–50.
Duarte, F. J. and Piper, J. A. (1982). Dispersion theory of multiple-prism beam expander for pulsed dye lasers. *Opt. Commun.* **43**, 303–307.
Duarte, F. J. and Piper, J. A. (1983). Generalized prism dispersion theory. *Am. J. Phys.* **51**, 1132–1134.

Appendix E: Complex Numbers

E.1 Complex Numbers

Here, we offer a brief and pragmatic introduction to complex numbers and some well-known trigonometric identities based on complex numbers.

The imaginary part of a complex number is represented by i. The number i has the basic property

$$i^2 = -1 \tag{E.1}$$

so that

$$i \cdot i = -1 \tag{E.2}$$

and

$$i \cdot (-i) = +1 \tag{E.3}$$

A complex number has a real and an imaginary part denoted by i. A complex number c is defined as

$$c = a + ib \tag{E.4}$$

where a and b are real. This complex number is depicted in Figure E.1. The complex conjugate of this number c is denoted by c^*:

$$c^* = a - ib \tag{E.5}$$

These two numbers can be multiplied as

$$cc^* = (a + ib)(a - ib) = a^2 + b^2 \tag{E.6}$$

and the magnitude of $c = a + ib$ is denoted by $|c|$:

$$|c| = |cc^*| = (a^2 + b^2)^{1/2} \tag{E.7}$$

and it represents the length of the vector $(a + ib)$ (see Figure E.1). Also,

$$|c|^2 = cc^* \tag{E.8}$$

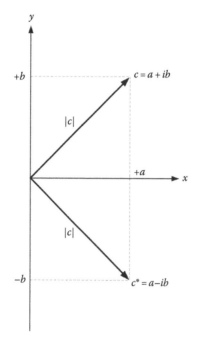

FIGURE E.1
Geometrical representation of a complex number.

Useful complex trigonometric identities include the Euler formulae

$$e^{i\theta} = \cos\theta + i\sin\theta \qquad\qquad\qquad (E.9)$$

$$e^{-i\theta} = \cos\theta - i\sin\theta \qquad\qquad\qquad (E.10)$$

which mean that

$$e^{i\pi} = -1 \qquad\qquad\qquad (E.11)$$

$$e^{-i\pi} = +1 \qquad\qquad\qquad (E.12)$$

$$e^{i\pi/2} = +i \qquad\qquad\qquad (E.13)$$

$$e^{-i\pi/2} = -i \qquad\qquad\qquad (E.14)$$

Also,

$$\cos\theta = \frac{1}{2}(e^{i\theta} + e^{-i\theta}) \tag{E.15}$$

$$\sin\theta = \frac{1}{2i}(e^{i\theta} - e^{-i\theta}) \tag{E.16}$$

$$\cosh\theta = \frac{1}{2}(e^{\theta} + e^{-\theta}) \tag{E.17}$$

$$\sinh\theta = \frac{1}{2}(e^{\theta} - e^{-\theta}) \tag{E.18}$$

Appendix F: Trigonometric Identities

F.1 Trigonometric Identities

Here, we list a number of useful trigonometric identities applied in various places throughout this book:

$$\sin^2\phi + \cos^2\phi = 1 \tag{F.1}$$

$$\sin(-\phi) = -\sin\phi \tag{F.2}$$

$$\cos(-\phi) = \cos\phi \tag{F.3}$$

$$\sin(\phi + \psi) = \sin\phi\cos\psi + \cos\phi\sin\psi \tag{F.4}$$

$$\sin(\phi - \psi) = \sin\phi\cos\psi - \cos\phi\sin\psi \tag{F.5}$$

$$\cos(\phi + \psi) = \cos\phi\cos\psi - \sin\phi\sin\psi \tag{F.6}$$

$$\cos(\phi - \psi) = \cos\phi\cos\psi + \sin\phi\sin\psi \tag{F.7}$$

$$\sin 2\phi = 2\sin\phi\cos\phi \tag{F.8}$$

$$\cos 2\phi = \cos^2\phi - \sin^2\phi \tag{F.9}$$

$$\cos 2\phi = 1 - 2\sin^2\phi \tag{F.10}$$

$$\cos 2\phi = 2\cos^2\phi - 1 \tag{F.11}$$

$$2\sin^2\phi = 1 - \cos 2\phi \tag{F.12}$$

$$2\cos^2\phi = 1 + \cos 2\phi \tag{F.13}$$

$$\cos\phi + \cos\psi = 2\cos(\phi - \psi)\cos(\phi + \psi) \tag{F.14}$$

$$\cos\phi - \cos\psi = -2\sin(\phi - \psi)\sin(\phi + \psi) \tag{F.15}$$

$$-\cos\phi + \cos\psi = 2\sin(\phi - \psi)\sin(\phi + \psi) \tag{F.16}$$

$$-\cos\phi - \cos\psi = -2\cos(\phi - \psi)\cos(\phi + \psi) \tag{F.17}$$

$$\sin\phi + \sin\psi = 2\cos(\phi - \psi)\sin(\phi + \psi) \tag{F.18}$$

$$\sin\phi - \sin\psi = 2\sin(\phi - \psi)\cos(\phi + \psi) \tag{F.19}$$

$$-\sin\phi + \sin\psi = -2\sin(\phi - \psi)\cos(\phi + \psi) \tag{F.20}$$

$$-\sin\phi - \sin\psi = -2\cos(\phi - \psi)\sin(\phi + \psi) \tag{F.21}$$

$$\sin\phi\sin\psi = \frac{1}{2}\left(\cos(\phi - \psi) - \cos(\phi + \psi)\right) \tag{F.22}$$

$$\cos\phi\cos\psi = \frac{1}{2}\left(\cos(\phi - \psi) + \cos(\phi + \psi)\right) \tag{F.23}$$

$$\sin\phi\cos\psi = \frac{1}{2}\left(\sin(\phi + \psi) + \sin(\phi - \psi)\right) \tag{F.24}$$

$$\cos\phi\sin\psi = \frac{1}{2}\left(\sin(\phi + \psi) - \sin(\phi - \psi)\right) \tag{F.25}$$

Appendix G: Calculus Basics

G.1 Calculus Basics

Here, we provide a brief and pragmatic survey of some useful well-known calculus mechanics and rules. A good reference on calculus is Flanders et al. (1970).

G.1.1 Differentiation Product Rule

$$\frac{d(fg)}{dx} = f\frac{dg}{dx} + g\frac{df}{dx} \tag{G.1}$$

Example: Differentiate the product xe^{ikx}:

$$\frac{d(xe^{ikx})}{dx} = x(ik)e^{ikx} + e^{ikx} = e^{ikx}(1 + ikx)$$

G.1.2 Differentiation Quotient Rule

$$\frac{d(f/g)}{dx} = g^{-2}\left(g\frac{df}{dx} - f\frac{dg}{dx} \right) \tag{G.2}$$

G.1.3 Differentiation Power Rule

If n is an integer,

$$\frac{d(f^n)}{dx} = nf^{n-1}\frac{df}{dx} \tag{G.3}$$

Example: Differentiate $(x^2 + 1)^2$:

$$\frac{d(x^2+1)^2}{dx} = 4x(x^2+1)$$

G.1.4 Differentiation Chain Rule

If y and x are functions of t,

$$\frac{dy}{dt} = \frac{dy}{dx}\frac{dx}{dt} \tag{G.4}$$

Example: Differentiate the function $y = e^{t^2+2t+1}$ Set $y = e^x$ and $x = t^2 + 2t + 1$. Then apply the chain rule:

$$\frac{dy}{dt} = e^x(2t+2) = e^{t^2+2t+1}(2t+2)$$

G.1.5 Integration by Parts

$$\int f \, dg = fg - \int g \, df \tag{G.5}$$

Example: Integrate by parts $\int xe^{ikx}dx$. Set $f = x$, $df = dx$, $dg = e^{ikx}dx$, and $g = e^{ikx}/ik$.

Then apply Equation G.5:

$$\int xe^{ikx}dx = \frac{x}{ik}e^{ikx} - \int \frac{e^{ikx}}{ik}dx = \frac{e^{ikx}}{ik}\left(x - \frac{1}{ik}\right) + C$$

where C is a constant. Differentiation of $F(x) = (e^{ikx}/ik)(x - (1/ik)) + C$, using the product rule leads back to xe^{ikx}.

Reference

Flanders, H., Korfhage, R. R., and Price, J. J. (1970). *Calculus*. Academic Press, New York.

Appendix H: Poincaré's Space

H.1 Poincaré's Space

A useful tool in polarization notation is derived from *Poincaré's sphere* (Poincaré, 1892).

This sphere, depicted in Figure H.1, has three axes 1, 2, and 3. Axis 2 is analogous to the usual Cartesian axis x, axis 3 is analogous to the usual Cartesian axis y, and axis 1 is analogous to the usual Cartesian axis z, that is,

$$1 \rightarrow z$$

$$2 \rightarrow x$$

$$3 \rightarrow y$$

Adopting the notation of Robson (1974), the radius of the sphere is denoted by I. The angular displacement in planes 1–2 is 2ψ and the angular displacement between planes 1–2 and axis 3 is denoted by 2χ. In this system, the points P_1, P_2, P_3 are given by

$$P_1 = I \cos 2\chi \cos 2\psi \qquad (\text{H.1})$$

$$P_2 = I \cos 2\chi \sin 2\psi \qquad (\text{H.2})$$

$$P_3 = I \sin 2\chi \qquad (\text{H.3})$$

These are known as the *Stokes parameters*. In addition to Poincaré's sphere, the polarization space described here is also known as Bloch's sphere (Pelliccia et al., 2003).

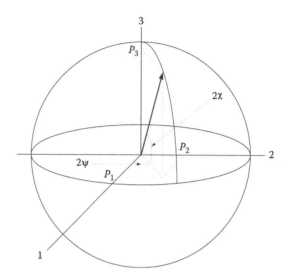

FIGURE H.1
Poincaré's sphere.

References

Pelliccia, D., Schettini, V., Sciarrino, F., Sias, C., and De Martini, F. (2003). Contextual realization of the universal quantum cloning machine and of the universal-NOT gate by quantum-injected optical parametric amplification. *Phys. Rev. A* **68**, 042306.
Poincaré, H. (1892). *Théorie Mathématique de la Lumière*, Vol. 2. Corré, Paris, France.
Robson, B. A. (1974). *The Theory of Polarization Phenomena*. Clarendon Press, Oxford, U.K.

Appendix I: N-Slit Interferometric Calculations

I.1 Introduction

In this appendix, we attempt to further explain, in plain language, how to proceed with numerical calculations based on the generalized 1D interferometric equation (Duarte, 1991, 1993). This explanation, provided in addition to the description given in Chapter 4, is intended to help in the construction of logical pathways and flowcharts for computer-based numerical calculations.

I.2 Interferometric Equation

The physics is described by the probability amplitude equation

$$\langle x \,|\, s \rangle = \sum_{j=1}^{N} \langle x \,|\, j \rangle \langle j \,|\, s \rangle \tag{I.1}$$

where

$$\langle j \,|\, s \rangle = \Psi(r_{j,s}) e^{-i\theta_j} \tag{I.2}$$

$$\langle x \,|\, j \rangle = \Psi(r_{x,j}) e^{-i\phi_j} \tag{I.3}$$

Multiplying by the complex conjugate and rearranging yields the generalized interferometric equation

$$|\langle x \,|\, s \rangle|^2 = \sum_{j=1}^{N} \Psi(r_j)^2 + 2 \sum_{j=1}^{N} \Psi(r_j) \left(\sum_{m=j+1}^{N} \Psi(r_m) \cos(\Omega_m - \Omega_j) \right) \tag{I.4}$$

Key in these calculations is the phase angle

$$\Omega_j = (\theta_j + \phi_j) \tag{I.5}$$

which is related to the wavelength and the exact geometry of the interferometer.

A good simplifying assumption is to set

$$\langle j \,|\, s \rangle = \Psi(r_{j,s})e^{-i\theta_j} = 1 \tag{I.6}$$

which represents uniform illumination of the grating. In a more advanced version of the program, this assumption is not used and the illumination can be varied at will. Hence, the only wave function needed is

$$\langle x \,|\, j \rangle = \Psi(r_{x,j})e^{-i\phi_j} \tag{I.7}$$

where the amplitude can take the form of a Gaussian or a similar mathematical representation.

I.3 Geometry

In these calculations, it is very important to get the geometry represented exactly and correctly. The usual geometrical and angular approximations are not allowed. The geometrical configuration, illustrated in Figure I.1, includes the following:

1. w is the slit width.
2. ϕ_j is the jth phase angle.
3. h is the center to center slit distance.
4. $D_{(x|j)}$ is the distance from the N-slit array (j) to the interference plane (x).
5. Δx_i is the distance from the center of the ith slit to the reference position at the interference.

Thus, the phase angles are given by exact geometrical expressions, such as

$$(\phi_1 + \phi_2) = \left(\frac{1\pi}{\lambda}\right)\left[\frac{\Delta x_1}{\left(d^2 + \Delta x_1^2\right)^{1/2}}\right] \tag{I.8}$$

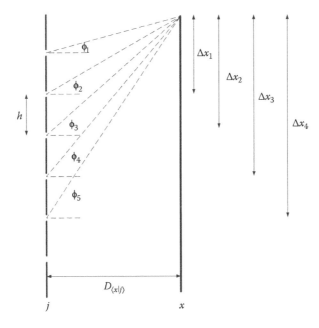

FIGURE I.1
Detailed view of the N-slit array (*j*) and the interferometric plane (*x*). See the text for a description of the quantities involved.

$$(\phi_1 + \phi_3) = \left(\frac{2\pi h}{\lambda}\right)\left[\frac{\Delta x_2}{\left(d^2 + \Delta x_2^2\right)^{1/2}}\right] \tag{I.9}$$

$$(\phi_1 + \phi_4) = \left(\frac{3\pi h}{\lambda}\right)\left[\frac{\Delta x_3}{\left(d^2 + \Delta x_3^2\right)^{1/2}}\right] \tag{I.10}$$

$$(\phi_2 + \phi_3) = \left(\frac{\pi h}{\lambda}\right)\left[\frac{\Delta x_2}{\left(d^2 + \Delta x_2^2\right)^{1/2}}\right] \tag{I.11}$$

$$(\phi_2 + \phi_4) = \left(\frac{2\pi h}{\lambda}\right)\left[\frac{\Delta x_3}{\left(d^2 + \Delta x_3^2\right)^{1/2}}\right] \tag{I.12}$$

$$(\phi_3 + \phi_4) = \left(\frac{\pi h}{\lambda}\right)\left[\frac{\Delta x_3}{\left(d^2 + \Delta x_3^2\right)^{1/2}}\right] \tag{I.13}$$

where λ is the wavelength of the laser.

Note that in this example, both the slit width (w) and the separation of the slits (h) are assumed to be constant. As shown in Duarte (1993), that is a special case as these parameters can be numerically varied (allowing for dimensional errors), thus introducing an element of variability in the calculations. Also note that in Chapter 2 of *Tunable Laser Optics* (Duarte, 2003), an alternative and equivalent description (also included in Chapter 4) of the exact geometrical equations is given.

References

Duarte, F. J. (1991). Dispersive dye lasers. In *High Power Dye Lasers* (Duarte, F. J., ed.). Springer-Verlag, Berlin, Germany, Chapter 2.

Duarte, F. J. (1993). On a generalized interference equation and interferometric measurements. *Opt. Commun.* **103**, 8–14.

Duarte, F. J. (2003). *Tunable Laser Optics*. Elsevier-Academic, New York.

Appendix J: N-Slit Interferometric Calculations—Numerical Approach

J.1 Introduction

The first numerical calculations representing the interferometric equation

$$|\langle x|s\rangle|^2 = \sum_{j=1}^{N}\Psi(r_j)^2 + 2\sum_{j=1}^{N}\Psi(r_j)\left(\sum_{m=j+1}^{N}\Psi(r_m)\cos(\Omega_m - \Omega_j)\right) \qquad (J.1)$$

and its associated geometry was performed using Fortran IV (Duarte and Paine 1989; Duarte 1991, 1993). Then the program was transitioned to Visual Fortran (Duarte, 2002). MATLAB® versions of the calculations have been developed more recently Duarte et al. (2013). Here, we provide a MATLAB version of a simple program that deals with the experimental situation related to the basic probability amplitude

$$\langle x|s\rangle = \sum_{j=1}^{N}\langle x|j\rangle\langle j|s\rangle \qquad (J.2)$$

In Duarte et al. (2013), the numerical calculations were performed using a MATLAB version of the program that allows cascade calculations from one interferometric plane to another interferometric plane as previously demonstrated in Duarte (1993), while using Fortran IV.

J.2 Program

This is a sample program, written in MATLAB language. It is a simplified version of the type of program used by Duarte et al. (2013). This program, applicable to a uniform array of N slits, assumes a slit separation equal to the slit width w and exact dimensions for the slit width and slit separation. The program uses the following input data:

1. Number of slits N
2. Slit width w in micrometers (μm)
3. Intra-interferometric distance $D_{\langle x|j\rangle}$ in meters (m)
4. Wavelength λ in nanometers (nm)

```
01%NSLITSGENA.m
02 clear all
03 datestr(now)
04 N = input('Number of slits N = ');
05 w = input('Slit width in micrometers w = ');
06 d = input('Slit to screen distance in meters d = ');
07 lambda = input('Wavelength in nanometers lambda = ');
08 d = d*1e6;
09 lambda = lambda/1e3;
10 NP = 40030;% Screen size in micrometers.
11 s = 2*pi/lambda;%wavenumber
12 j0 = 20015-w*(N-0.5);% left side left aperture position
13 j01 = j0+1;
14 j0n = j0+w;
15 for ii = 1:NP, % Screen
16     for j = j01:j0n,%Field due to left slit on screen.
17     r = sqrt((ii-j)^2+d^2);
18     E(j) = 1e3*(i*d/lambda)*(exp(i*s*r))/r^2;
19     end
20             ii10 = ii/1000-round(ii/1000);
21 if ii10 = =0
22     iactual = [ii NP]
23 else
24 end
25             EF(ii) = sum(E);
26             E12(ii) = sum(E)*conj(sum(E));
27             E1(ii) = abs(sum(E));%(module)
28             E2(ii) = angle(sum(E));%(angle)
29 end
30 for k = 1:N
31         for ii = 1:NP
32             EF1(k,ii+2*w*(k-1)) = EF(ii);
33             FI2(k,ii+2*w*(k-1)) = E12(ii);
34             FIR(k,ii+2*w*(k-1)) = E1(ii);
35             fir(k,ii+2*w*(k-1)) = E2(ii);
36                 end
37 end
```

```
38  HQ = sum(FI2);
39  IIMAX = NP+(N-1)*2*w;
40  H = (1:IIMAX)*0;
41  for j = 1:N
42      for m = (j+1):N,
43  H = H+2*FIR(j,:).*FIR(m,:).*cos(fir(m,:)-fir(j,:));
44  end
45  end
46          for j = 1:N
47  FIRx(j,:) = FIR(j,:).*cos(fir(j,:));%REAL PART
48  FIRy(j,:) = FIR(j,:).*sin(fir(j,:));%IMAGINARY PART
49      end
50      FIREX = sum(FIRx);
51      FIREY = sum(FIRy);
52      Efield1(1,:) = FIREX;
53      Efield1(2,:) = FIREY;
54      save Efield1 Efield1 -ascii
55          ESQ = H+HQ;
56              plot(ESQ,'r')
57                  hold on
58  datestr(now)
```

The specifics of this program differ from the descriptive approach disclosed in Appendix I. Here, *ii* is the position on the interferometric plane (screen) and *j* the position at the slit array. Also, on line 18, and associated definitions, a particular wave function amplitude is assumed. It should be noted that the formalism, defined in Equation J.1, tolerates various forms of wave function amplitudes. In this regard, the search for the simplest and most suitable wave function amplitude, as determined by comparisons with measured interferograms, is ongoing.

References

Duarte, F. J. (1991). Dispersive dye lasers. In *High Power Dye Lasers* (Duarte, F. J., ed.). Springer, Berlin, Germany, Chapter 2.

Duarte, F. J. (1993). On a generalized interference equation and interferometric measurements. *Opt. Commun.* **103**, 8–14.

Duarte, F. J. (2002). Secure interferometric communications in free space. *Opt. Commun.* **205**, 313–319.

Duarte, F. J. and Paine, D. J. (1989). Quantum mechanical description of *N*-slit interference phenomena, in *Proceedings of the International Conference on Lasers'88*, (Sze, R. C. and Duarte, F. J., eds.). STS Press, McLean, VA, pp. 42–27.

Duarte, F. J., Taylor, T. S., Black, A. M., and Olivares, I. E. (2013). Diffractive patterns superimposed over propagating *N*-slit interferograms. *J. Mod. Opt.* **60**, 136–140.

Appendix K: Physical Constants and Optical Quantities

K.1 Fundamental Physical Constants

Physical constants useful in optics are listed in Table K.1. The values of these constants are those listed by the *National Institute of Science and Technology* (NIST) available at the time of publication.

The format and context of the tables included here is adapted from Duarte (2003).

K.2 Conversion Quantities

Conversion quantities often used in optics are listed in Table K.2. The conversion values for the electron volt and the atomic mass unit are the values listed by NIST available at the time of publication.

TABLE K.1

Fundamental Physical Constants

Name	Symbol	Value	Units
Boltzmann's constant	k_B	$1.3806488 \times 10^{-23}$	$J\,K^{-1}$
Elementary charge	e	$1.6021\,76\,565 \times 10^{-19}$	C
Newtonian constant of gravitation	G	6.67384×10^{-11}	$m^3\,kg^{-1}\,s^{-2}$
Magnetic constant[a,b]	μ_0	$4\pi \times 10^{-7}$	$N\,A^{-2}$
Electric constant[c]	ε_0	$8.854187817 \times 10^{-12}$	$F\,m^{-1}$
Planck's constant	h	$6.62606957 \times 10^{-34}$	$J\,s$
Speed of light in vacuum	c	2.99792458×10^{8}	$m\,s^{-1}$

[a] Also known as permeability of vacuum.
[b] $\pi = 3.141592654\ldots$
[c] Also known as permittivity of vacuum.

TABLE K.2

Conversion Quantities

Name	Symbol	Value	Units
Electron volt	eV	$1.602176565 \times 10^{-19}$	J
Atomic mass unit	u	$1.660538921 \times 10^{-27}$	kg
Frequency	ν		Hz = s^{-1}
Linewidth	$\Delta \nu = c / \Delta x$		Hz
Linewidth	$\Delta \lambda = \lambda^2 / \Delta x$		m
Wavelength	$\lambda = c / \nu$		m
Wave number	$k = 2\pi / \lambda$		m^{-1}
1 reciprocal cm	1 cm^{-1}	2.99792458×10^{1}	GHz

K.3 Units of Optical Quantities

Units of optical quantities used throughout this book are listed in Table K.3.

TABLE K.3

Units of Optical Quantities

Name	Symbol	Units[a]
Angular dispersion	$\nabla_\lambda \phi$	m^{-1}
Angular frequency	$\omega = 2\pi \nu$	Hz
Beam divergence	$\Delta \theta$	rad
Beam magnification	M	Dimensionless
Beam waist	w	m
Cross section	σ	m^2
Diffraction limited $\Delta \theta$	$\Delta \theta = \lambda / \pi w$	rad
Energy	E	J
Frequency	ν	Hz
Intensity	I	J s^{-1} m^{-2}
Laser linewidth	$\Delta \nu$	Hz
Laser linewidth	$\Delta \lambda$	m
Power	P	W = J s^{-1}
Rayleigh length	$L_R = \pi w^2 / \lambda$	m
Refractive index	n	Dimensionless
Time	t	s
Wavelength	λ	m
Wave number	$k = 2\pi / \lambda$	m^{-1}
Wave number	$k = \omega / c$	m^{-1}

[a] Quantities like I and σ are also used in cgs units.

Reference

Duarte, F. J. (2003). *Tunable Laser Optics*. Elsevier-Academic, New York.

Index

Milton Keynes UK
Ingram Content Group UK Ltd.
UKHW031139141024
449569UK00024B/1215